Transformation Thermotics and Extended Theories

Liu-Jun Xu · Ji-Ping Huang

Transformation Thermotics and Extended Theories

Inside and Outside Metamaterials

 Springer

Liu-Jun Xu
Department of Physics
Fudan University
Shanghai, China

Ji-Ping Huang
Department of Physics
Fudan University
Shanghai, China

National Natural Science Foundation of China
The Science and Technology Commission of Shanghai Municipality

ISBN 978-981-19-5910-3 ISBN 978-981-19-5908-0 (eBook)
https://doi.org/10.1007/978-981-19-5908-0

This Springer imprint is published by the registered company Springer Nature Singapore Pte Ltd.
The registered company address is: 152 Beach Road, #21-01/04 Gateway East, Singapore 189721, Singapore

Contents

Chapter 1
Preface

1.1 Traditional Thermodynamics Versus Theoretical Thermotics

What you do not know always determines what you know. Unfortunately, what you know often hinders you from knowing what you do not know yet. In this sense, it is valuable for inheritance and innovation to systematize the existing scattered knowledge. We believe now is the time to present "theoretical thermotics" as a new discipline with systematic knowledge constructed by transformation thermotics and its extended theories. See Fig. 1.1. Here, transformation thermotics is known to originate from transformation optics [7], but the latter always handles wave systems rather than diffusion systems (that serve as a focus of transformation thermotics).

If you want to be a big tree, compare yourself with other big trees, rather than grass. Let us compare "theoretical thermotics" with "traditional thermodynamics". As shown in Table 1.1, theoretical thermotics distinctly differs from traditional thermodynamics. Certainly, as one of the most fundamental theoretical frameworks for describing nature, traditional thermodynamics must also work for all the artificial systems studied by theoretical thermotics. Nevertheless, theoretical thermotics has its purposes, systems, and frameworks, thus distinguishing it from traditional thermodynamics (Table 1.1).

Though the word "thermotics" is not commonly used, I choose it for the new discipline, "theoretical thermotics". Here, "thermotics" can always be translated into "heat transfer (heat transfer theory)" and sometimes into "thermodynamics". But, the reason why I do not choose to use "theoretical heat transfer" is two-folded: I hope to add new concepts (say, those from condensed matter physics, optics, statistical physics, etc.) to "thermotics", which goes beyond traditional heat transfer; I do not hope the existing knowledge of conventional "heat transfer" affects the understanding of the connotation of "theoretical thermotics". These two reasons also hold for another name, "theoretical thermodynamics". If a name can be easily followed without confusion, work can be accomplished, one of Confucius's sayings. Anyway, the future name is up to others, but what we can do now is up to us.

© The Author(s) 2023
L.-J. Xu and J.-P. Huang, *Transformation Thermotics and Extended Theories*,
https://doi.org/10.1007/978-981-19-5908-0_1

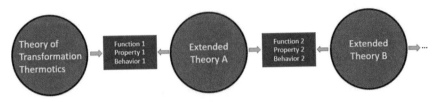

Fig. 1.1 The new discipline of "theoretical thermotics" is constructed by the theory of transformation thermotics and its extended theories with different levels. All these theories are connected with functions/properties/behaviors. For example, "theory of transformation thermotics for steady state" [1, 2] ⟹ "thermal cloaking (Function 1)" [1, 2] ⟸ "scattering cancellation (Extended Theory A)" [3, 4] ⟹ "thermal camouflage (Function 2)" [5] ⟸ "effective medium theory (Extended Theory B)" [6] ⟹ ···

Table 1.1 Traditional thermodynamics versus theoretical thermotics. Here, the phrase "passive description" means that people cannot change the heat phenomena of natural systems but understand them according to the four thermodynamic laws. In contrast, the phrase "active control" represents that people can change the heat phenomena at will by designing artificial systems based on transformation thermotics and its extended theories. These theories also make theoretical thermotics different from the existing heat transfer theory (which is much more familiar to engineering thermophysicists than physicists). Adapted from Ref. [8]

	Main purpose	Key systems	Theoretical framework
Traditional thermodynamics	Passive description	Natural systems	The four laws of thermodynamics
Theoretical thermotics	Active control	Artificial systems	Transformation thermotics and extended theories

For transformation thermotics, the starting point of theoretical thermotics, its foundations could be summarized as "four properties" in the following.

A. Invariance: Thermal equations have form invariance. Many thermal equations, including those describing heat conduction, have the same form in different coordinate systems;

B. Anisotropy: Thermophysical quantities can be anisotropic. The physical properties can be anisotropic, which are described by anisotropic thermophysical quantities like thermal conductivity;

C. Inhomogeneity: Thermophysical quantities can be inhomogeneous. The physical properties can be non-uniformly distributed in space, which are described by inhomogeneous thermophysical quantities like thermal conductivity;

D. Effectiveness: Thermophysical quantities have effective properties. The thermophysical quantities described in B and C above can be equivalent to the composite of isotropic homogeneous materials.

Based on the above A, we can deduce B and C, and the prior existence of B and C also ensures the necessity of A's existence. Therefore, A, B, and C lie in the same column, supporting and guarding each other. More importantly, B and C make the existence of D indispensable. Otherwise, the experiment cannot easily verify the theoretical prediction based on A-C, thus blocking the engineering application.

The four foundations (A–D) help construct the whole discipline of theoretical thermotics by starting from transformation thermotics.

1.2 Theoretical Thermotics Meets Metamaterials: Inside Versus Outside Metamaterials

Theoretical thermotics is interdisciplinary with three first-class disciplines, i.e., physics, engineering thermophysics, and materials science; see Fig. 1.2. On the other hand, the mature discipline of metamaterials is also interdisciplinary with many disciplines, say, optics, electromagnetics, acoustics, classical mechanics, quantum mechanics, etc. When theoretical thermotics meets metamaterials, what will happen? They give birth to a new direction, thermal metamaterials [9]; see Fig. 1.3. The first monograph on thermal metamaterials was published in 2020 [10]. Thanks to Ref. [9], the name "thermal metamaterials" was first used to cover the five works on thermal cloaks for controlling thermal conduction [1, 2, 11–13]. The first monograph on thermal cloaks was published in 2022 [14]. The connotation of thermal metamaterials has been extended significantly from thermal conduction to convection and radiation. As a result, so far, theoretical thermotics has been studied and developed from pure science to technology and engineering; see Fig. 1.4. The biennial International Conference on Thermodynamics and Thermal Metamaterials has been organized since 2020 to promote the development; see Fig. 1.5.

The key factor for treating an artificial structural material as a metamaterial is that the construction unit should have a characteristic length. The concept of effective media helps to understand the novel properties associated with metamaterials. For example, the characteristic length of electromagnetic metamaterials is the incident wavelength, that of thermal conduction metamaterials is the diffusion length, that of thermal convection metamaterials is the migration length of fluids, and that of thermal radiation metamaterials is the radiation wavelength.

Metamaterials can be classified in diverse ways: wave metamaterials versus diffusion metamaterials, programmable metamaterials versus unprogrammable metamaterials, bulk metamaterials versus metasurfaces, and so on. Figure 1.3 displays that theoretical thermotics can be classified as "inside metamaterials" and "outside metamaterials". Currently, the part of "inside metamaterials" has received much attention [8, 10, 20–25]. In the meantime, the part of "outside metamaterials" is rapidly developing as well [26–28].

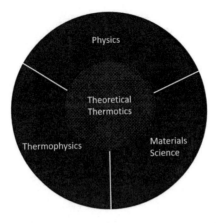

Fig. 1.2 Theoretical thermotics is interdisciplinary with three first-class disciplines, namely, physics (thermodynamics and statistical physics), engineering thermophysics (heat transfer), and materials science (material thermodynamics). A huge number of articles have appeared in the professional journals corresponding to these three disciplines (such as Physical Review Letters, Physical Review E, Physical Review Applied, and Applied Physics Letters for physics; International Journal of Heat and Mass Transfer for engineering thermophysics; and Advanced Materials for materials science), besides those interdisciplinary journals (say, Science, Nature, and Proceedings of the National Academy of Sciences of the United States of America)

Fig. 1.3 Theoretical thermotics (an interdisciplinary subject) meets metamaterials (another interdisciplinary subject), yielding a new central branch of thermal metamaterials. Metamaterials have a characteristic length larger or much larger than the construction unit

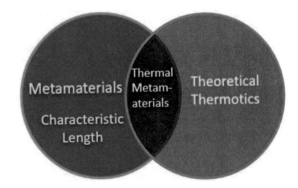

1.3 Acknowledgment and Some Additional Notes

Liu-Jun Xu, the first author of this monograph, would like to thank Prof. Ji-Ping Huang for involving him in this book writing. Supervised by Prof. Huang, Liu-Jun came into contact with and immersed in theoretical thermotics, making his five-year Ph.D. career fulfilling and rewarding. Liu-Jun also appreciates Prof. Cheng-Wei Qiu's careful guidance when Liu-Jun spent one year at the National University of Singapore. He has received a doctoral degree from the Department of Physics, Fudan University, Shanghai, China, in June 2022. (Notes: Liu-Jun Xu wrote this paragraph in the third person.)

Fig. 1.4 Theoretical thermotics contains the research on the whole chain from science (from zero to one) to technology (from virtual to real) and engineering (from useless to useful). This book focuses on the part of science. The parts of technology and engineering exist in Ref. [15–19]

The content of this book mainly comes from the articles published by my group. We add "Exercise and Solution" because we hope this book could be a monograph for experts to read and a textbook for newcomers to practice (so that they could engage in this new field as soon as possible). Incidentally, each chapter in the book has its symbols to facilitate reading. In this sense, to read this book, the reader may start with any chapter.

I am also grateful to my family members, especially my wife (Yan-Jiao Zhao) and my two daughters (Ji-Yan Huang with the nickname of Qian-Qian and Ji-Yang Huang with the nickname of Yue-Yue), for bringing me great happiness. Qian-Qian also helped polish Figs. 1.1, 1.2, 1.3 and 1.4 in this preface. I have stayed at home or the residential area due to COVID-19 between April 1, 2022 and May 31, 2022.

Fig. 1.5 Group photo: 2020 International Conference on Thermodynamics and Thermal Metamaterials, held on August 7–9, 2020, in Zoom (Online)

When writing this preface, I refer to my previous book Ref. [10].

Last, we acknowledge financial support from the National Natural Science Foundation of China under Grants No. 11725521 and No. 12035004 and the Science and Technology Commission of Shanghai Municipality under Grant No. 20JC1414700.

Shanghai, China   Ji-Ping Huang
June 17, 2022

References

1. Fan, C.Z., Gao, Y., Huang, J.P.: Shaped graded materials with an apparent negative thermal conductivity. Appl. Phys. Lett. **92**, 251907 (2008)
2. Chen, T.Y., Weng, C.N., Chen, J.S.: Cloak for curvilinearly anisotropic media in conduction. Appl. Phys. Lett. **93**, 114103 (2008)
3. Xu, H.Y., Shi, X.H., Gao, F., Sun, H.D., Zhang, B.L.: Ultrathin three-dimensional thermal cloak. Phys. Rev. Lett. **112**, 054301 (2014)

4. Han, T.C., Bai, X., Gao, D.L., Thong, J.T.L., Li, B.W., Qiu, C.-W.: Experimental demonstration of a bilayer thermal cloak. Phys. Rev. Lett. **112**, 054302 (2014)
5. Han, T., Bai, X., Thong, J.T.L., Li, B., Qiu, C.-W.: Full control and manipulation of heat signatures: cloaking, camouflage and thermal metamaterials. Adv. Mater. **26**, 1731–1734 (2014)
6. Dai, G.L., Shang, J., Huang, J.P.: Theory of transformation thermal convection for creeping flow in porous media: cloaking, concentrating, and camouflage. Phys. Rev. E **97**, 022129 (2018)
7. Pendry, J.B., Schurig, D., Smith, D.R.: Controlling electromagnetic fields. Science **312**, 1780–1782 (2006)
8. Yang, S., Wang, J., Dai, G.L., Yang, F.B., Huang, J.P.: Controlling macroscopic heat transfer with thermal metamaterials: theory, experiment and application. Phys. Rep. **908**, 1–65 (2021)
9. Maldovan, M.: Sound and heat revolutions in phononics. Nature **503**, 209–217 (2013)
10. Huang, J.-P.: Theoretical Thermotics: Transformation Thermotics and Extended Theories for Thermal Metamaterials. Springer, Singapore (2020)
11. Guenneau, S., Amra, C., Veynante, D.: Transformation thermodynamics: cloaking and concentrating heat flux. Opt. Express **20**, 8207–8218 (2012)
12. Narayana, S., Sato, Y.: Heat flux manipulation with engineered thermal materials. Phys. Rev. Lett. **108**, 214303 (2012)
13. Schittny, R., Kadic, M., Guenneau, S., Wegener, M.: Experiments on transformation thermodynamics: molding the flow of heat. Phys. Rev. Lett. **110**, 195901 (2013)
14. Yeung, W.-S., Yang, R.-J.: Introduction to Thermal Cloaking: Theory and Analysis in Conduction and Convection. Springer, Singapore (2022)
15. Dede, E.M., Schmalenberg, P., Nomura, T., Ishigaki, M.: Design of anisotropic thermal conductivity in multilayer printed circuit boards. IEEE Trans. Compon. Pack. Manuf. Technol. **5**, 1763–1774 (2015)
16. Dede, E.M., Zhou, F., Schmalenberg, P., Nomura, T.: Thermal metamaterials for heat flow control in electronics. J. Electron. Packag. **140**, 010904 (2018)
17. Kim, J.C., Ren, Z., Yuksel, A., Dede, E.M., Bandaru, P.R., Oh, D., Lee, J.: Recent advances in thermal metamaterials and their future applications for electronics packaging. J. Electron. Packag. **143**, 010801 (2021)
18. Zhai, Y., Ma, Y.G., David, S.N., Zhao, D.L., Lou, R.N., Tan, G., Yang, R.G., Yin, X.B.: Scalable-manufactured randomized glass-polymer hybrid metamaterial for daytime radiative cooling. Science **355**, 1062–1066 (2017)
19. Huang, J.P.: Technologies for Controlling Thermal Energy: Design, Simulation and Experiment based on Thermal Metamaterial Theories including Transformation Thermotics (in Chinese). Higher Education Press, Beijing (2022)
20. Li, Y., Li, W., Han, T., Zheng, X., Li, J., Li, B., Fan, S., Qiu, C.-W.: Transforming heat transfer with thermal metamaterials and devices. Nat. Rev. Mater. **6**, 488–507 (2021)
21. Ma, Y.G., Liu, Y.C., Raza, M., Wang, Y.D., He, S.L.: Experimental demonstration of a multiphysics cloak: manipulating heat flux and electric current simultaneously. Phys. Rev. Lett. **113**, 205501 (2014)
22. Leonhardt, U.: Cloaking of heat. Nature **498**, 440–441 (2013)
23. Wegener, M.: Metamaterials beyond optics. Science **342**, 939–940 (2013)
24. Ball, P.: Against the flow. Nat. Mater. **11**, 566–566 (2012)
25. Xu, L.J., Xu, G.Q., Huang, J.P., Qiu, C.-W.: Diffusive Fizeau drag in spatiotemporal thermal metamaterials. Phys. Rev. Lett. **128**, 145901 (2022)
26. Li, Y., Peng, Y.-G., Han, L., Miri, M.-A., Li, W., Xiao, M., Zhu, X.-F., Zhao, J., Alù, A., Fan, S., Qiu, C.-W.: Anti-parity-time symmetry in diffusive systems. Science **364**, 170–713 (2019)
27. Xu, L., Wang, J., Dai, G., Yang, S., Yang, F., Wang, G., Huang, J.: Geometric phase, effective conductivity enhancement, and invisibility cloak in thermal convection-conduction. Int. J. Heat Mass Transf. **165**, 120659 (2021)
28. Xu, G., Li, W., Zhou, X., Li, H., Li, Y., Fan, S., Zhang, S., Christodoulides, D.N., Qiu, C.-W.: Observation of Weyl exceptional rings in thermal diffusion. Proc. Natl. Acad. Sci. U.S.A. **119**, e2110018119 (2022)

Chapter 2
Introduction

Abstract In this chapter, we present the background and organization of this book.

Keywords Theoretical thermotics · Characteristic lengths · Metamaterials and beyond

2.1 Theoretical Thermotics

Theoretical thermotics originates from the theory of transformation thermotics [1, 2]. With the artificial heat regulation development, the connotation of theoretical thermotics has been greatly extended, not limited to those theories for designing thermal cloaks, concentrators, and rotators. Therefore, theoretical thermotics is the summarization of "transformation thermotics and extended theories". For clarity, we mainly divide theoretical thermotics into three levels according to the historical development.

The first level (LV1) is those transformation-related theories for designing cloaking, concentrating, rotating, etc. Since the theory of transformation thermotics was proposed for controlling steady and passive heat conduction in 2008 [1, 2], extended transformation theories have been developed successively from steady and passive to transient and active heat conduction [3]. Then, temperature-dependent (nonlinear) thermal conductivities were considered for developing nonlinear transformation thermotics [4]. These coordinate transformations were all time-independent, making it challenging to deal with time-dependent coordinate transformations. Thus, spatiotemporal coordinate transformations were discussed [5]. Beyond conduction, convection is also a primary heat transfer mode, so researchers developed a transformation theory for convection control [6]. Nevertheless, it was still challenging to guide convective velocities directly. Therefore, the Darcy law in porous media was introduced to transform convection and ensure feasibility [7, 8]. Another convective model with creeping flows was also explored [9]. The last basic heat transfer scheme is radiation, and researchers also proposed a transformation theory to regulate the radiation described by the Rosseland diffusion approximation [10]. With these efforts, conduction, convection, and radiation can be unified in the transfor-

© The Author(s) 2023
L.-J. Xu and J.-P. Huang, *Transformation Thermotics and Extended Theories*,
https://doi.org/10.1007/978-981-19-5908-0_2

mation framework [11]. Besides, heat transfer may also be accompanied by other physical processes, such as electric transport. Therefore, a transformation theory was put forward to regulate thermal and electric fields simultaneously [12]. Researchers further studied the coupling between thermal and electric fields, i.e., the thermo-electric effect, and proposed a transformation theory [13]. Therefore, most thermal phenomena can be manipulated by transformation theories.

The second level (LV2) is other theories for designing functions predicted by the transformation theory. Although the transformation theory is powerful, it still has some limitations. For example, the parameters for thermal cloaking should be anisotropic, inhomogeneous, and even singular. Thus, other theories beyond the transformation theory were proposed. We take thermal cloaking as an example. A bilayer scheme was proposed [14–16] to remove anisotropic and inhomogeneous parameters. Then, an active scheme was developed to remove all parametric require-ments because only active temperature control was required [17]. Furthermore, a dipole-based scheme was considered to simplify the active temperature control [18]. Besides these analytical theories, topology optimization is an indispensable method that largely simplifies the design [19, 20]. These theories and schemes are distinctly different from the transformation theory, but they are still applied to design functions predicted by the transformation theory.

The third level (LV3) is other theories for designing functions not predicted by the transformation theory. With the development of theoretical thermotics, many phenomena and functions beyond the predictions of transformation thermotics were revealed, such as the anti-parity-time symmetry in diffusive systems [21, 22], diffu-sive geometric phases [23, 24], thermal wave nonreciprocity [25–29], thermal edge states [30–34], and thermal skin effects [35, 36]. These emerging theories may guide the future development of theoretical thermotics.

2.2 Characteristic Length

Compared with traditional thermodynamics (A in Fig. 2.1), theoretical thermotics focuses on the active control of heat based on transformation thermotics and extended theories (B in Fig. 2.1). Since theoretical thermotics also designs artificial structures for heat regulation, what is the relation between theoretical thermotics and the emerg-ing field of metamaterials? The answer is the characteristic length.

Metamaterials generally refer to those artificial structures with a structural unit size (much) smaller than the characteristic length (C in Fig. 2.1). In this way, an artificial structure has novel parameters that do not exist in nature or chemical com-pounds according to effective media, such as negative permittivities. Electromagnetic metamaterials (C2 in Fig. 2.1) originate from the research on negative refractive index [37–39]. Then, the metamaterial research was extended to other wave systems (C3 in Fig. 2.1), such as acoustics [40, 41] and elastodynamics [42, 43]. In 2008, transformation thermotics and thermal cloaking were proposed [1, 2], extending the metamaterial physics from wave to diffusion systems (right part of C1 in Fig. 2.1).

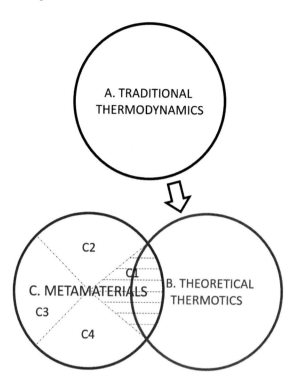

Fig. 2.1 Theoretical thermotics meets metamaterials. A: Focus on the passive description of heat based on the four laws of thermodynamics. B: Focus on the active control of heat based on transformation thermotics and extended theories. C: Artificial structures with characteristic lengths have novel properties. C1: Diffusive systems. Right area: thermotics (thermal metamaterials or metamaterial-based devices, which are interdisciplinary with physics, thermophysics, and materials science); left area: other diffusions (particle/plasma diffusion, etc.). C2: Wave systems (electromagnetic/optical waves). C3: Wave systems (acoustic/elastodynamic waves, etc.). C4: Systems other than diffusion and wave systems

Therefore, thermal metamaterials are an interdisciplinary product of metamaterials and theoretical thermotics [44, 45]. Certainly, characteristic lengths should be available to distinguish thermal metamaterials from other thermal materials. For heat conduction, the characteristic length is the thermal diffusion length $L = \sqrt{\kappa t/\rho C}$, where κ, t, ρ, and C are thermal conductivity, time, density, and heat capacity, respectively. The characteristic length for thermal convection is the geometric length of fluid migration. For thermal radiation, the characteristic length is the wavelength of electromagnetic waves. Beyond heat transfer, metamaterials were also studied in other diffusive systems, such as mass diffusion [46] and light diffusion [47] (left part of C1 in Fig. 2.1). Besides, there are also metamaterials beyond wave and diffusion (C4 in Fig. 2.1), such as origami metamaterials [48] and robotic metamaterials [49].

2.3 Book Organization

We divide this book into two parts according to characteristic lengths, i.e., inside and outside metamaterials. Those theories with a characteristic length (much) larger than the structural unit size belong to Part I (inside metamaterials). The others belong to Part II (outside metamaterials). See Fig. 2.2.

In Part I (inside metamaterials), we introduce fourteen theories, classified into three levels for logical clarity. We start from the transformation theory, the foundation of theoretical thermotics (LV1), for dealing with complex thermal materials (Chap. 3) and thermoelectric materials (Chap. 4). Although the transformation theory is powerful, the required complicated parameters make practical fabrications challenging. Therefore, we introduce other theories beyond the transformation theory (i.e., effective medium theory) but still design functions predicted by the transformation theory (LV2), such as cloaks (Chap. 5–7), concentrators (Chap. 8), rotators (Chap. 9), sensors (Chaps. 10–12), and metasurfaces (Chap. 13). Based on LV1 and LV2, we develop the wavelike diffusion theory for designing functions not predicted by the transformation theory (LV3), such as advectionlike behavior (Chap. 14), diffusive Fizeau drag (Chap. 15), and thermal refraction effect (Chap. 16).

In Part II (outside metamaterials), we propose six theories, starting from the invisibility function. Thermal invisibility can be realized with metamaterials designed by the transformation theory or the effective medium theory, see Part I. A natural question is whether it can achieve thermal invisibility without metamaterials. The first theory (active dipole theory, Chap. 17) in Part II answers this question by demonstrating that thermal invisibility can be realized by an active dipole (without metamaterials, LV2). Based on nonlinear thermal conductivity (Chap. 18) and complex thermal conductivity (Chap. 19), we then develop theories for achieving functions not predicted by the transformation theory (LV3). With these foundations, we explore the topology-related approach for uncovering three-port thermal nonre-

Fig. 2.2 Book framework

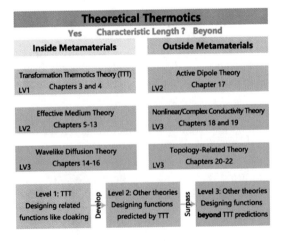

ciprocity (Chap. 20), thermal geometric phases (Chap. 21), and thermal edge states (Chap. 22).

Finally, we summarize this book and make an outlook in Chap. 23.

The research paradigms of theoretical thermotics can be extended to other diffusive systems. To provide more insights, we add an appendix to introduce other diffusion systems, including particle diffusion (Appendix A) and plasma diffusion (Appendix B).

References

1. Fan, C.Z., Gao, Y., Huang, J.P.: Shaped graded materials with an apparent negative thermal conductivity. Appl. Phys. Lett. **92**, 251907 (2008)
2. Chen, T.Y., Weng, C.-N., Chen, J.-S.: Cloak for curvilinearly anisotropic media in conduction. Appl. Phys. Lett. **93**, 114103 (2008)
3. Guenneau, S., Amra, C., Veynante, D.: Transformation thermodynamics: cloaking and concentrating heat flux. Opt. Express **20**, 8207 (2012)
4. Li, Y., Shen, X.Y., Wu, Z.H., Huang, J.Y., Chen, Y.X., Ni, Y.S., Huang, J.P.: Temperature-dependent transformation thermotics: from switchable thermal cloaks to macroscopic thermal diodes. Phys. Rev. Lett. **115**, 195503 (2015)
5. García-Meca, C., Barceló, C.: Dynamically tunable transformation thermodynamics. J. Opt. **18**, 044026 (2016)
6. Guenneau, S., Petiteau, D., Zerrad, M., Amra, C., Puvirajesinghe, T.: Transformed Fourier and Fick equations for the control of heat and mass diffusion. AIP Adv. **5**, 053404 (2015)
7. Dai, G.L., Shang, J., Huang, J.P.: Theory of transformation thermal convection for creeping flow in porous media: cloaking, concentrating, and camouflage. Phys. Rev. E **97**, 022129 (2018)
8. Dai, G.L., Huang, J.P.: A transient regime for transforming thermal convection: cloaking, concentrating, and rotating creeping flow and heat flux. J. Appl. Phys. **124**, 235103 (2018)
9. Wang, B., Shih, T.-M., Huang, J.P.: Transformation heat transfer and thermo-hydrodynamic cloaks for creeping flows: manipulating heat fluxes and fluid flows simultaneously. Appl. Therm. Eng. **190**, 116726 (2021)
10. Xu, L.J., Dai, G.L., Huang, J.P.: Transformation multithermotics: controlling radiation and conduction simultaneously. Phys. Rev. Appl. **13**, 024063 (2020)
11. Xu, L.J., Yang, S., Dai, G.L., Huang, J.P.: Transformation omnithermotics: simultaneous manipulation of three basic modes of heat transfer. ES Energy Environ. **7**, 65 (2020)
12. Li, J.Y., Gao, Y., Huang, J.P.: A bifunctional cloak using transformation media. J. Appl. Phys. **108**, 074504 (2010)
13. Stedman, T., Woods, L.M.: Cloaking of thermoelectric transport. Sci. Rep. **7**, 6988 (2017)
14. Xu, H.Y., Shi, X.H., Gao, F., Sun, H.D., Zhang, B.L.: Ultrathin three-dimensional thermal cloak. Phys. Rev. Lett. **112**, 054301 (2014)
15. Han, T.C., Bai, X., Gao, D.L., Thong, J.T.L., Li, B.W., Qiu, C.-W.: Experimental demonstration of a bilayer thermal cloak. Phys. Rev. Lett. **112**, 054302 (2014)
16. Ma, Y.G., Liu, Y.C., Raza, M., Wang, Y.D., He, S.L.: Experimental demonstration of a multiphysics cloak: Manipulating heat flux and electric current simultaneously. Phys. Rev. Lett. **113**, 205501 (2014)
17. Nguyen, D.M., Xu, H.Y., Zhang, Y.M., Zhang, B.L.: Active thermal cloak. Appl. Phys. Lett. **107**, 121901 (2015)
18. Xu, L.J., Yang, S., Huang, J.P.: Dipole-assisted thermotics: experimental demonstration of dipole-driven thermal invisibility. Phys. Rev. E **100**, 062108 (2019)
19. Dede, E.M., Nomura, T., Lee, J.: Thermal-composite design optimization for heat flux shielding, focusing, and reversal. Struct. Multidiscip. Optim. **49**, 59 (2014)

20. Fujii, G., Akimoto, Y., Takahashi, M.: Exploring optimal topology of thermal cloaks by CMA-ES. Appl. Phys. Lett. **112**, 061108 (2018)
21. Li, Y., Peng, Y.-G., Han, L., Miri, M.-A., Li, W., Xiao, M., Zhu, X.-F., Zhao, J.L., Alù, A., Fan, S.H., Qiu, C.-W.: Anti-parity-time symmetry in diffusive systems. Science **364**, 170 (2019)
22. Cao, P.C., Li, Y., Peng, Y.G., Qiu, C.W., Zhu, X.F.: High-order exceptional points in diffusive systems: robust APT symmetry against perturbation and phase oscillation at APT symmetry breaking. ES Energy Environ. **7**, 48 (2020)
23. Xu, L.J., Wang, J., Dai, G.L., Yang, S., Yang, F., Wang, G., Huang, J.P.: Geometric phase, effective conductivity enhancement, and invisibility cloak in thermal convection-conduction. Int. J. Heat Mass Transf. **165**, 120659 (2021)
24. Xu, L.J., Dai, G.L., Wang, G., Huang, J.P.: Geometric phase and bilayer cloak in macroscopic particle-diffusion systems. Phys. Rev. E **102**, 032140 (2020)
25. Xu, L.J., Huang, J.P., Ouyang, X.P.: Tunable thermal wave nonreciprocity by spatiotemporal modulation. Phys. Rev. E **103**, 032128 (2021)
26. Ordonez-Miranda, J., Guo, Y.Y., Alvarado-Gil, J.J., Volz, S., Nomura, M.: Thermal-wave diode. Phys. Rev. Appl. **16**, L041002 (2021)
27. Xu, L.J., Huang, J.P., Ouyang, X.P.: Nonreciprocity and isolation induced by an angular momentum bias in convection-diffusion systems. Appl. Phys. Lett. **118**, 221902 (2021)
28. Shimokusu, T.J., Zhu, Q., Rivera, N., Wehmeyer, G.: Time-periodic thermal rectification in heterojunction thermal diodes. Int. J. Heat Mass Transf. **182**, 122035 (2022)
29. Xu, L.J., Xu, G.Q., Huang, J.P., Qiu, C.-W.: Diffusive Fizeau drag in spatiotemporal thermal metamaterials. Phys. Rev. Lett. **128**, 145901 (2022)
30. Yoshida, T., Hatsugai, Y.: Bulk-edge correspondence of classical diffusion phenomena. Sci. Rep. **11**, 888 (2021)
31. Xu, L.J., Huang, J.P.: Robust one-way edge state in convection-diffusion systems. EPL **134**, 60001 (2021)
32. Xu, G.Q., Li, Y., Li, W., Fan, S.H., Qiu, C.-W.: Configurable phase transitions in a topological thermal material. Phys. Rev. Lett. **127**, 105901 (2021)
33. Makino, S., Fukui, T., Yoshida, T., Hatsugai, Y.: Edge states of a diffusion equation in one dimension: Rapid heat conduction to the heat bath. Phys. Rev. E **105**, 024137 (2022)
34. Xu, G.Q., Yang, Y.H., Zhou, X., Chen, H.S., Alù, A., Qiu, C.-W.: Diffusive topological transport in spatiotemporal thermal lattices. Nat. Phys. **18**, 450 (2022)
35. Cao, P.-C., Li, Y., Peng, Y.-G., Qi, M.H., Huang, W.-X., Li, P.-Q., Zhu, X.-F.: Diffusive skin effect and topological heat funneling. Commun. Phys. **4**, 230 (2021)
36. Cao, P.-C., Peng, Y.-G., Li, Y., Zhu, X.-F.: Phase-locking diffusive skin effect. Chin. Phys. Lett. **39**, 057801 (2022)
37. Veselago, V.G.: The electrodynamics of substances with simultaneously negative values of ϵ and μ. Sov. Phys. Usp. **10**, 509 (1968)
38. Pendry, J.B., Holden, A.J., Stewart, W.J., Youngs, I.: Extremely low frequency plasmons in metallic mesostructures. Phys. Rev. Lett. **76**, 4773 (1996)
39. Pendry, J.B., Holden, A.J., Robbins, D.J., Stewart, W.J.: Magnetism from conductors and enhanced nonlinear phenomena. IEEE Trans. Microw. Theory Tech. **47**, 2075 (1999)
40. Liu, Z.Y., Zhang, X.X., Mao, Y.W., Zhu, Y.Y., Yang, Z.Y., Chan, C.T., Sheng, P.: Locally resonant sonic materials. Science **289**, 1734 (2000)
41. Cummer, S.A., Schurig, D.: One path to acoustic cloaking. New J. Phys. **9**, 45 (2007)
42. Farhat, M., Guenneau, S., Enoch, S.: Ultrabroadband elastic cloaking in thin plates. Phys. Rev. Lett. **103**, 024301 (2009)
43. Stenger, N., Wilhelm, M., Wegener, M.: Experiments on elastic cloaking in thin plates. Phys. Rev. Lett. **108**, 014301 (2012)
44. Narayana, S., Sato, Y.: Heat flux manipulation with engineered thermal materials. Phys. Rev. Lett. **108**, 214303 (2012)
45. Maldovan, M.: Sound and heat revolutions in phononics. Nature **503**, 209 (2013)
46. Guenneau, S., Puvirajesinghe, T.M.: Fick's second law transformed: one path to cloaking in mass diffusion. J. R. Soc. Interface **10**, 20130106 (2013)

47. Schittny, R., Kadic, M., Bückmann, T., Wegener, M.: Invisibility cloaking in a diffusive light scattering medium. Science **345**, 427 (2014)
48. Silverberg, J.L., Evans, A.A., McLeod, L., Hayward, R.C., Hull, T., Santangelo, C.D., Cohen, I.: Using origami design principles to fold reprogrammable mechanical metamaterials. Science **345**, 647 (2014)
49. Brandenbourger, M., Locsin, X., Lerner, E., Coulais, C.: Non-reciprocal robotic metamaterials. Nat. Commun. **10**, 4608 (2019)

Part I
Inside Metamaterials

Chapter 3
Theory for Thermal Wave Control: Transformation Complex Thermotics

Abstract In this chapter, we develop a transformation theory for controlling wave-like temperature fields (called thermal waves herein) in conduction and advection. We first unify these two basic heat transfer modes by coining a complex thermal conductivity whose real and imaginary parts are related to conduction and advection. Consequently, the conduction-advection process supporting thermal waves is described by a complex conduction equation, thus called complex thermotics. We then propose the principle for transforming complex thermal conductivities. We further design three metamaterials to control thermal waves with cloaking, concentrating, and rotating functions. Experimental suggestions are also provided based on porous media.

Keywords Transformation complex thermotics · Thermal waves · Porous media

3.1 Opening Remarks

Conduction and advection are ubiquitous, with crucial parameters of thermal conductivities and advection velocities, respectively. Therefore, these two heat transfer modes are generally considered independent, challenging their simultaneous manipulation. Recently, transformation theories have been proposed to control conduction and advection simultaneously, yielding practical applications such as cloaking, concentrating, and rotating [1–3]. These theories apply to constant-temperature boundary conditions but are not necessarily appropriate for periodic boundary conditions supporting thermal waves.

To solve the problem, we resort to a complex thermal conductivity $\kappa = \sigma + i\tau$, where σ and τ are two real numbers [4]. The κ can be well understood with the complex plane shown in Fig. 3.1. We consider thermal waves with rightward advection velocities. The temperature profiles in the right ($\sigma > 0$) and left ($\sigma < 0$) half-planes have loss and gain of heat energy, respectively. The motion of the temperature profiles in the upper ($\tau > 0$) and lower ($\tau < 0$) half-planes is rightward and leftward, respectively. Therefore, the conduction-advection process supporting thermal waves can be described by a complex conduction equation, thus called complex thermotics

© The Author(s) 2023
L.-J. Xu and J.-P. Huang, *Transformation Thermotics and Extended Theories*,
https://doi.org/10.1007/978-981-19-5908-0_3

Fig. 3.1 Connotation of the complex thermal conductivity $\kappa = \sigma + i\tau$. Red curves denote thermal waves. Advection velocities are rightward. The arrow in the center of each temperature profile indicates loss or gain. Adapted from Ref. [6]

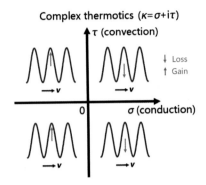

herein. In other words, advection can be regarded as a complex form of conduction. Transforming complex materials was also realized in wave systems with a similar idea [5], where gain/loss leads to non-Hermitian dielectrics.

We further study the complex conduction equation and propose the theory of transformation complex thermotics, linking spatial transformations and material transformations. We first prove the form-invariance of the complex conduction equation under coordinate transformations and derive the principle for transforming complex thermal conductivities. The present theory further allows us to cloak, concentrate, and rotate thermal waves as three model applications. Specifically, cloaking can hide an obstacle without distorting the thermal waves in the background; concentrating can enhance the density of thermal waves; rotating can control the direction of thermal waves. We further provide experimental suggestions based on porous media whose effective parameters can be calculated by weighted average.

3.2   Theoretical Foundation

Complex thermotics can be described by a complex conduction equation,

$$\rho C \frac{\partial T}{\partial t} + \nabla \cdot (-\kappa \nabla T) = 0, \tag{3.1}$$

where ρ, C, κ, T, and t are density, heat capacity, complex thermal conductivity, temperature, and time, respectively. The complex thermal conductivity κ can be expressed as [4]

$$\kappa = \sigma + i\tau = \sigma + i\frac{\rho C v \cdot \beta}{\beta^2}, \tag{3.2}$$

where v is advection velocity, and β is wave vector. By applying a wavelike temperature field [7, 8] described by $T = A_0 e^{i(\beta \cdot r - \omega t)} + T_0$, we can derive the dispersion relation of complex thermotics,

$$\omega = \boldsymbol{v} \cdot \boldsymbol{\beta} - \mathrm{i}\frac{\sigma \beta^2}{\rho C}, \tag{3.3}$$

where A_0, $\boldsymbol{r}$, ω, and T_0 are the amplitude, position vector, angular frequency, and reference temperature of the wavelike temperature field, respectively. The wavelike temperature field can also be called a thermal wave because it mathematically corresponds to a plane wave. Note that thermal waves herein have a distinct mechanism from those thermal-relaxation-related heat waves [9, 10]. Equation (3.2) is a mathematical skill to unify conduction and advection. Due to the feature of thermal waves (say, $\nabla T = \mathrm{i}\boldsymbol{\beta}T$), we can derive $\mathrm{i}\tau \cdot \nabla T = -\tau \boldsymbol{\beta}T$ which just corresponds to an advection term.

We then prove that the complex conduction equation (Eq. (3.1)) is form-invariant under the spatial transformation from a curvilinear space X to a physical space X'. For this purpose, we rewrite Eq. (3.1) as

$$\rho C \frac{\partial T}{\partial t} + \nabla \cdot (-\sigma \nabla T) + \nabla \cdot (\tau \boldsymbol{\beta}T) = 0. \tag{3.4}$$

We suppose $\boldsymbol{u} = \tau \boldsymbol{\beta}$ and write down the component form of Eq. (3.4) in the curvilinear space with a contravariant basis $\left(\boldsymbol{g}^1, \boldsymbol{g}^2, \boldsymbol{g}^3\right)$ and contravariant components $\left(x^1, x^2, x^3\right)$,

$$\sqrt{g}\rho C \partial_t T + \partial_j \left(-\sqrt{g}\sigma^{jk}\partial_k T\right) + \partial_j \left(\sqrt{g}u^j T\right) = 0, \tag{3.5}$$

where g is the determinant of the matrix $\boldsymbol{g}_j \cdot \boldsymbol{g}_k$ with $(\boldsymbol{g}_1, \boldsymbol{g}_2, \boldsymbol{g}_3)$ being a covariant basis, and j (or k) takes 1, 2 or 3. Equation (3.5) is expressed in the curvilinear space, and then we rewrite it in the physical space with Cartesian coordinates $\left(x^{1'}, x^{2'}, x^{3'}\right)$,

$$\sqrt{g}\rho C \partial_t T + \partial_{j'} \frac{\partial x^{j'}}{\partial x^j} \left(-\sqrt{g}\sigma^{jk}\frac{\partial x^{k'}}{\partial x^k}\partial_{k'} T\right) + \partial_{j'} \frac{\partial x^{j'}}{\partial x^j} \left(\sqrt{g}u^j T\right) = 0, \tag{3.6}$$

where $\partial x^{j'}/\partial x^j$ and $\partial x^{k'}/\partial x^k$ are the components of the Jacobian transformation matrix $\tilde{J}$, and $\sqrt{g} = 1/\det \tilde{J}$. We turn the spatial transformation into the transformation of materials or vectors, so Eq. (3.6) becomes

$$\frac{\rho C}{\det \tilde{J}} \partial_t T + \partial_{j'} \left(-\frac{\frac{\partial x^{j'}}{\partial x^j}\sigma^{jk}\frac{\partial x^{k'}}{\partial x^k}}{\det \tilde{J}}\partial_{k'} T\right) + \partial_{j'} \left(\frac{\frac{\partial x^{j'}}{\partial x^j}u^j}{\det \tilde{J}}T\right) = 0. \tag{3.7}$$

The transformation rule can be derived,

$$(\rho C)' = \frac{\rho C}{\det \tilde{J}}, \tag{3.8a}$$

$$\sigma' = \frac{\tilde{J}\sigma\tilde{J}^\dagger}{\det \tilde{J}}, \tag{3.8b}$$

$$u' = \frac{\tilde{J}u}{\det \tilde{J}}, \tag{3.8c}$$

where $\tilde{J}^\dagger$ represents the transpose of $\tilde{J}$. Since $u = \tau\beta$, Eq. (3.8c) becomes

$$(\tau\beta)' = \frac{\tilde{J}(\tau\beta)}{\det \tilde{J}}. \tag{3.9}$$

We do not transform the wave vector, i.e., $\beta' = \beta$, so Eq. (3.9) turns into

$$\tau' = \frac{\tilde{J}\tau}{\det \tilde{J}}. \tag{3.10}$$

Therefore, the principle for transforming complex thermal conductivities can be summarized as

$$(\rho C)' = \frac{\rho C}{\det \tilde{J}}, \tag{3.11a}$$

$$\sigma' = \frac{\tilde{J}\sigma\tilde{J}^\dagger}{\det \tilde{J}}, \tag{3.11b}$$

$$\tau' = \frac{\tilde{J}\tau}{\det \tilde{J}}. \tag{3.11c}$$

Equation (3.11) is the first key result, acting as the foundation of transformation complex thermotics. Physically, Eqs. (3.11a) and (3.11b) agree with the result given by the theory of transformation thermotics for conduction [11, 12]. A crucial point is to show that Eq. (3.11c) does not violate physical laws either. For this purpose, we substitute the expression of τ (Eq. (3.2)) into Eq. (3.11c), thus yielding

$$\left(\frac{\rho C v \cdot \beta}{\beta^2}\right)' = \frac{\tilde{J}\left(\frac{\rho C v \cdot \beta}{\beta^2}\right)}{\det \tilde{J}}. \tag{3.12}$$

With Eq. (3.11a) and $\beta' = \beta$, Eq. (3.12) can be reduced to

$$v' = \tilde{J}v, \tag{3.13}$$

which also agrees with the theory of transformation thermotics for advection [1–3]. Therefore, we may briefly summarize two conclusions: (I) complex thermotics indicates that the real and imaginary parts of a complex thermal conductivity (Eq. (3.2)) are related to conduction (featuring dissipation) and advection (featuring propagation), respectively; and (II) the governing equation of complex thermotics (Eq. (3.1)) is form-invariant under coordinate transformations.

3.3 Model Application

The form-invariance of the complex conduction equation (Eq. (3.1)) allows us to cloak, concentrate, and rotate thermal waves. A schematic diagram of cloaking is shown in Fig. 3.2a. The left and right ends are set with periodic boundary conditions, say, $T_L = T_R$. The upper and lower boundaries are insulated. We consider the case with $\boldsymbol{v}//\boldsymbol{\beta}$ where the imaginary part of κ appears, as calculated by Eq. (3.2). We take on the wave vector $\beta = 2\pi m/W$ with $m = 10$, and the time period of the thermal wave is $t_0 = 20$ s according to Eq. (3.3). We set the initial wavelike temperature field as $T = 40 \sin(\beta x) + 323$ K (Fig. 3.2b). When there is an obstacle without motion in the center, the thermal wave is distorted (Fig. 3.2c and d). Different from the schemes with analytical design [13–15] and topological optimization [16–19], we apply the present theory of transformation complex thermotics to design thermal cloaking. The coordinate transformation can be expressed as $r = ar' + b$ and $\theta = \theta'$, where (r, θ) denote cylindrical coordinates in the physical space, $a = (r_2 - r_1)/r_2$, and $b = r_1$. Here, r_1 and r_2 are the inner and outer radii of the shell, respectively. The Jacobian transformation matrix $\tilde{J}$ can be calculated as $\tilde{J} = \text{diag}[a, ar/(r-b)]$. We design the cloak according to Eq. (3.11). The initial wavelike temperature field in the cloak turns into $T = 40 \sin\{\beta[(r-b)x/(ar)]\} + 323$ K (Fig. 3.2e). Here, the wave vector β is not transformed indeed, and only the coordinate x becomes $(r-b)x/(ar)$. The obstacle does not distort the thermal wave in the background, so the cloaking effect is achieved (Fig. 3.2f and g). Since the dispersion relation (Eq. (3.3)) indicates that the decay rate $(-\text{Im}(\omega))$ is in direct proportion to thermal conductivity, the temperature of the obstacle (with a high thermal conductivity of $120 \,\text{W m}^{-1} \,\text{K}^{-1}$) decays quickly and becomes a constant. Meanwhile, the thermal wave has energy loss due to the positive real part of κ, and propagates rightwards along x axis due to the positive imaginary part of κ. After propagating for one period (20 s), the thermal wave approximately gains a phase difference of 2π, thus going back to the initial position (Fig. 3.2e and g).

With the similar method for cloaking, we can also design concentrating and rotating. The transformation of concentrating is $r = cr'$ for $0 < r' < r_m, r = dr' + f$ for $r_m < r' < r_2$, and $\theta = \theta'$. Here, $c = r_1/r_m$, $d = (r_2 - r_1)/(r_2 - r_m)$, $f = (r_1 - r_m)r_2/(r_2 - r_m)$, and r_m is an intermediate radius between r_1 and r_2. The concentrating effect is determined by the parameter $1/c = r_m/r_1$, whose maximum value is r_2/r_1. Therefore, increasing the value of r_m can enhance the thermal gradient inside the concentrator. We can derive the Jacobian matrix $\tilde{J}$ in the core as $\tilde{J} = \text{diag}[c, c]$, and that for the shell as $\tilde{J} = \text{diag}[d, dr/(r-f)]$. The initial wavelike temperature field in the core turns into $T = 40 \sin[\beta(x/c)] + 323$ K, and that in the shell becomes $T = 40 \sin\{\beta[(r-f)x/(dr)]\} + 323$ K (Fig. 3.3a). The thermal waves

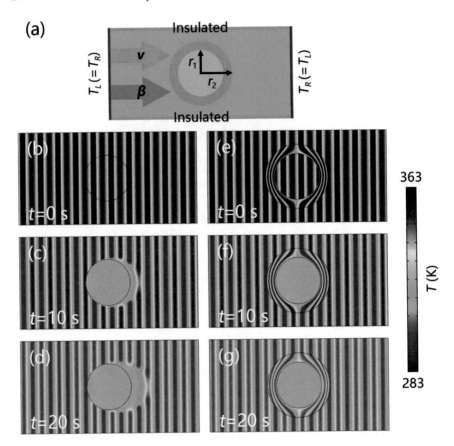

Fig. 3.2 a Schematic diagram of cloaking. **b–d** Simulations with an obstacle in the center. **e–g** Simulations with an obstacle coated by a cloak. Background parameters: $W = 20$ cm, $H = 10$ cm, $\rho = 1000$ kg/m^3, $C = 4200$ J kg^{-1} K^{-1}, $\sigma = 0.6$ W m^{-1} K^{-1}, and $v = 0.1$ cm/s. The obstacle is without motion, and has only a different parameter of $\sigma = 120$ W m^{-1} K^{-1} from background parameters. Cloaking parameters: the product of density and heat capacity is $\rho C (r - b)/(a^2 r)$; the real part of the complex thermal conductivity is diag$[(r - b)\sigma/r, r\sigma/(r - b)]$; and the velocity is $v[a\cos\theta, -ar\sin\theta/(r - b)]^\dagger$ with $r_1 = 2.5$ cm, $r_2 = 3.5$ cm, $a = 2/7$, and $b = 2.5$ cm. Adapted from Ref. [6]

at $t = 10$ s and $t = 20$ s are shown in Fig. 3.3b and c, respectively. The thermal wave in the center is concentrated indeed.

The transformation of rotating is $r = r'$, $\theta = \theta' + \theta_0$ for $0 < r' < r_1$, and $\theta = \theta' + h(r - r_2)$ for $r_1 < r' < r_2$. Here, $h = \theta_0/(r_1 - r_2)$, and θ_0 is rotating angle. We can derive the Jacobian matrix in the core as $\tilde{J} = $ diag$[1, 1]$, and that in the shell as $\tilde{J} = [(1, 0), (hr, 1)]$. The initial wavelike temperature field in the core turns into $T = 40\sin[\beta(x\cos\theta_0 + y\sin\theta_0)] + 323$ K, and that in the shell turns into $T = 40\sin\{\beta\{x\cos[h(r - r_2)] + y\sin[h(r - r_2)]\}\} + 323$ K (Fig. 3.3d). The thermal waves at $t = 10$ s and $t = 20$ s are shown in Fig. 3.3e and f, respectively. We

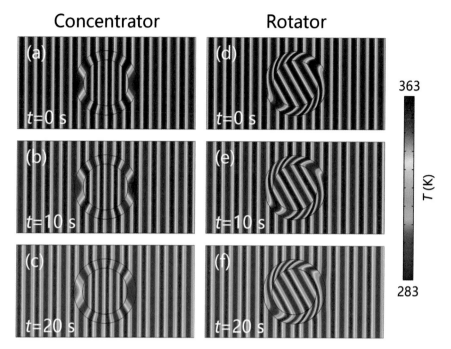

Fig. 3.3 Simulations of **a–c** concentrating and **d–f** rotating. The system sizes (W, H, r_1, and r_2) and background parameters (ρ, C, σ, and v) are the same as those for Fig. 3.2. Core parameters in **a–c**: $\rho C/c^2$, σ, and cv. Shell parameters in **a–c**: $\rho C \left(r - f\right)/\left(d^2 r\right)$, $\mathrm{diag}\left[(r - f)\sigma/r, r\sigma/(r - f)\right]$, and $v\left[d\cos\theta, -dr\sin\theta/(r - f)\right]^\dagger$ with $r_m = 3.2$ cm, $c = 25/32$, $d = 10/3$, and $f = -49/6$ cm. Core parameters in **d–f**: ρC, σ, and $v\left[\cos\theta_0, \sin\theta_0\right]^\dagger$. Shell parameters in **d–f**: ρC, $\sigma\left[(1, hr), \left(hr, h^2 r^2 + 1\right)\right]$, and $v\left[\cos\theta, hr\cos\theta - \sin\theta\right]^\dagger$ with $\theta_0 = \pi/6$ rad and $h = -\pi/6$ rad/cm. Adapted from Ref. [6]

can observe that the direction of thermal wave in the center is rotated by $\theta_0 = \pi/6$ anticlockwise.

Here, we only apply a single coordinate transformation to realize a single function. If one combines different coordinate transformations, it is possible to design devices with functions of cloaking-rotating [20] or concentrating-rotating [21]. Certainly, model applications are not limited to the above three devices, and many other applications can also be expected, such as thermal camouflage.

3.4 Experimental Suggestion

The transformation of τ (Eq. (3.11c)) is related to the transformation of v (Eq. (3.13)), which is mathematically easy but experimentally difficult. Meanwhile, we should transform the density and heat capacity of moving media, which is also experimen-

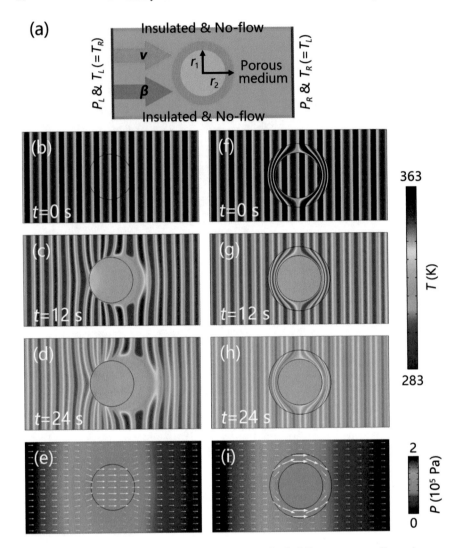

Fig. 3.4 a Schematic diagram of cloaking in porous media. **b–d** Temperature profiles and **e** pressure distribution with an obstacle located in the center. **f–h** Temperature profiles and **i** pressure distribution with the same obstacle coated by a cloak. White arrows in **e** and **i** denote advection velocities. $P_L = 2 \times 10^5$ Pa and $P_R = 0$ Pa. The fluid is still water with $\rho_f = 1000$ kg/m^3, $C_f = 4200$ J kg^{-1} K^{-1}, $\sigma_f = 0.6$ W m^{-1} K^{-1}, and $\xi = 10^{-3}$ Pa s. The background solid is stone with parameters $\rho_s = 4000$ kg/m^3, $C_s = 840$ J kg^{-1} K^{-1}, $\sigma_s = 2$ W m^{-1} K^{-1}, $\eta = 10^{-12}$ m^2, and $\phi = 0.8$. The obstacle has only different parameters of $\sigma = 400$ W m^{-1} K^{-1} and $\eta = 2 \times 10^{-10}$ m^2 from the background solid. The parameters in the shell are transformed as Eq. (3.17). Adapted from Ref. [6]

tally difficult. Fortunately, many fluid models can help [22–31]. Here, we utilize porous media [22] to proceed. Then, we should extend transformation complex thermotics from pure materials to composite materials. The porous medium is composed of solid and fluid with solid porosity of ϕ. We denote the density and heat capacity of the solid (or fluid) as ρ_s (or ρ_f) and C_s (or C_f), respectively. The effective density (ρ) and heat capacity (C) of the porous medium can be derived from the weighted average of the solid and fluid, say, $\rho C = \phi \rho_f C_f + (1 - \phi) \rho_s C_s$. Similar to Eq. (3.2), the complex thermal conductivities of the solid and fluid can be expressed as

$$\kappa_s = \sigma_s + i\tau_s = \sigma_s + i\frac{\rho_s C_s \boldsymbol{v}_s \cdot \boldsymbol{\beta}}{\beta^2}, \tag{3.14a}$$

$$\kappa_f = \sigma_f + i\tau_f = \sigma_f + i\frac{\rho_f C_f \boldsymbol{v}_f \cdot \boldsymbol{\beta}}{\beta^2}, \tag{3.14b}$$

where $\boldsymbol{v}_s$ and $\boldsymbol{v}_f$ are the velocities of the solid and fluid, respectively. The imaginary part of Eq. (3.14a) generally vanishes ($\tau_s = 0$) when the solid does not move ($\boldsymbol{v}_s = 0$). It is reasonable to handle the real parts of Eq. (3.14) with the method of weighted average, thus yielding the real part of the effective complex thermal conductivity as $\sigma = \phi \sigma_f + (1 - \phi) \sigma_s$ [32]. The next question is how to handle the imaginary parts of Eq. (3.14). We know that the imaginary part τ of the effective complex thermal conductivity is related to propagation, which has vector property to some extent. Therefore, it is also physical to use the method of weighted average to derive the effective imaginary part, say, $\tau = \phi \tau_f + (1 - \phi) \tau_s$. Therefore, the effective complex thermal conductivity κ of the porous medium can be expressed as

$$\kappa = \sigma + i\tau = \phi \sigma_f + (1 - \phi) \sigma_s + i\left[\phi \tau_f + (1 - \phi) \tau_s\right] \equiv \phi \kappa_f + (1 - \phi) \kappa_s. \tag{3.15}$$

Equation (3.15) is the second key result, describing the effective complex thermal conductivity of composite materials. By substituting Eq. (3.15) into Eq. (3.1), we can obtain the dispersion relation in porous media,

$$\omega = \frac{\phi \rho_f C_f}{\rho C} \boldsymbol{v}_f \cdot \boldsymbol{\beta} + \frac{(1 - \phi) \rho_s C_s}{\rho C} \boldsymbol{v}_s \cdot \boldsymbol{\beta} - i\frac{\sigma \beta^2}{\rho C}. \tag{3.16}$$

When $\phi = 1$, the porous medium becomes pure fluid, and Eq. (3.16) is reduced to Eq. (3.3) naturally.

With the understanding of Eq. (3.15), we can still use the result of Eq. (3.11), but it is not enough. We should consider the Darcy law and mass conservation. The Darcy law indicates that the origin of advection velocity is pressure difference, say, $\boldsymbol{v} = -(\eta/\xi) \nabla P$ where η is permeability, ξ is dynamic viscosity, and P denotes pressure. Since the pressure field is stable, density does not change with time and mass conservation is satisfied naturally. With these two physical conditions, we can obtain the transformation rule in porous media,

$$\left(\rho_f C_f\right)' = \rho_f C_f, \tag{3.17a}$$

$$\sigma'_f = \sigma_f, \tag{3.17b}$$

$$(\rho_s C_s)' = \frac{(\rho C)' - \phi \rho_f C_f}{1 - \phi}, \tag{3.17c}$$

$$\sigma'_s = \frac{\sigma' - \phi \sigma_f}{1 - \phi}, \tag{3.17d}$$

$$\eta' = \frac{\tilde{J}\eta\tilde{J}^\dagger}{\det \tilde{J}}, \tag{3.17e}$$

where $(\rho C)'$ and σ' are given by Eqs. (3.11a) and (3.11b), respectively. Equation (3.17) is the third key result, revealing the theory of transformation complex thermotics in porous media. We only transform the parameters of solids and avoid transforming advection velocities and moving fluids directly. Therefore, the physical problems for experiments have been solved, and the remaining problems are to find practical materials with anisotropic and inhomogeneous thermal conductivities and permeabilities, which have been widely studied based on multilayered structures [33–42].

Figure 3.4a shows the schematic diagram of our experimental suggestion. We use two modules: the heat transfer in porous media and the Darcy law. The left and right boundaries are also set at high pressure (P_L) and low pressure (P_R). We take the wave vector $\beta = 2\pi m / W$ with $m = 10$, and the time period of the thermal wave is $t_0 = 24$ s according to Eq. (3.16) with $v_s = 0$. The initial wavelike temperature field (Fig. 3.4b and f) are the same as those in Fig. 3.2b and e. If there does not exist a cloak coating the obstacle, the thermal wave (Fig. 3.4c and d) and the pressure field (Fig. 3.4e) are strongly distorted. In contrast, a cloak can avoid the distortion of the thermal wave (Fig. 3.4g and h) and the pressure field (Fig. 3.4i). The thermal wave in Fig. 3.4h also has energy loss because of the positive real part of κ. After propagating for one period (24 s), the thermal wave in Fig. 3.4h approximately gains a phase difference of 2π, thus being at the same position as Fig. 3.4f. We also provide experimental suggestions for concentrating and rotating, whose parameters are designed according to Eq. (3.17). The simulation results are shown in Fig. 3.5a–d and e–h, respectively. The concentrating and rotating effects are achieved indeed with a porous media. Therefore, the predictions of Eqs. (3.15)–(3.17) are physical, confirming the validity of transformation complex thermotics in composite materials.

3.5   Conclusion

We have coined a complex thermal conductivity κ and a complex conduction equation (say, complex thermotics) to unify conduction and advection. The real and imaginary parts of κ correspond to conduction and advection, respectively. We have also

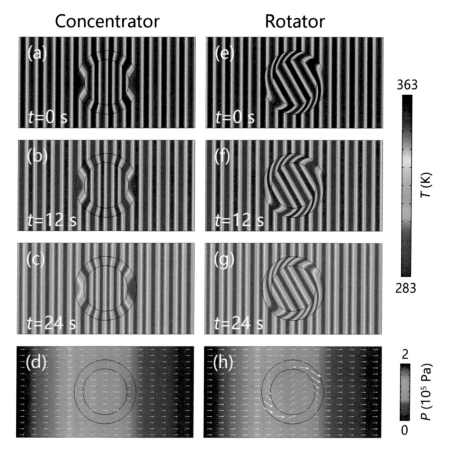

Fig. 3.5 Simulations of **a–d** concentrating and **e–g** rotating in porous media. The system sizes and background parameters are the same as those for Fig. 3.4. Other parameters are designed with Eq. (3.17). Adapted from Ref. [6]

proved the form-invariance of the complex conduction equation under coordinate transformations and derived the principle for transforming complex thermal conductivities. The current theory allows us to control thermal waves flexibly. Three practical devices have been designed with cloaking, concentrating, and rotating functions. Experimental suggestions are also provided, with the method of weighted average to derive the effective complex thermal conductivities of composite materials such as porous media.

3.6   Exercise and Solution

Exercise

1. Prove that a complex conduction equation can also describe the conduction-advection process in porous media.

Solution

1. The conductive energy density E_1 is

$$E_1 = -\sigma \nabla T. \tag{3.18}$$

The advection energy density induced by moving fluids E_2 is

$$E_2 = \phi \rho_f C_f \boldsymbol{v}_f T, \tag{3.19}$$

and that induced by moving solids E_3 is

$$E_3 = (1 - \phi) \rho_s C_s \boldsymbol{v}_s T. \tag{3.20}$$

Therefore, the total energy E that flows into the closed surface Σ from time t_1 to t_2 is

$$E = -\int_{t_1}^{t_2} \oiint_{\Sigma} \boldsymbol{n} \cdot (E_1 + E_2 + E_3) \, dSdt = -\int_{t_1}^{t_2} \iiint_{\Omega} \nabla \cdot (E_1 + E_2 + E_3) \, dVdt, \tag{3.21}$$

where Ω is the region enclosed by the surface Σ, $\boldsymbol{n}$ is unit normal vector, dS is the surface element, and dV is the volume element.

On the other hand, the absorbed energy E' can also be derived from the thermodynamic formula

$$E' = \iiint_{\Omega} [\rho C T (t_2) - \rho C T (t_1)] \, dV = \int_{t_1}^{t_2} \iiint_{\Omega} \rho C \left(\frac{\partial T}{\partial t}\right) dVdt. \tag{3.22}$$

According to the law of energy conservation, there must be $E = E'$, and we can derive the energy equation of the conduction-convection process in a porous medium as

$$\rho C \frac{\partial T}{\partial t} + \nabla \cdot \left[-\sigma \nabla T + \phi \rho_f C_f \boldsymbol{v}_f T + (1 - \phi) \rho_s C_s \boldsymbol{v}_s T\right] = 0. \tag{3.23}$$

With the effective complex thermal conductivity of the porous media (Eq. (3.15)), Eq. (3.23) can be reduced to

$$\rho C \frac{\partial T}{\partial t} + \nabla \cdot (-\kappa \nabla T) = 0. \tag{3.24}$$

Therefore, the conduction-advection process in porous media can be still described by a complex conduction equation where ρC and κ are the weighted average of solids and fluids.

References

1. Guenneau, S., Petiteau, D., Zerrad, M., Amra, C., Puvirajesinghe, T.: Transformed Fourier and Fick equations for the control of heat and mass diffusion. AIP Adv. **5**, 053404 (2015)
2. Dai, G.L., Shang, J., Huang, J.P.: Theory of transformation thermal convection for creeping flow in porous media: cloaking, concentrating, and camouflage. Phys. Rev. E **97**, 022129 (2018)
3. Xu, L.J., Yang, S., Dai, G.L., Huang, J.P.: Transformation omnithermotics: simultaneous manipulation of three basic modes of heat transfer. ES Energy Environ. **7**, 65 (2020)
4. Xu, L.J., Huang, J.P.: Negative thermal transport in conduction and convection. Chin. Phys. Lett. **37**, 080502 (2020)
5. Krešiá, I., Makris, K.G., Leonhardt, U., Rotter, S.: Transforming space with non-Hermitian dielectrics. Phys. Rev. Lett. **128**, 183901 (2022)
6. Xu, L.J., Huang, J.P.: Controlling thermal waves with transformation complex thermotics. Int. J. Heat Mass Transf. **159**, 120133 (2020)
7. Li, Y., Peng, Y.-G., Han, L., Miri, M.-A., Li, W., Xiao, M., Zhu, X.-F., Zhao, J.L., Alù, A., Fan, S.H., Qiu, C.-W.: Anti-parity-time symmetry in diffusive systems. Science **364**, 170 (2019)
8. Cao, P.C., Li, Y., Peng, Y.G., Qiu, C.-W., Zhu, X.-F.: High-Order exceptional points in diffusive systems: robust APT symmetry against perturbation and phase oscillation at APT symmetry breaking. ES Energy Environ. **7**, 48 (2020)
9. Farhat, M., Chen, P.-Y., Bagci, H., Amra, C., Guenneau, S., Alù, A.: Thermal invisibility based on scattering cancellation and mantle cloaking. Sci. Rep. **5**, 9876 (2015)
10. Farhat, M., Guenneau, S., Chen, P.-Y., Alù, A., Salama, K.N.: Scattering cancellation-based cloaking for the Maxwell-Cattaneo heat waves. Phys. Rev. Appl. **11**, 044089 (2019)
11. Fan, C.Z., Gao, Y., Huang, J.P.: Shaped graded materials with an apparent negative thermal conductivity. Appl. Phys. Lett. **92**, 251907 (2008)
12. Chen, T.Y., Weng, C.-N., Chen, J.-S.: Cloak for curvilinearly anisotropic media in conduction. Appl. Phys. Lett. **93**, 114109 (2008)
13. Liu, Y.X., Guo, W.L., Han, T.C.: Arbitrarily polygonal transient thermal cloaks with natural bulk materials in bilayer configurations. Int. J. Heat Mass Transf. **115**, 1 (2017)
14. Guo, J., Qu, Z.G.: Thermal cloak with adaptive heat source to proactively manipulate temperature field in heat conduction process. Int. J. Heat Mass Transf. **127**, 1212 (2018)
15. Qin, J., Luo, W., Yang, P., Wang, B., Deng, T., Han, T.C.: Experimental demonstration of irregular thermal carpet cloaks with natural bulk material. Int. J. Heat Mass Transf. **141**, 487 (2019)
16. Dede, E., Nomura, T., Lee, J.: Thermal-composite design optimization for heat flux shielding, focusing, and reversal. Struct. Multidiscip. Optim. **49**, 59 (2014)
17. Fujii, G., Akimoto, Y., Takahashi, M.: Exploring optimal topology of thermal cloaks by CMA-ES. Appl. Phys. Lett. **112**, 061108 (2018)
18. Fujii, G., Akimoto, Y.: Topology-optimized thermal carpet cloak expressed by an immersed boundary level-set method via a covariance matrix adaptation evolution strategy. Int. J. Heat Mass Transf. **137**, 1312 (2019)
19. Fujii, G., Akimoto, Y.: Optimizing the structural topology of bifunctional invisible cloak manipulating heat flux and direct current. Appl. Phys. Lett. **115**, 174101 (2019)
20. Zhou, L.L., Huang, S.Y., Wang, M., Hu, R., Luo, X.B.: While rotating while cloaking. Phys. Lett. A **383**, 759 (2019)
21. Tsai, Y.-L., Li, J.Y., Chen, T.Y.: Simultaneous focusing and rotation of a bifunctional thermal metamaterial with constant anisotropic conductivity. J. Appl. Phys. **126**, 095103 (2019)

22. Urzhumov, Y.A., Smith, D.R.: Fluid flow control with transformation media. Phys. Rev. Lett. **107**, 074501 (2011)
23. Urzhumov, Y.A., Smith, D.R.: Flow stabilization with active hydrodynamic cloaks. Phys. Rev. E **86**, 056313 (2012)
24. Bowen, P.T., Urzhumov, Y.A., Smith, D.R.: Wake control with permeable multilayer structures: the spherical symmetry case. Phys. Rev. E **92**, 063030 (2015)
25. Park, J., Youn, J.R., Song, Y.S.: Hydrodynamic metamaterial cloak for drag-free flow. Phys. Rev. Lett. **123**, 074502 (2019)
26. Park, J., Youn, J.R., Song, Y.S.: Fluid-flow rotator based on hydrodynamic metamaterial. Phys. Rev. Appl. **12**, 061002 (2019)
27. Wang, Z.Y., Li, C.Y., Zatianina, R., Zhang, P., Zhang, Y.Q.: Carpet cloak for water waves. Phys. Rev. E **96**, 053107 (2017)
28. Li, C.Y., Xu, L., Zhu, L.L., Zou, S.Y., Liu, Q.H., Wang, Z.Y., Chen, H.Y.: Concentrators for water waves. Phys. Rev. Lett. **121**, 104501 (2018)
29. Zou, S.Y., Xu, Y.D., Zatianina, R., Li, C.Y., Liang, X., Zhu, L.L., Zhang, Y.Q., Liu, G.H., Liu, Q.H., Chen, H.Y., Wang, Z.Y.: Broadband waveguide cloak for water waves. Phys. Rev. Lett. **123**, 074501 (2019)
30. Li, Y., Zhu, K.-J., Peng, Y.-G., Li, W., Yang, T.Z., Xu, H.-X., Chen, H., Zhu, X.-F., Fan, S.H., Qiu, C.-W.: Thermal meta-device in analogue of zero-index photonics. Nat. Mater. **18**, 48 (2019)
31. Xu, L.J., Huang, J.P.: Chameleonlike metashells in microfluidics: a passive approach to adaptive responses. Sci. China-Phys. Mech. Astron. **63**, 228711 (2020)
32. Bear, J., Corapcioglu, M.Y.: Fundamentals of Transport Phenomena in Porous Media. Springer, Netherlands (1984)
33. Vemuri, K.P., Bandaru, P.R.: Geometrical considerations in the control and manipulation of conductive heat flux in multilayered thermal metamaterials. Appl. Phys. Lett. **103**, 133111 (2013)
34. Yang, T.Z., Vemuri, K.P., Bandaru, P.R.: Experimental evidence for the bending of heat flux in a thermal metamaterial. Appl. Phys. Lett. **105**, 083908 (2014)
35. Vemuri, K.P., Canbazoglu, F.M., Bandaru, P.R.: Guiding conductive heat flux through thermal metamaterials. Appl. Phys. Lett. **105**, 193904 (2014)
36. Shang, J., Wang, R.Z., Xin, C., Dai, G.L., Huang, J.P.: Macroscopic networks of thermal conduction: Failure tolerance and switching processes. Int. J. Heat Mass Transf. **121**, 321 (2018)
37. Xu, L.J., Yang, S., Huang, J.P.: Thermal theory for heterogeneously architected structure: fundamentals and application. Phys. Rev. E **98**, 052128 (2018)
38. Xu, L.J., Yang, S., Huang, J.P.: Thermal transparency induced by periodic interparticle interaction. Phys. Rev. Appl. **11**, 034056 (2019)
39. Li, J.X., Li, Y., Li, T.L., Wang, W.Y., Li, L.Q., Qiu, C.-W.: Doublet thermal metadevice. Phys. Rev. Appl. **11**, 044021 (2019)
40. Xu, L.J., Huang, J.P.: Metamaterials for manipulating thermal radiation: Transparency, cloak, and expander. Phys. Rev. Appl. **12**, 044048 (2019)
41. Dai, G.L., Huang, J.P.: Nonlinear thermal conductivity of periodic composites. Int. J. Heat Mass Transf. **147**, 118917 (2020)
42. Huang, J.P.: Theoretical Thermotics: Transformation Thermotics and Extended Theories for Thermal Metamaterials. Springer, Singapore (2020)

Chapter 4
Theory for Thermoelectric Effect Control: Transformation Nonlinear Thermoelectricity

Abstract Temperature-dependent (nonlinear) transformation thermotics provides a powerful tool for designing multifunctional, switchable, or intelligent metamaterials in diffusion systems. However, its extension to multiphysics remains studied, in which the temperature dependence of intrinsic parameters is ubiquitous. Here, we theoretically establish a temperature-dependent transformation method for controlling multiphysics. Taking thermoelectric transport as a typical case, we prove the form invariance of its temperature-dependent governing equations and formulate the corresponding transformation rules. Our finite-element simulations demonstrate robust thermoelectric cloaking, concentrating, and rotating performance in temperature-dependent backgrounds. We further design two practical applications with temperature-dependent transformation: an ambient-responsive cloak-concentrator thermoelectric device that can switch between cloaking and concentrating; an improved thermoelectric cloak with nearly-thermostat performance inside. Our theoretical frameworks and application designs may provide guidance for efficiently controlling temperature-related multiphysics and enlighten subsequent intelligent multiphysical metamaterial research.

Keywords Transformation nonlinear thermoelectricity · Thermoelectric coupling · Nonlinear parameters

4.1 Opening Remarks

Recent advances in metamaterials and metadevices for controlling diffusion systems have witnessed a development tendency of adaptability, adjustability, and integration [1–5]. The nonlinear transformation thermotics [6–8], evolving from the linear transformation theory [9–13], provides a definite method to exactly map the diffusive single-field distribution to a required one in temperature-dependent backgrounds. On the basis of it, metamaterial research for manipulating diffusive flows achieves enhanced convertibility [14–17] and intellectualization [18–21].

However, in practical applications, it is important to consider how to manipulate multiphysics, which is ubiquitous in nature, industry, and daily life. Until now, almost

L.-J. Xu and J.-P. Huang, *Transformation Thermotics and Extended Theories*,
https://doi.org/10.1007/978-981-19-5908-0_4

all efforts in controlling multiphysics have been confined to linear medium [22–30]. This approximation may not only deviate from practical situations to some extent but also limit the advancement of manipulating multiple fields. Referring to thermo-electric(TE) effects [31–33], temperature-dependent transport processes have been investigated due to the electron-phonon coupling mechanism [34–36] or strong inter-action in quantum-dot systems [37, 38]. At the macroscopic level, the nonlinearity of material is often embodied in temperature-dependent thermal conductivities, elec-trical conductivities, and Seeback coefficients [39], which may introduce better TE performance beyond linear response to temperature or voltage bias [40]. In detail, the thermal conductivity κ may have a power-law form T^n (n is a real number) with different experiential values of n for different conditions or materials, which induce different electrical conductivities according to the Wiedemann-Franz law [41]. The Seeback coefficient S is usually directly proportional to T for metals and some semi-conductors [42]. Although nonlinear transformation thermotics can be extended to decoupled multiphysics readily due to the form similarity of independent governing equations, it needs to be further studied if the nonlinear transformation theory still works in regulating coupled multiphysical fields like thermoelectricity.

Inspired by nonlinear transformation thermotics [6–8], we consider the temperature-dependent TE transport where material properties and/or spatial trans-formation operations are temperature dependent. In this way, functions of passive devices may become flexible and automatically adapt to changes in environments. Our study represents an example of applying the temperature-dependent transforma-tion theory to design intelligent multiphysical metamaterials and metadevices, which can be generalized to other multiphysics.

4.2 Theoretical Foundation

We consider a nonlinear TE coupling transport process as a typical temperature-dependent multiphysics case. First, the nonlinearity indicates the temperature depen-dence of electrical conductivity, which has been adequately studied. The general form of a temperature-dependent electrical conductivity tensor can be written as $\sigma(T)$. On the other hand, according to the Wiedemann-Franz law [43], a considerable amount of materials with electron domination in heat conduction will thus have temperature-dependent thermal conductivity tensors $\kappa(T)$. In addition, a nonlinear Seeback coef-ficient tensor is given as $S(T)$ without loss of generality. When the temperature and voltage biases are applied on the TE medium simultaneously, the coupled heat and electrical currents will be induced by each other separately besides their respective independent transport. Thus, the constitutive relations of electric current density $\boldsymbol{J}$ and heat current density $\boldsymbol{J}_Q$ can be described as [44, 45]

$$\begin{aligned} \boldsymbol{J} &= -\sigma(T)\nabla\mu - \sigma(T)S(T)\nabla T, \\ \boldsymbol{J}_Q &= -\kappa(T)\nabla T + TS^{\mathrm{tr}}(T)\boldsymbol{J}, \end{aligned} \tag{4.1}$$

where μ and T are position-related electrochemical potential and temperature, and $\boldsymbol{S}^{\mathrm{tr}}(T)$ is the transpose of $\boldsymbol{S}(T)$. Charge and heat flows are coupled by the Seebeck coefficient $\boldsymbol{S}(T)$. At the steady state with local equilibrium, the governing equations of TE transport are expressed as [44, 45]

$$\begin{aligned} \boldsymbol{\nabla} \cdot \boldsymbol{J} &= 0, \\ \boldsymbol{\nabla} \cdot \boldsymbol{J}_Q &= -\boldsymbol{\nabla}\mu \cdot \boldsymbol{J}. \end{aligned} \tag{4.2}$$

In contrast with single physics, TE coupling transport leads to the generation of a heat source term, namely, $-\boldsymbol{\nabla}\mu \cdot \boldsymbol{J}$, which can be interpreted as a Joule heating result. With the Onsager reciprocal requirement [46], electrical and thermal conductivity tensors should be symmetric. Thus, we can determine that $\boldsymbol{\sigma}(T) = \boldsymbol{\sigma}^{\mathrm{tr}}(T)$ and $\boldsymbol{\kappa}(T) = \boldsymbol{\kappa}^{\mathrm{tr}}(T)$. Substituting Eq. (4.1) into Eq. (4.2), the governing equations can be rewritten as

$$\boldsymbol{\nabla} \cdot [\boldsymbol{\sigma}(T)\boldsymbol{\nabla}\mu + \boldsymbol{\sigma}(T)\boldsymbol{S}(T)\boldsymbol{\nabla}T] = 0, \tag{4.3}$$

and

$$\begin{aligned} -\boldsymbol{\nabla} \cdot [\boldsymbol{\kappa}(T)\boldsymbol{\nabla}T + T\boldsymbol{S}^{\mathrm{tr}}(T)\boldsymbol{\sigma}(T)\boldsymbol{S}(T)\boldsymbol{\nabla}T + T\boldsymbol{S}^{\mathrm{tr}}(T)\boldsymbol{\sigma}(T)\boldsymbol{\nabla}\mu] \\ = \boldsymbol{\nabla}\mu \cdot [\boldsymbol{\sigma}(T)\boldsymbol{\nabla}\mu + \boldsymbol{\sigma}(T)\boldsymbol{S}(T)\boldsymbol{\nabla}T]. \end{aligned} \tag{4.4}$$

We are now in the position to prove that Eqs. (4.3) and (4.4) satisfy form invariance under arbitrary coordinate transformation, so that the transformation theory is still valid in the temperature-dependent TE transport process. In a curvilinear coordinate system with a set of contravariant bases $\{\boldsymbol{g}^i, \boldsymbol{g}^j, \boldsymbol{g}^k\}$, a group of covariant bases $\{\boldsymbol{g}_i, \boldsymbol{g}_j, \boldsymbol{g}_k\}$, and corresponding contravariant components (x^i, y^j, z^k), the component form of Eq. (4.3) can be expressed as

$$\partial_i[\sqrt{g}\sigma^{ij}(T)\partial_j\mu] + \partial_i[\sqrt{g}\sigma^{ij}(T)S_j^k(T)\partial_k T] = 0. \tag{4.5}$$

where g is the determinant of the matrix with components $g_{ij} = \boldsymbol{g}_i \cdot \boldsymbol{g}_j$. And the component form of Eq. (4.4) can be written as

$$\begin{aligned} \partial_j[\sqrt{g}\kappa^{jk}(T)\partial_k T + T\sqrt{g}(S^{\mathrm{tr}})_i^j(T)\sigma^{ij}(T)S_j^k(T)\partial_k T + T\sqrt{g}(S^{\mathrm{tr}})_i^j(T)\sigma^{ij}(T)\partial_j\mu] \\ = -\sqrt{g}(\partial_j\mu)[\sigma^{ij}(T)\partial_j\mu + \sigma^{ij}(T)S_j^k(T)\partial_k T]. \end{aligned} \tag{4.6}$$

where $(S^{\mathrm{tr}})_i^j(T)$ is the transpose of $S_i^j(T)$. Equations (4.5) and (4.6) have the same form under different coordinates. The only difference in diverse coordinate systems is the coefficient g. Here, g is not limited to position dependence and can be written as $g(T)$ if the coordinate transformation is temperature dependent. The theory for temperature-dependent transformation TE fields allows executing temperature-dependent coordinate transformations on temperature-dependent TE materials, and these two kinds of nonlinearity will be incorporated into transformed physical parameters.

For the simplicity of derivation on transformation rules, we first demonstrate the linear transformation. Now, consider a bijection $f : r \mapsto r'$, which is smooth enough from the pretransformed space to the transformed space in the three-dimensional Euclidean space. Due to the diffeomorphism between the pretransformed space (virtual space) with a chosen set of curvilinear coordinates $\{x, y, z\}$ and the transformed space (physical space) with another set of Cartesian coordinates $\{x', y', z'\}$, Eqs. (4.3) and (4.4) can be rewritten as

$$\nabla' \cdot [\sigma'(T')\nabla'\mu' + \sigma'(T')S'(T')\nabla T'] = 0, \tag{4.7}$$

and

$$-\nabla' \cdot [\kappa'(T')\nabla'T' + T'(S')^{\text{tr}}(T')\sigma'(T')S'(T')\nabla'T' + T'(S')^{\text{tr}}(T')\sigma'(T')\nabla'\mu']$$
$$= \nabla'\mu' \cdot [\sigma'(T')\nabla'\mu' + \sigma'(T)S'(T')\nabla'T']. \tag{4.8}$$

We can find that transformation rules given by Eqs. (4.3) and (4.4) are consistent with Eqs. (4.7) and (4.8). The transformed $\kappa'(T')$, $\sigma'(T')$, and $S'(T')$. can be expressed as

$$\kappa'(T'(r')) = \frac{A\kappa_0(T(f^{-1}(r')))A^{\text{tr}}}{\det A} \tag{4.9}$$

$$\sigma'(T'(r')) = \frac{A\sigma_0(T(f^{-1}(r')))A^{\text{tr}}}{\det A} \tag{4.10}$$

$$S'(T'(r')) = A^{-\text{tr}}S(T(f^{-1}(r')))A^{\text{tr}}. \tag{4.11}$$

The linear transformation on temperature-dependent TE background requires tailoring thermal conductivity, electrical conductivity, and Seebeck coefficient described in Eqs. (4.9)–(4.11).

We then return to the theory of nonlinear TE transformation to perform a temperature-dependent transformation on temperature-dependent TE backgrounds. Here, a temperature-dependent transformation means that the transformed operations are temperature-related, so the corresponding Jacobian matrixes become $A(T)$. We can see that the transformation rules of linear transformation can easily be generalized to nonlinear transformation by replacing r' with $r'(T)$ and replacing A with $A(T)$ in Eqs. (4.9)–(4.11).

4.3 Finite-Element Simulation

We now employ these rules to design TE metamaterials on temperature-dependent backgrounds. Equation (4.10) implies that the Seebeck coefficient remains invariant after coordinate transformation if the Seebeck coefficient before transformation is isotropic, it can be written as $S'(T) = S_0(T) = \gamma T$ (γ is constant). The temperature-dependent electrical conductivity and thermal conductivity satisfy the transformation rules in Eqs. (4.9) and (4.11). Here, we assign the background thermal conductivity a trivial scalar expression $\kappa_0(T) = \alpha + \beta T^n$ (α, β and n are constants). According to the Wiedemann-Franz law $\kappa/\sigma = LT$ (L is the Lorenz number) [43], the background electrical conductivity can be written as $\sigma_0(T) = (\alpha T^{-1} + \beta T^{n-1})/L$. The transformed material properties can then be expressed as

$$\kappa'(T) = \frac{A(\alpha + \beta T^n)A^{tr}}{\det A},$$

$$\sigma'(T) = \frac{A(\alpha + \beta T^n/(LT))A^{tr}}{\det A}, \qquad (4.12)$$

$$S'(T) = \gamma T.$$

Here, we consider cloaking, concentrating, and rotating functions in two-dimensional nonlinear backgrounds. The TE cloak keeps the central region free from heat flows and currents maintain constant temperatures and electric potentials without disturbing TE distributions outside, as shown in Fig. 4.1a. We can present the coordinate transformation relationship of the cloak in polar coordinates (r, θ) as

$$r' = r(r_2 - r_1)/r_2 + r_1,$$

$$\theta' = \theta, \qquad (4.13)$$

where $r \in [0, r_2]$ and $r' \in [r_1, r_2]$. The purpose of a TE concentrator is to collect more currents and heat flows in the central region to increase the local temperature gradient without disturbing the TE distribution outside, as shown in Fig. 4.1b. The detailed coordinate transformation can be given as

$$r'' = r_1 r/r_m \quad (r < r_m),$$

$$r'' = r(r_2 - r_1)/(r_2 - r_m) + r_2(r_1 - r_m)/(r_2 - r_m) \quad (r_m < r < r_2), \qquad (4.14)$$

$$\theta'' = \theta,$$

where r_m is the radius between r_1 and r_2. A TE rotator serves to rotate the currents and heat flows with angle θ_0 in the central circular region without disturbing the TE distributions outside, as indicated in Fig. 4.1c. The corresponding coordinate transformation can be described as

$$r''' = r,$$

$$\theta''' = \theta + \theta_0 \quad (r < r_1), \qquad (4.15)$$

$$\theta''' = \theta + \theta_0(r - r_2)/(r_1 - r_2) \quad (r_1 < r < r_2).$$

The Jacobian transformation matrix A can be expressed in polar coordinates as

$$A = \begin{bmatrix} \partial r^*/\partial r & \partial r^*/(r\partial\theta) \\ r^*\partial\theta^*/\partial r & r^*\partial\theta^*/(r\partial\theta) \end{bmatrix}, \qquad (4.16)$$

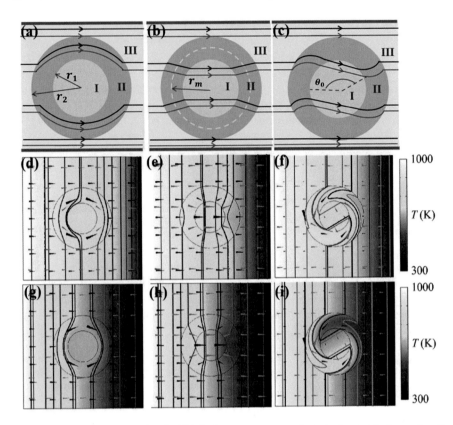

Fig. 4.1 a–c Schematic graphs of a TE cloak, concentrator, and rotator located in the center of a temperature-dependent background. Region I, II, and III represent the functional area, transformation layer, and background, respectively. **d–f** and **g–i** are simulation results of temperature-dependent and temperature-independent TE cloak, concentrator and rotator, separately. Background size is 8×8 cm. Inner radius of the transformed layer is $r_1 = 1$ cm and outer radius is $r_2 = 2$ cm. The virtual radius of concentrator is $r_m = 1.5$ cm, and the rotation angle of rotator is $\theta_0 = 120°$. The background thermal conductivity of **d–f** is $100 + 10T^3$ W/(m · K), electrical conductivity is $100/T + 10T^2$ S/m, and Seebeck coefficient is $S = 30T$ μV/K. The background thermal conductivity of **g–i** is 1000 W/(m · K), electrical conductivity is 100 S/m, and Seebeck coefficient is $S = 200$ μV/K. The left boundary is set as 1000 K and 0.01 V, and the right boundary is set as 300 K and 0 V. Upper and lower boundaries are thermally and electrically insulated. In **d–i**, color surfaces represent temperature distribution, black and blue arrows/lines denote thermal and electric flows/isothermal and isopotential. Adapted from Ref. [47]

where $r^* = r', r''$ or r''' and $\theta^* = \theta', \theta''$ or θ'''. Substituting Eqs. (4.13)–(4.15) into Eq. (4.16), we can obtain the corresponding Jacobian matrices of three metamaterials. In combination with Eq. (4.12), the transformed thermal conductivity and electrical conductivity of three metamaterials in the annulus region ($r_1 < r' < r_2$) can be expressed as

$$\kappa'(T) = (\alpha + \beta T^n)\, B^*,$$
$$\sigma'(T) = (\alpha + \beta T^n)/(LT)\, B^*. \tag{4.17}$$

where $B^* = B'$, B'', and B''' corresponding to TE cloaks, concentrators and rotators. They can be written separately

$$B' = \mathrm{diag}[(r' - r_1)/r', r'/(r' - r_1)],$$

$$B'' = \mathrm{diag}\left[\frac{(r_2 - r_m)r'' - (r_1 - r_m)r_2}{(r_2 - r_m)r''}, \frac{(r_2 - r_m)r''}{(r_2 - r_m)r'' - (r_1 - r_m)r_2}\right], \tag{4.18}$$

$$B''' = \left(\left[1, \frac{\theta_0 r'''}{r_1 - r_2}\right], \left[\frac{\theta_0 r'''}{r_1 - r_2}, \left(\frac{\theta_0 r'''}{r_1 - r_2}\right)^2 + 1\right]\right).$$

For TE cloaks, concentrators and rotators, the electrical conductivities and thermal conductivities in the center circular region with radius r_1 are the same as the backgrounds, and they can be written as

$$\kappa^*(T) = (\alpha + \beta T^n)\, \mathrm{diag}[1, 1],$$
$$\sigma^*(T) = (\alpha + \beta T^n)/(LT)\, \mathrm{diag}[1, 1]. \tag{4.19}$$

We then execute finite-element simulations of the designed temperature-dependent and temperature-independent TE cloak, concentrator, and rotator with the commercial software COMSOL Multiphysics. We use the steady TE-effect module in the two-dimensional system to simulate the temperature and potential distributions of coupled TE fields; the results are shown in the second and third panels of Fig. 4.1, respectively. In Fig. 4.1d–i, temperature or potential distributions in backgrounds are inhomogeneous under horizontal external thermal and electrical fields due to the temperature-dependent parameters, but the cloaking, concentrating, or rotating functionalities are still valid. To further verify the robustness of the proposed meta-materials, we subject them to different temperature boundary conditions; see Fig. 4.2. We retain the electrical boundary conditions and fix the right boundary at 300 K. With increasing temperatures up to 1500 K at the left boundary, the nonlinearity effect gradually emerges, which can be seen from the isothermal lines. However, cloaking, concentrating, and rotating still function effectively. For clarity, the references with pure backgrounds are also displayed for comparison. Furthermore, we plot the simulation data of the central lines horizontally crossing the center of metamaterials in Fig. 4.3. The results are echoed well in region III (outside metamaterials), indicating no distortion in backgrounds. The relations between temperatures (or potentials) and positions in region III are not linear but tend to be nonlinear with increasing high-temperature boundary conditions. In particular, voltages at 0.06–0.08 m show negative differentials due to the coupled TE effects.

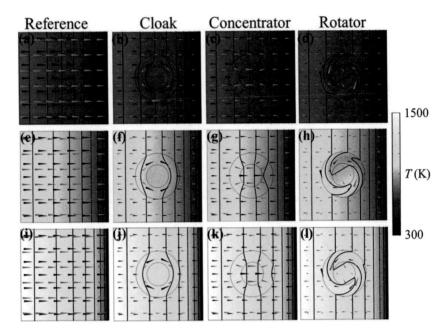

Fig. 4.2 Simulation results of the temperature-dependent TE cloak, concentrator, and rotator under different temperature boundary conditions. The parameter settings are the same as those in Fig. 4.1. The right boundaries are set at 300 K thermally, with electrical grounding. The left boundaries are separately set at 700 K [**a–d**], 1100 K [**e–h**], 1500 K [**i–l**], and 0.01 V. Black and blue arrows indicate magnitudes and directions of heat and electric flows, respectively. References in the first column are bare backgrounds without any internal structures. Adapted from Ref. [47]

4.4 Model Application

A. Ambient-responsive TE cloak-concentrator. Based on the proposed temperature-dependent transformation TE field theory, we further design an ambient-responsive TE cloak-concentrator device as a practical application. Due to the temperature-dependent features, cloaking and concentrating functionalities function under different environmental temperature regions, resulting in a switchable TE cloak-concentrator. Here, we skip the original constitutive parameters and only consider the parameters after transformation operation. The emphasis of achieving TE cloak-concentrator is to make transformed thermal conductivity and electrical conductivity be corresponding to different functions under different temperatures. Thus we consider temperature-related coordinate transformation to realize it. If we carefully check the coordinate transformation relationship in Eqs. (4.13) and (4.14), we can find that Eq. (4.14) has the same form as Eq. (4.13) when $r_m = 0$. Thus, a temperature-dependent function can be constructed by replacing r_m in Eq. (4.14) with $r_m^*(T)$, for which the coordinate transformation relationship corresponds to cloak at $r_m^*(T) = 0$ and concentrator at $r_m^*(T) = r_m$. Equation (4.13) can be rewritten as

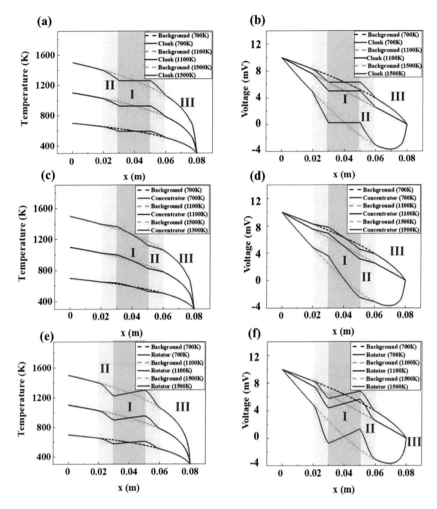

Fig. 4.3 Quantitative comparison between backgrounds assembled with metamaterials and bare backgrounds. The data are extracted at center lines along the horizontal direction from the simulation results in Fig. 4.2. Three panels denote the cloak, concentrator, and rotator separately by row. The left and right columns are temperature and voltage data, respectively. Regions I, II, and III are the corresponding regions in Fig. 4.1. Adapted from Ref. [47]

$$r_m^*(T) = \frac{r_m}{1 + \exp[\eta(T - T_C)]}. \tag{4.20}$$

Here, T_C is a critical temperature around which $r_m^*(T)$ can be distinguished by 0 or r_m, as schematically shown in Fig. 4.4a. η is a scaling coefficient for ensuring the step change around T_C. The coordinate transformation of the shell region can be rewritten as

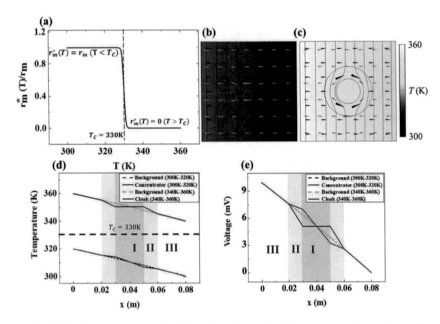

Fig. 4.4 TE cloak-concentrator with different functions under different temperature regions. **a** The curve of $r_m^*(T)/r_m$ with temperature, and $T_C = 300$ K, $\eta = 2.5$ K^{-1} in this expression of Eq. (4.20). **b–c** Simulation results of the TE cloak-concentrator. It exhibits concentration under the temperature region 300–320 K and cloaking under the temperature region 340–360 K. The higher temperature and voltage 0.01 V are set at the left boundaries, and the lower temperature and electrical grounding are set at the right boundaries. **d** and **e** are the temperature and voltage curves of the center lines extracted from simulation results in **b** and **c**. Regions I, II, and III are the corresponding regions in Fig. 4.1. Adapted from Ref. [47]

$$r' = r\frac{r_2 - r_1}{r_2 - r_m^*(T)} + r_2\frac{r_1 - r_m^*(T)}{r_2 - r_m^*(T)},$$

$$\theta' = \theta.$$
(4.21)

where the transformed coordinates are temperature-dependent, which also meets the requirements of nonlinear transformations. The expressions of transformed thermal and electrical conductivities can be obtained by replacing r_m in Eq. (4.18) with $r_m^*(T)$, so expressions can be transformed into TE cloaks when the environment temperature is higher than T_C and into TE concentrators when the environment temperature is lower than T_C.

We present our results more intuitively by finite-element simulation. The simulation results are shown in Fig. 4.4b and c, in accordance with the expected effects. The device shows the concentrating effect within the temperature region of 300–320 K and cloaking from 340–360 K. That is, it can automatically transfer the function from TE cloaking (or concentrating) to TE concentrating (or cloaking) when the temperature of the environment changes. This direct result of ambient-responsive

TE parameters is impossible in linear transport processes. Additionally, we extract the data of temperature and voltage on the central line horizontally crossing the center in different temperature regions in Fig. 4.4d and e. Background temperature and potential data in region III coincide between with- and without-device cases, indicating that the background is not influenced. Cloaking and concentrating features are obvious in region I when comparing the slope with pure backgrounds.

The thermal conductivity and electrical conductivity of a TE cloak-concentrator are anisotropic and can achieve abrupt change in different temperature ranges. It is difficult for actual natural materials to meet this requirement. Shape memory alloy may be a candidate to achieve the switch of function [6, 14] due to its sharp deform with changing temperatures. Thus, We can employ composites of shape memory alloy and TE materials on the transformation layer to realize the ambient-responsive TE cloak-concentrator.

B. Improved TE cloak. An improved TE cloak that can maintain a nearly constant temperature internally is designed. This is different from the existing TE cloak [29], in which the temperature inside relies on the boundary conditions. Four individual components compose the cloak, which is placed in a temperature-dependent background depicted as region III, as shown in Fig. 4.5a. A linear transformation is executed to region II, where the fixed thermal and electrical conductivities follow Eq. (4.17). In regions IV and V, the nonlinear transformation is achieved by two symmetrical equations: Eq. (4.20) and

$$r_m^*(T) = \frac{r_m \exp[\eta(T - T_C)]}{1 + \exp[\eta(T - T_C)]}. \tag{4.22}$$

It is clear that Eqs. (4.20) and (4.22) exhibit opposite behaviors around T_C. We obtain expressions of transformed thermal and electrical conductivities by substituting them into Eq. (4.19). This operation may make the internal temperature approach T_C [14, 16]. The temperature and voltage distributions of this cloak are shown in Fig. 4.5d and f. We obtain simultaneous near-thermostat performance inside the cloak, while the distributions of temperature and potential remain unchanged outside the cloak. For comparison, we also demonstrate the conventional TE cloaks under the same boundary conditions in Fig. 4.5c and e. Furthermore, for quantitative verification, we extract temperature data of center lines from simulation results; see Fig. 4.5b. In region I, the temperature tends to T_C, marked as a yellow dashed line. Under changed high or low boundary temperatures, the designed cloak exhibits robustness in that the internal temperatures deviate to T_C, compared with conventional cloaks indicated by solid lines in Fig. 4.5b. In addition, in region III, good fitting of conventional and improved cloaks is visually concluded. Thus, the designed cloak can improve the thermostat performance internally without loss of concealing functionalities.

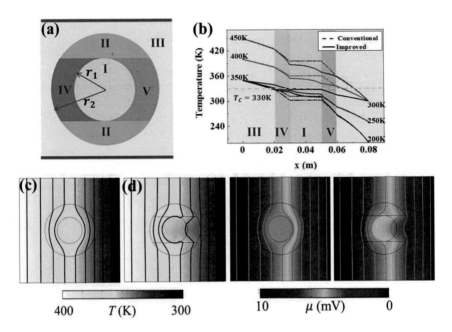

Fig. 4.5 Improved thermoelectric cloak. **a** Constituent structure. **c** and **e** show the temperature and potential distributions of a conventional TE cloak, respectively. **d** and **f** show the temperature and potential distributions of the improved TE cloak with near-thermostat functionality inside. The temperature 400 K and voltage 0.01 V are set at the left boundaries, and the temperature 300 K and electrical grounding are set at the right boundaries. **b** denotes temperature curves of the center lines extracted from the conventional cloak and improved cloaks. Different colors indicate different temperature conditions. Adapted from Ref. [47]

4.5   Discussion

We have verified that the form invariance of temperature-dependent TE governing equations remains valid under a spatial coordinate transformation. The temperature-dependent transformation emphasizes the transformation operation, and backgrounds can depend on temperature simultaneously, usually leading to temperature-dependent design parameters. Recent studies in pyroelectricity [39–42] have shown that temperature dependence of TE materials is ubiquitous, especially on small scales, due to the corresponding large bias of temperature or voltage. From this perspective, the proposed temperature-dependent transformation multiphysics generally applies to controlling TE fields with naturally existing materials.

As two representative applications of temperature-dependent TE transformation theory, ambient-responsive TE cloak-concentrator devices and improved TE cloaks are constructed by executing nonlinear transformation operations in nonlinear backgrounds. For the former, switching between cloaking and concentrating can be achieved under different ambient temperature regions. The devices can avoid thermal or electrical damage caused by high ambient temperatures, and heat and electric flows can be used effectively under low temperatures. For the latter, the desired temperature T_C can be approximately achieved inside the cloak. The thermostat effects of improved TE cloaks are not rigorous due to the intrinsic limits of transformed layers. That is, the zero thermal or electrical conductivities can be reached only at r_1, which limits physical spaces with respect to capturing specific temperatures [14, 16]. However, the simulation results show obvious improvement in internal temperature preservation compared with conventional cloaks. Although another solid scheme may resort to the bilayer design [14, 16], here we verify that the transformed layers may also play the role of temperature trapper, which may be of benefit in some situations where the transformation method is employed.

4.6 Conclusion

In summary, the transformation theory on the nonlinear multiphysical backgrounds is established, so linear and nonlinear transformations can be performed on the background. Three nonlinear metamaterials with functions of cloaking, concentrating, and rotating are demonstrated, confirming the theory. For practical applications, two temperature-responsive multiphysical devices are designed whose functionalities exceed those of their linear analogies. Our theory and design can be extended from thermoelectricity to other fields of multiphysics such as thermo-optics or thermomagnetics. For example, enhancement of magnetic field on a metal-coated superconductor thermomagnetic system will generate Joule heating sources [48]. The governing equation of the thermomagnetic effect in the superconductor satisfies the form invariance under a coordinate transformation. Thus, the transformation theory can be used to control such a thermomagnetic field. Various multiphysical intelligent metamaterials can be expected, which may facilitate multiple flow guidance and benefit the development of self-adaptation in metamaterial design.

4.7 Exercise and Solution

Exercise

1. The ambient-responsive thermoelectric cloak-concentrator designed with temperature-dependent transformation thermoelectricity can realize functional switching at different ambient temperatures. How to realize the functional switch of cloak and concentrator through a coordinate transformation? Please derive the transformed material parameters.

Solution

1. Temperature-dependent transformation thermoelectricity requires that the background material parameters or the coordinate transformations are temperature-dependent. The temperature-dependent background material parameters can help heat transfer, and the temperature-dependent coordinate transformation can realize function switching. Here, we use both temperature-dependent background material parameters and coordinate transformations. The transformed material properties based on temperature-dependent transformation thermoelectricity can be expressed as

$$\kappa'(T) = \frac{A(\alpha + \beta T)A^{tr}}{\det A},$$

$$\sigma'(T) = \frac{A(\beta/L + \alpha/LT)A^{tr}}{\det A}, \qquad (4.23)$$

$$S' = S_0.$$

The Jacobian transformation matrix A in polar coordinates can be expressed as

$$A = \begin{bmatrix} \partial r^*/\partial r & \partial r^*/(r\partial\theta) \\ r^*\partial\theta^*/\partial r & r^*\partial\theta^*/(r\partial\theta) \end{bmatrix}, \qquad (4.24)$$

Therefore, to obtain the transformed material parameters, the coordinate transformation must be obtained first. Consider the virtual radius of the concentrator to be temperature dependent, it can be written as

$$r_m^*(T) = \frac{r_m}{1 + \exp^{\beta(T - T_C)}}. \qquad (4.25)$$

Here, T_C is a critical temperature around which the coordinate transformation is cloak or concentrator, and β is a scaling coefficient. The coordinate transformation relationship corresponds to cloak at $r_m^*(T) = 0$ and concentrator at $r_m^*(T) = r_m$. The coordinate transformation of concentrator can be rewritten as

$$r' = r\frac{r_2 - r_1}{r_2 - r_m^*(T)} + r_2\frac{r_1 - r_m^*(T)}{r_2 - r_m^*(T)}, \qquad (4.26)$$

$$\theta' = \theta.$$

Substitute Eq. (4.26) into Eqs. (4.24) and (4.23), the expressions of transformation electrical conductivity and thermal conductivity in polar coordinates can be expressed as

$$\kappa'(T) = (\alpha + \beta T) \begin{bmatrix} \dfrac{(r_2 - r_m^*(T))r' - (r_1 - r_m^*(T))r_2}{(r_2 - r_m^*(T))r'} & 0 \\ 0 & \dfrac{(r_2 - r_m^*(T))r'}{(r_2 - r_m^*(T))r' - (r_1 - r_m^*(T))r_2} \end{bmatrix},$$

$$\sigma'(T) = \frac{\alpha + \beta T}{LT} \begin{bmatrix} \dfrac{(r_2 - r_m^*(T))r' - (r_1 - r_m^*(T))r_2}{(r_2 - r_m^*(T))r'} & 0 \\ 0 & \dfrac{(r_2 - r_m^*(T))r'}{(r_2 - r_m^*(T))r' - (r_1 - r_m^*(T))r_2} \end{bmatrix}.$$

(4.27)

Equation (4.27) is transformed to transformation parameters of the thermoelectric cloak when the environment temperature is higher than T_C, and to transformation parameters of the thermoelectric concentrator when the environment temperature is lower than T_C. Thus, we use Eq. (4.27) to realize the function switching under different ambient temperatures.

References

1. Li, Y., Li, W., Han, T., Zheng, X., Li, J., Li, B., Fan, S.H., Qiu, C.-W.: Transforming heat transfer with thermal metamaterials and devices. Nat. Rev. Mater. **6**, 488 (2021)
2. Yang, S., Wang, J., Dai, G.L., Yang, F.B., Huang, J.P.: Controlling macroscopic heat transfer with thermal metamaterials: theory, experiment and application. Phys. Rep. **908**, 1 (2021)
3. Sklan, S.R., Li, B.W.: Thermal metamaterials: functions and prospects. Natl. Sci. Rev. **5**, 138 (2018)
4. Wang, J., Dai, G., Huang, J.: Thermal metamaterial: fundamental, application, and outlook. Iscience **23**, 101637 (2020)
5. Dai, G.L.: Designing nonlinear thermal devices and metamaterials under the Fourier law: a route to nonlinear thermotics. Front. Phys. **16**, 53301 (2021)
6. Li, Y., Shen, X.Y., Wu, Z.H., Huang, J.Y., Chen, Y.X., Ni, Y.S., Huang, J.P.: Temperature-dependent transformation thermotics: from switchable thermal cloaks to macroscopic thermal diodes. Phys. Rev. Lett. **115**, 195503 (2015)
7. Li, Y., Shen, X.Y., Huang, J.P., Ni, Y.S.: Temperature-dependent transformation thermotics for unsteady states: switchable concentrator for transient heat flow. Phys. Lett. A **380**, 1641 (2016)
8. Sklan, S.R., Li, B.W.: A unified approach to nonlinear transformation materials. Sci. Rep. **8**, 4436 (2018)
9. Fan, C.Z., Gao, Y., Huang, J.P.: Shaped graded materials with an apparent negative thermal conductivity. Appl. Phys. Lett. **92**, 251907 (2008)
10. Chen, T.Y., Weng, C.-N., Chen, J.-S.: Cloak for curvilinearly anisotropic media in conduction. Appl. Phys. Lett. **93**, 114103 (2008)
11. Guenneau, S., Amra, C., Veynante, D.: Transformation thermodynamics: cloaking and concentrating heat flux. Opt. Express **20**, 8207 (2012)
12. Narayana, S., Sato, Y.: Heat flux manipulation with engineered thermal materials. Phys. Rev. Lett. **108**, 214303 (2012)
13. Schittny, R., Kadic, M., Guenneau, S., Wegener, M.: Experiments on transformation thermodynamics: molding the flow of heat. Phys. Rev. Lett. **110**, 195901 (2013)
14. Shen, X.Y., Li, Y., Jiang, C.R., Huang, J.P.: Temperature trapping: energy-free maintenance of constant temperatures as ambient temperature gradients change. Phys. Rev. Lett. **117**, 055501 (2016)
15. Shen, X.Y., Li, Y., Jiang, C.R., Ni, Y.S., Huang, J.P.: Thermal cloak-concentrator. Appl. Phys. Lett. **106**, 031907 (2016)

16. Wang, J., Shang, J., Huang, J.: Negative energy consumption of thermostats at ambient temperature: electricity generation with zero energy maintenance. Phys. Rev. Appl. **11**, 024053 (2019)
17. Su, C., Xu, L.J., Huang, J.P.: Nonlinear thermal conductivities of core-shell metamaterials: rigorous theory and intelligent application. EPL **130**, 34001 (2020)
18. Wehmeyer, G., Yabuki, T., Monachon, C., Wu, J.Q., Dames, C.: Thermal diodes, regulators, and switches: physical mechanisms and potential applications. Appl. Phys. Rev. **4**, 041304 (2017)
19. Wang, J., Dai, G.L., Yang, F.B., Huang, J.P.: Designing bistability or multistability in macroscopic diffusive systems. Phys. Rev. E **101**, 022119 (2020)
20. Li, Y., Li, J.X., Qi, M.H., Qiu, C.-W., Chen, H.S.: Diffusive nonreciprocity and thermal diode. Phys. Rev. B **103**, 014307 (2021)
21. Xu, L.J., Huang, J.P., Jiang, T., Zhang, L., Huang, J.P.: Thermally invisible sensors. EPL **132**, 14002 (2020)
22. Li, J.Y., Gao, Y., Huang, J.P.: A bifunctional cloak using transformation media. J. Appl. Phys. **108**, 074504 (2010)
23. Moccia, M., Castaldi, G., Savo, S., Sato, Y., Galdi, V.: Independent manipulation of heat and electrical current via bifunctional metamaterials. Phys. Rev. X **5**, 021025 (2014)
24. Ma, Y., Liu, Y., Raza, M., Wang, Y., He, S.: Experimental demonstration of a multiphysics cloak: manipulating heat flux and electric current simultaneously. Phys. Rev. Lett. **113**, 205501 (2014)
25. Raza, M., Liu, Y., Ma, Y.: A multi-cloak bifunctional device. J. Appl. Phys. **117**, 024502 (2015)
26. Lan, C., Bi, K., Fu, X., Li, B., Zhou, J.: Bifunctional metamaterials with simultaneous and independent manipulation of thermal and electric fields. Opt. Express **24**, 23072 (2016)
27. Lan, C., Bi, K., Fu, X., Gao, Z., Li, B., Zhou, J.: Achieving bifunctional cloak via combination of passive and active schemes. Appl. Phys. Lett. **109**, 201903 (2016)
28. Fujii, G., Akimoto, Y.: Optimizing the structural topology of bifunctional invisible cloak manipulating heat flux and direct current. Appl. Phys. Lett. **115**, 174101 (2019)
29. Stedman, T., Woods, L.M.: Cloaking of thermoelectric transport. Sci. Rep. **7**, 6988 (2017)
30. Shi, W.C., Stedman, T., Woods, L.M.: Transformation optics for thermoelectric flow. J. Phys. Energy **1**, 025002 (2019)
31. He, J., Tritt, T.M.: Advances in thermoelectric materials research: looking back and moving forward. Science **357**, 9997 (2017)
32. Zhou, X., Yan, Y., Lu, X., Zhu, H., Han, X., Chen, G., Ren, Z.: Routes for high-performance thermoelectric materials. Mater. Today **21**, 974 (2018)
33. Zevalkink, A., Smiadak, D.M., Blackburn, J.L., Ferguson, A.J., Chabinyc, M.L., Delaire, O., Wang, J., Kovnir, K., Martin, J., Schelhas, L.T., Sparks, T.D., Kang, S.D., Dylla, M.T., Snyder, G.J., Ortiz, B.R., Toberer, E.S.: A practical field guide to thermoelectrics: fundamentals, synthesis, and characterization. Appl. Phys. Rev. **5**, 021303 (2018)
34. Melnick, C., Kaviany, M.: From thermoelectricity to phonoelectricity. Appl. Phys. Rev. **6**, 021305 (2019)
35. Karki, D.B., Kiselev, M.N.: Nonlinear Seebeck effect of SU (N) Kondo impurity. Phys. Rev. B **100**, 125426 (2019)
36. Zebarjadi, M., Esfarjani, K., Shakouri, A.: Nonlinear Peltier effect in semiconductors. Appl. Phys. Lett. **91**, 122104 (2007)
37. Mazal, Y., Meir, Y., Dubi, Y.: Nonmonotonic thermoelectric currents and energy harvesting in interacting double quantum dots. Phys. Rev. B **99**, 075433 (2019)
38. Lavagna, M., Talbo, V., Duong, T.Q., Crepieux, A.: Level anticrossing effect in single-level or multilevel double quantum dots: electrical conductance, zero-frequency charge susceptibility, and Seebeck coefficient. Phys. Rev. B **102**, 125426 (2020)
39. Ishida, A.: Formula for energy conversion efficiency of thermoelectric generator taking temperature dependent thermoelectric parameters into account. J. Appl. Phys. **128**, 135105 (2020)
40. Pourkiaei, S.M., Ahmadi, M.H., Sadeghzadeh, M., Moosavi, S., Pourfayaz, F., Chen, L.G., Yazdi, M.A.P., Kumar, R.: Thermoelectric cooler and thermoelectric generator devices: a review of present and potential applications, modeling and materials. Energy **186**, 115849 (2019)

41. Ziabari, A., Zebarjadi, M., Vashaee, D., Shakouri, A.: Nanoscale solid-state cooling: a review. Rep. Prog. Phys. **79**, 095901 (2016)
42. Snyder, G.J., Toberer, E.S.: Complex thermoelectric materials. Nat. Mater. **7**, 105 (2008)
43. Tian, Z., Lee, S., Chen, G.: Heat transfer in thermoelectric materials and devices. J. Heat Transf. **135**, 061605 (2013)
44. Tritt, T.M., Subramanian, M.A.: Thermoelectric materials, phenomena, and applications: a bird's eye view. Mater. Res. Soc. Bull. **31**, 188 (2006)
45. Domenicali, C.A.: Irreversible thermodynamics of thermoelectricity. Rev. Mod. Phys. **26**, 237 (1954)
46. Onsager, L.: Reciprocal relations in irreversible processes I. Phys. Rev. **37**, 405 (1931)
47. Lei, M., Wang, J., Dai, G.L., Tan, P., Huang, J.P.: Temperature-dependent transformation multiphysics and ambient-adaptive multiphysical metamaterials. EPL **135**, 54003 (2021)
48. Vestgarden, J.I., Johansen, T.H., Galperin, Y.M.: Nucleation and propagation of thermomagnetic avalanches in thin-film superconductors. Low Temp. Phys. **44**, 460 (2018)

Chapter 5
Theory for Zero-Index Conductive Cloaks: Constant-Temperature Scheme

Abstract In this chapter, we propose an exact approach to an effectively infinite thermal conductivity with a constant-temperature boundary condition, which an external thermostatic sink can easily realize. Since (effectively) infinite thermal conductivity corresponds to zero refractive indexes in photonics, it has direct applications in designing zero-index thermal metamaterials. Therefore, we experimentally demonstrate zero-index thermal cloaks, which can work in highly conductive backgrounds with simple structures. These results provide insights into thermal management with effectively infinite thermal conductivities.

Keywords Effectively infinite thermal conductivity · Constant-temperature boundary conditions · Zero-index thermal cloaks

5.1 Opening Remarks

Thermal conductivity plays a crucial role in heat transfer, and extreme (zero and infinite) thermal conductivities are always a research focus due to their excellent properties. For low thermal conductivities, a recent study reported that the thermal conductivity of ceramic aerogel could be as low as 0.0024 W m^{-1} K^{-1} [1]. For high thermal conductivities, there is still a long way ahead. Although many materials have high thermal conductivity, such as boron nitride with 600 W m^{-1} K^{-1} [2], carbon nanotube with 2300 W m^{-1} K^{-1} [3], and graphene with 5300 W m^{-1} K^{-1} [4], they are still far from infinite thermal conductivities.

A recent study reported that the effective thermal conductivity of moving fluids could approximately tend to infinity [5]. Such an effectively infinite thermal conductivity requires the velocity of moving fluids to be also infinite, which cannot be exactly realized. To go further, we propose an exact approach to effectively infinite thermal conductivities with simple structures. By applying a constant-temperature boundary condition to an object with a finite thermal conductivity, the object can effectively have infinite thermal conductivity. Meanwhile, an external thermostatic sink can easily realize the constant-temperature boundary condition, which is beneficial for practical applications.

© The Author(s) 2023
L.-J. Xu and J.-P. Huang, *Transformation Thermotics and Extended Theories*,
https://doi.org/10.1007/978-981-19-5908-0_5

Since (effectively) infinite thermal conductivities are in analogue of zero refractive indexes in photonics [6–11], they can be used to design zero-index thermal metamaterials. We take thermal cloaking [12–21] as an example, which can be realized by transformation thermotics [12, 13] or scattering cancellation [15–17]. Here, we use infinite thermal conductivity to realize zero-index thermal cloaks, which can work in highly conductive backgrounds with simple structures. Specifically, if the previous bilayer scheme [15–17] is applied to a highly conductive background (such as copper, $400\,\text{W m}^{-1}\,\text{K}^{-1}$), the thermal conductivity of the inner shell is zero, and that of the outer shell should be larger than $400\,\text{W m}^{-1}\,\text{K}^{-1}$. However, few common materials have thermal conductivities higher than $400\,\text{W m}^{-1}\,\text{K}^{-1}$ [5]. Although some rare materials like diamond have high thermal conductivities, the cost and difficulty of practical applications also increase. In contrast, if the zero-index scheme is applied, the core with a constant-temperature boundary condition can effectively have infinite thermal conductivity. Therefore, the thermal conductivity of the outer shell can be smaller than $400\,\text{W m}^{-1}\,\text{K}^{-1}$, and many common materials such as aluminum can be applied. Therefore, the zero-index scheme is free from the thermal conductivities of backgrounds.

5.2 Thermal Zero Index Connotation

The Fourier law describes thermal conduction, namely $\boldsymbol{J} = -\kappa \nabla T$, where $\boldsymbol{J}$ is the heat flux, κ is the thermal conductivity, and T denotes temperature. To understand the temperature field effect of infinite thermal conductivity (i.e., zero-index thermal conductivity), we put a two-dimensional elliptical particle (with thermal conductivity $\kappa_p = \infty$, actually set as 10^{10} W m^{-1} K^{-1}) in the background (with thermal conductivity κ_b) and apply a horizontal thermal field $\boldsymbol{K}_0$. Consequently, the isotherms are all repelled, and the black arrows (denoting the directions of heat fluxes) are always perpendicular to the exterior boundary of the particle (Fig. 5.1a). The particle is isothermal, and a brief proof is as follows. We denote the temperature distribution of the particle as T_p. By solving the Laplace equation $\nabla \cdot (-\kappa \nabla T) = 0$, we can derive T_p as

$$T_p = \frac{-\kappa_b}{L_{p1}\kappa_p + (1 - L_{p1})\kappa_b} K_0 x_1 + T_0, \tag{5.1}$$

where $K_0 = |\boldsymbol{K}_0|$, T_0 is the reference temperature, and (x_1, x_2, x_3) denote the Cartesian coordinates. L_{p1} is the shape factor of the particle along x_1 axis, which will be discussed later. Equation (5.1) indicates that whatever value L_{p1} takes on, if $\kappa_p = \infty$, T_p is always a constant T_0. Physically, since heat fluxes ($\boldsymbol{J} = -\kappa \nabla T$) do not diverge, a direct conclusion from $\kappa = \infty$ is $\nabla T = 0$. In other words, a finite thermal conductivity with a constant-temperature boundary condition is equivalent to an infinite thermal conductivity. For comparison, we reset κ_p to a finite value ($\kappa_p < \infty$, actually set as 0.026 W m^{-1} K^{-1}) and apply a constant-temperature boundary condition on the boundary of the particle (Fig. 5.1b). As a result, the temperature profile and direc-

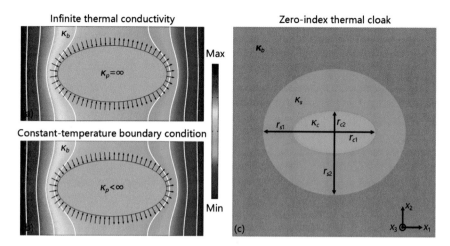

Fig. 5.1 **a** Temperature profile with an elliptical particle ($\kappa_p = \infty$, actually set as 10^{10} W m^{-1} K^{-1}) embedded in the background ($\kappa_b = 400$ W m^{-1} K^{-1}). **b** Temperature profile with a common particle ($\kappa_p < \infty$, actually set as 0.026 W m^{-1} K^{-1}) and a constant-temperature boundary condition with temperature (Max+Min)/2 embedded in the same background. Here, Max and Min denote the temperatures of the left and right boundaries, respectively. Rainbow surfaces denote temperature distributions, and white lines represent isotherms. **c** Schematic diagram of the zero-index thermal cloak. A constant-temperature boundary condition is applied to the core boundary, so the core has an effectively infinite thermal conductivity. Adapted from Ref. [22]

tions of heat fluxes are the same as those in Fig. 5.1(a), thus achieving an effectively infinite thermal conductivity with a constant-temperature boundary condition. Note that such an equivalence is only exact for temperature distributions.

5.3 Zero-Index Thermal Cloak

Zero-index metamaterials have been widely explored to manipulate electromagnetic waves due to their excellent properties [6–11]. We know that the directions of heat fluxes are always perpendicular to the exterior boundary of the particle with an (effectively) infinite thermal conductivity (Fig. 5.1a and b). This phenomenon follows zero refractive indexes in photonics, where electromagnetic waves travel outward vertically from materials with zero refractive indexes. Therefore, (effectively) infinite thermal conductivities can be directly used to design zero-index thermal metamaterials.

Zero-index thermal cloaks are a typical example of zero-index thermal metamaterials, which can be realized by introducing thermal convection [5]. Such a scheme requires the velocity of moving fluids to be infinite, which cannot be exactly realized, thus called near-zero-index thermal cloaks. In contrast, the present approach can realize exact-zero-index thermal cloaks with simple structures because only an external thermostatic sink is required to realize a constant-temperature boundary condition. In a word, thermal zero-index parameters indicate that thermal conductivities are (effectively) infinite. We apply a constant-temperature boundary condition on the core to realize an effectively infinite thermal conductivity, so the present thermal cloaks are also called zero-index thermal cloaks.

Zero-index thermal cloaks are essentially a core-shell structure (Fig. 5.1c). We denote the thermal conductivities of the core and shell as κ_c and κ_s, respectively. The subscript c (or s) represents the core (or shell) throughout this chapter. For generality, we consider an ellipsoidal case in three dimensions. The semi axes of the core and shell along x_i axis ($i = 1, 2, 3$) are denoted as r_{ci} and r_{si}, respectively. The effective thermal conductivity of such a core-shell structure (denoted as κ_e) is anisotropic, and the component along x_i axis (denoted as κ_{ei}) can be calculated by

$$\kappa_{ei} = \kappa_s \frac{L_{ci}\kappa_c + (1 - L_{ci})\kappa_s + f(1 - L_{si})(\kappa_c - \kappa_s)}{L_{ci}\kappa_c + (1 - L_{ci})\kappa_s - fL_{si}(\kappa_c - \kappa_s)}, \tag{5.2}$$

where $f = r_{c1}r_{c2}r_{c3}/(r_{s1}r_{s2}r_{s3})$ is core fraction. L_{ci} and L_{si} are, respectively, the shape factors of the core and shell along x_i axis, which can be calculated by

$$L_{wi} = \frac{r_{w1}r_{w2}r_{w3}}{2} \int_0^\infty \frac{du}{(u + r_{wi}^2)\sqrt{(u + r_{w1}^2)(u + r_{w2}^2)(u + r_{w3}^2)}}, \tag{5.3}$$

where the subscript w can take c or s, representing the shape factor of the core or shell. Note that only when the core-shell structure is concentric or confocal, can Eq. (5.2) predict the effective thermal conductivities exactly.

When a constant-temperature boundary condition is applied, the thermal conductivity of the core turns to infinity, namely $\kappa_c = \infty$. Then, Eq. (5.2) becomes

$$\kappa_{ei} = \kappa_s \frac{L_{ci} + f(1 - L_{si})}{L_{ci} - fL_{si}}. \tag{5.4}$$

Equation (5.4) can also be applied to two dimensions as long as we take $r_{w3} = \infty$ and $f = r_{c1}r_{c2}/(r_{s1}r_{s2})$. Then, Eq. (5.3) can be reduced to $L_{w1} = r_{w2}/(r_{w1} + r_{w2})$, $L_{w2} = r_{w1}/(r_{w1} + r_{w2})$, and $L_{w3} = 0$. As an intrinsic property, $L_{w1} + L_{w2} + L_{w3} = 1$ is always valid no matter in two or three dimensions.

5.4 Finite-Element Simulation

We perform simulations with COMSOL Multiphysics to confirm these theoretical analyses. Without loss of generality, we discuss two two-dimensional cases, including a circular one and an elliptical one. Figure 5.2a and b show the circular case where the thermal conductivities of the core and shell are $\kappa_c = 0.026$ and

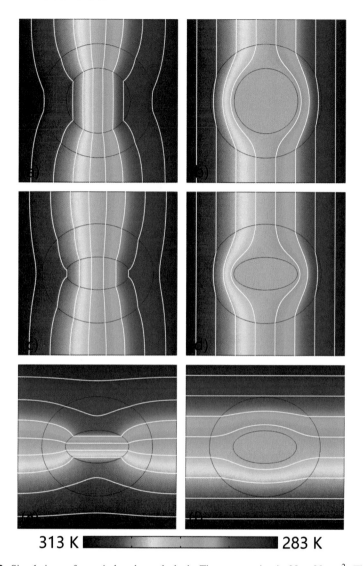

313 K ████████████ 283 K

Fig. 5.2 Simulations of zero-index thermal cloak. The system size is 20×20 cm^2. The thermal conductivities of the core and shell are $\kappa_c = 0.026$ and $\kappa_s = 203$ W m^{-1} K^{-1}, respectively. The thermal conductivities of the background in **a–b** and **c–f** are $\kappa_b = 400$ and $\kappa_b = \mathrm{diag}\,(358, 270)$ W m^{-1} K^{-1}, respectively. The inner and outer radii of the shell in **a** and **b** are $r_{c1} = r_{c2} = 4$ and $r_{s1} = r_{s2} = 7$ cm, respectively. The inner and outer semiaxes of the elliptical shell in **c–f** are $r_{c1} = 4$, $r_{c2} = 2$, $r_{s1} = 7$, and $r_{s2} = 6$ cm, respectively. The left and right columns show the temperature profiles without and with a constant-temperature boundary condition, respectively. The constant-temperature boundary condition is set at 298 K. The high and low temperatures are set at 313 and 283 K, respectively. The other boundaries are insulated. Adapted from Ref. [22]

$\kappa_s = 203$ W m^{-1} K^{-1}, respectively. The thermal conductivity of the background is set as $\kappa_b = \kappa_e = 400$ W m^{-1} K^{-1} which is derived from Eq. (5.4). When a constant-temperature boundary condition is not applied, the isotherms are contracted due to the smaller effective thermal conductivity of the core-shell structure (Fig. 5.2a). However, if we apply a constant temperature boundary condition to the boundary of the core, thermal cloaking can be achieved because the core has an effectively infinite thermal conductivity (Fig. 5.2b).

We further discuss the elliptical case with the same thermal conductivities of the core-shell structure, namely $\kappa_c = 0.026$ and $\kappa_s = 203$ W m^{-1} K^{-1}. Unlike the circular case, the effective thermal conductivity of the elliptical core-shell structure is anisotropic. Therefore, we set the thermal conductivity of the background as (expressed in the Cartesian coordinates) $\kappa_b = \kappa_e = \mathrm{diag}\,(358,\ 270)$ W m^{-1} K^{-1}, which is also derived from Eq. (5.4). In the presence of a horizontal thermal field, the smaller effective thermal conductivity of the core-shell structure makes the isotherms contracted (Fig. 5.2c), whereas a constant-temperature boundary condition helps us achieve thermal cloaking (Fig. 5.2d). The results are similar if the system is in the presence of a vertical thermal field (Fig. 5.2e and f).

5.5 Laboratory Experiment

For experimental demonstration, we fabricate six samples to confirm the six simulations in Fig. 5.2. We use integrated fabrication technology, indicating that the samples have no weld joints. The three samples without a constant-temperature boundary condition are presented in Fig. 5.3a, c, and e. Air holes are drilled on the copper plate to realize the designed thermal conductivities of the shell and background. Another three samples with a constant-temperature boundary condition are presented in Fig. 5.3b, d, and f. By immersing the central hollow cylinders in an external thermostatic sink with medium temperature, a constant-temperature boundary condition can be obtained, and effectively infinite thermal conductivity is achieved. Compared with other active schemes [23–25], our scheme does not require complicated temperature settings. These six samples' upper and lower surfaces are covered with transparent and foamed plastic to reduce environmental interferences. The sample photos of Fig. 5.3b, d, and f with top view are presented in Fig. 5.3g–i, respectively.

Then, we use the Flir E60 infrared camera to detect temperature distributions. The measured results corresponding to the six samples in Fig. 5.3 are presented in the left two columns of Fig. 5.4. We also perform finite-element simulations according to these six samples, and corresponding results are shown in the third and fourth columns of Fig. 5.4. For quantitative analyses, we plot the temperature distributions at $x_1 = -8$ cm for the first two rows and $x_2 = -8$ cm for the last row (the origin is in the center of each simulation). The experiments and simulations agree well with each other (Fig. 5.4m–o), thus confirming the feasibility of realizing zero-index thermal cloaks with effectively infinite thermal conductivities.

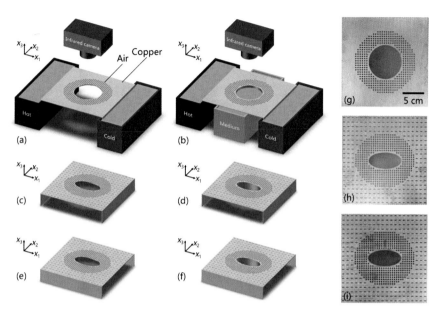

Fig. 5.3 Schematic diagrams of six samples and experimental setup. The thermal conductivities of air and copper are 0.026 and 400 W m^{-1} K^{-1}, respectively. The size of each sample is $20 \times 20 \times 4$ cm^3 with a copper thickness of 2 mm. The central air hole in **a** and **b** has a radius of 4 cm, and that in **c–f** has semiaxes of $r_{c1} = 4$ and $r_{c2} = 2$ cm. The effective shell radius in **a** and **b** is 7 cm, and the effective shell semiaxes in **c–f** are $r_{s1} = 7$ and $r_{s2} = 6$ cm. The air holes in the shell regions in **a–f** have the same radius of 1.6 mm, thus making the effective thermal conductivity of the shells to be 203 W m^{-1} K^{-1}. The air holes in the background regions in **c–f** have a major semi axis of 2.9 mm and a minor one of 0.8 mm, thus making the effective thermal conductivity of the backgrounds to be diag (358, 270) W m^{-1} K^{-1}. The distance between air holes in the shell region is 5 mm, and that in the background region is 10 mm. The temperatures of the hot, medium and cold sources are set at 313, 298, and 283 K, respectively. **g–i** Sample photos of **b**, **d**, and **f**, respectively. Adapted from Ref. [22]

The cloaking effect is also robust under more complicated conditions such as different directions of external fields, point heat sources, and three dimensions. Furthermore, thermal cloaking can be extended to other functions such as thermal camouflaging. Nevertheless, the scheme is applicable for only stable states because the temperature of a constant-temperature boundary condition is fixed.

5.6 Conclusion

We have shown that an effectively infinite thermal conductivity can be precisely achieved by applying a constant-temperature boundary condition to a common material. Meanwhile, an external thermostatic sink can easily realize the constant-temperature boundary condition. The current approach has direct applications in

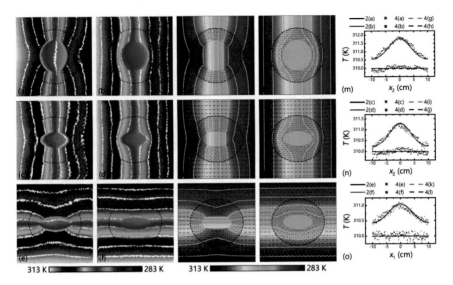

Fig. 5.4 Measured results (left two columns) and simulated results (the third and fourth columns) of the six samples in Fig. 5.3. Dashed lines are plotted for the convenience of comparison. (m) and (n) show the temperature distributions at $x_1 = -8$ cm (the origin is in the center of each simulation), and (o) shows the temperature distributions at $x_2 = -8$ cm. Each line corresponds to a figure shown in the legend. Adapted from Ref. [22]

designing zero-index thermal cloaks, which can work in highly conductive backgrounds with simple structures. These features, such as accuracy and simplicity, benefit practical applications. This work applies a constant-temperature boundary condition to realize effectively infinite thermal conductivity, which is expected to design more zero-index thermal metamaterials.

5.7 Exercise and Solution

Exercise

1. Derive Eq. (5.2).

Solution

1. Suppose the semiaxes of the core and shell along x_i to be r_{ci} and r_{si}, respectively. The conversion between the Cartesian coordinates and ellipsoidal (or elliptical) coordinates can be expressed as

$$\sum_i \frac{x_i^2}{\rho_j + r_{ci}^2} = 1, \tag{5.5}$$

with $j = 1$, 2 for two dimensions and $j = 1$, 2, 3 for three dimensions. $\rho_1 \left(> -r_{ci}^2 \right)$ represents the boundary of an ellipse or an ellipsoid. For example, $\rho_1 = \rho_c \, (= 0)$ and $\rho_1 = \rho_s$ can represent the inner and outer boundaries of the shell, respectively. In the presence of a thermal field along x_i, the heat conduction equation can be expressed as

$$\frac{\partial}{\partial \rho_1} \left(g \left(\rho_1 \right) \frac{\partial T}{\partial \rho_1} \right) + \frac{g \left(\rho_1 \right)}{\rho_1 + r_{ci}^2} \frac{\partial T}{\partial \rho_1} = 0, \tag{5.6}$$

with $g \left(\rho_1 \right) = \prod_i \left(\rho_1 + r_{ci}^2 \right)^{1/2}$. Accordingly, the temperatures of the core T_{ci}, shell T_{si}, and matrix T_{si} can be expressed as

$$T_{ci} = A_{ci} x_i, \tag{5.7a}$$
$$T_{si} = \left(A_{si} + B_{si} \phi_i \left(\rho_1 \right) \right) x_i, \tag{5.7b}$$
$$T_{mi} = \left(A_{mi} + B_{mi} \phi_i \left(\rho_1 \right) \right) x_i, \tag{5.7c}$$

with $\phi_i \left(\rho_1 \right) = \int_{\rho_c}^{\rho_1} \left(\left(\rho_1 + r_{ci}^2 \right) g \left(\rho_1 \right) \right)^{-1} d\rho_1$. A_{ci}, A_{si}, B_{si}, and B_{mi} are determined by the following boundary conditions,

$$T_{ci} \left(\rho_1 = \rho_c \right) = T_{si} \left(\rho_1 = \rho_c \right), \tag{5.8a}$$
$$T_{mi} \left(\rho_1 = \rho_s \right) = T_{si} \left(\rho_1 = \rho_s \right), \tag{5.8b}$$
$$\kappa_c \frac{\partial T_{ci}}{\partial \rho_1} \left(\rho_1 = \rho_c \right) = \kappa_s \frac{\partial T_{si}}{\partial \rho_1} \left(\rho_1 = \rho_c \right), \tag{5.8c}$$
$$\kappa_m \frac{\partial T_{mi}}{\partial \rho_1} \left(\rho_1 = \rho_s \right) = \kappa_s \frac{\partial T_{si}}{\partial \rho_1} \left(\rho_1 = \rho_s \right). \tag{5.8d}$$

We also need the following two mathematical skills,

$$\frac{\partial x_i}{\partial \rho_1} = \frac{x_i}{2 \left(\rho_1 + r_{ci}^2 \right)}, \tag{5.9a}$$

$$\frac{\partial}{\partial \rho_1} \left(\phi_i \left(\rho_1 \right) x_i \right) = \frac{x_i}{2 \left(\rho_1 + r_{ci}^2 \right)} \phi_i \left(\rho_1 \right) + \frac{x_i}{\left(\rho_1 + r_{ci}^2 \right) g \left(\rho_1 \right)}$$

$$= \frac{x_i}{2 \left(\rho_1 + r_{ci}^2 \right)} \left(\phi_i \left(\rho_1 \right) + \frac{2}{g \left(\rho_1 \right)} \right). \tag{5.9b}$$

Based on Eqs. (5.7) and (5.9), Eq. (5.8) can be written as

$$A_{ci} = A_{si} + B_{si} \phi_i \left(\rho_c \right), \tag{5.10a}$$
$$A_{mi} + B_{mi} \phi_i \left(\rho_s \right) = A_{si} + B_{si} \phi_i \left(\rho_s \right), \tag{5.10b}$$
$$\kappa_c A_{ci} = \kappa_s \left(A_{si} + B_{si} \phi_i \left(\rho_c \right) + \frac{2 B_{si}}{g \left(\rho_c \right)} \right), \tag{5.10c}$$

$$\kappa_m \left(A_{mi} + B_{mi} \phi_i \left(\rho_s \right) + \frac{2 B_{mi}}{g \left(\rho_s \right)} \right) = \kappa_s \left(A_{si} + B_{si} \phi_i \left(\rho_s \right) + \frac{2 B_{si}}{g \left(\rho_s \right)} \right). \quad (5.10d)$$

To further simplify $\phi_i \left(\rho_c \right)$ and $\phi_i \left(\rho_s \right)$, we define the shape factors of the core and shell along x_i as L_{ci} and L_{si}, respectively. They can be written as

$$L_{ci} = \frac{g \left(\rho_c \right)}{2} \int_{\rho_c}^{\infty} \frac{d\rho_1}{\left(\rho_1 + r_{ci}^2 \right) g \left(\rho_1 \right)}, \quad (5.11a)$$

$$L_{si} = \frac{g \left(\rho_s \right)}{2} \int_{\rho_s}^{\infty} \frac{d\rho_1}{\left(\rho_1 + r_{ci}^2 \right) g \left(\rho_1 \right)}, \quad (5.11b)$$

with $g \left(\rho_c \right) = \prod_i r_{ci}$, $g \left(\rho_s \right) = \prod_i r_{si}$, and $\sum_i L_{ci} = \sum_i L_{si} = 1$. For two dimensions, the shape factors can be further reduced to

$$L_{c1} = \frac{r_{c2}}{r_{c1} + r_{c2}}, \quad (5.12a)$$

$$L_{c2} = \frac{r_{c1}}{r_{c1} + r_{c2}}, \quad (5.12b)$$

$$L_{s1} = \frac{r_{s2}}{r_{s1} + r_{s2}}, \quad (5.12c)$$

$$L_{s2} = \frac{r_{s1}}{r_{s1} + r_{s2}}. \quad (5.12d)$$

Based on Eq. (5.11), we can derive

$$\phi_i \left(\rho_c \right) = \int_{\rho_c}^{\rho_c} \frac{d\rho_1}{\left(\rho_1 + r_{ci}^2 \right) g \left(\rho_1 \right)} = 0, \quad (5.13a)$$

$$\phi_i \left(\rho_s \right) = \left(\int_{\rho_c}^{\infty} - \int_{\rho_s}^{\infty} \right) \frac{d\rho_1}{\left(\rho_1 + r_{ci}^2 \right) g \left(\rho_1 \right)} = \frac{2 L_{ci}}{g \left(\rho_c \right)} - \frac{2 L_{si}}{g \left(\rho_s \right)}. \quad (5.13b)$$

With Eq. (5.10), we can derive A_{ci}, A_{si}, B_{si}, and B_{mi}. By setting $B_{mi} = 0$, we can further derive the effective thermal conductivity of the core-shell structure,

$$\kappa_e = \kappa_m = \kappa_s \frac{L_{ci} \kappa_c + (1 - L_{ci}) \kappa_s + (1 - L_{si}) (\kappa_c - \kappa_s) f}{L_{ci} \kappa_c + (1 - L_{ci}) \kappa_s - L_{si} (\kappa_c - \kappa_s) f}, \quad (5.14)$$

with $f = g \left(\rho_c \right) / g \left(\rho_s \right) = \prod_i r_{ci} / r_{si}$, indicating area fraction for two dimensions and volume fraction for three dimensions.

References

1. Xu, X., Zhang, Q.Q., Hao, M.L., Hu, Y., Lin, Z.Y., Peng, L.L., Wang, T., Ren, X.X., Wang, C., Zhao, Z.P., Wan, C.Z., Fei, H.L., Wang, L., Zhu, J., Sun, H.T., Chen, W.L., Du, T., Deng, B.W., Cheng, G.J., Shakir, I., Dames, C., Fisher, T.S., Zhang, X., Li, H., Huang, Y., Duan, X.F.: Double-negative-index ceramic aerogels for thermal superinsulation. Science **363**, 723 (2019)
2. Lindsay, L., Broido, D.A.: Enhanced thermal conductivity and isotope effect in single-layer hexagonal boron nitride. Phys. Rev. B **84**, 155421 (2011)
3. Pop, E., Mann, D., Wang, Q., Goodson, K., Dai, H.J.: Thermal conductance of an individual single-wall carbon nanotube above room temperature. Nano Lett. **6**, 96 (2006)
4. Balandin, A.A., Ghosh, S., Bao, W.Z., Calizo, I., Teweldebrhan, D., Miao, F., Lau, C.N.: Superior thermal conductivity of single-layer graphene. Nano Lett. **8**, 902 (2008)
5. Li, Y., Zhu, K.-J., Peng, Y.-G., Li, W., Yang, T.Z., Xu, H.-X., Chen, H., Zhu, X.-F., Fan, S.H., Qiu, C.-W.: Thermal meta-device in analogue of zero-index photonics. Nat. Mater. **18**, 48 (2019)
6. Huang, X.Q., Lai, Y., Hang, Z.H., Zheng, H.H., Chan, C.T.: Dirac cones induced by accidental degeneracy in photonic crystals and zero-refractive-index materials. Nat. Mater. **10**, 582 (2011)
7. Maas, R., Parsons, J., Engheta, N., Polman, A.: Experimental realization of an epsilon-near-zero metamaterial at visible wavelengths. Nat. Photonics **7**, 907 (2013)
8. Li, Y., Kita, S., Munoz, P., Reshef, O., Vulis, D.I., Yin, M., Loncar, M., Mazur, E.: On-chip zero-index metamaterials. Nat. Photonics **9**, 738 (2015)
9. Alam, M.Z., Leon, I.D., Boyd, R.W.: Large optical nonlinearity of indium tin oxide in its epsilon-near-zero region. Science **352**, 795 (2016)
10. Alam, M.Z., Schulz, S.A., Upham, J., Leon, I.D., Boyd, R.W.: Large optical nonlinearity of nanoantennas coupled to an epsilon-near-zero material. Nat. Photonics **12**, 79 (2018)
11. Chu, H.C., Li, Q., Liu, B.B., Luo, J., Sun, S.L., Hang, Z.H., Zhou, L., Lai, Y.: A hybrid invisibility cloak based on integration of transparent metasurfaces and zero-index materials. Light-Sci. Appl. **7**, 50 (2018)
12. Fan, C.Z., Gao, Y., Huang, J.P.: Shaped graded materials with an apparent negative thermal conductivity. Appl. Phys. Lett. **92**, 251907 (2008)
13. Chen, T.Y., Weng, C.-N., Chen, J.-S.: Cloak for curvilinearly anisotropic media in conduction. Appl. Phys. Lett. **93**, 114103 (2008)
14. Narayana, S., Sato, Y.: Heat flux manipulation with engineered thermal materials. Phys. Rev. Lett. **108**, 214303 (2012)
15. Xu, H.Y., Shi, X.H., Gao, F., Sun, H.D., Zhang, B.L.: Ultrathin three-dimensional thermal cloak. Phys. Rev. Lett. **112**, 054301 (2014)
16. Han, T.C., Bai, X., Gao, D.L., Thong, J.T.L., Li, B.W., Qiu, C.-W.: Experimental demonstration of a bilayer thermal cloak. Phys. Rev. Lett. **112**, 054302 (2014)
17. Ma, Y.G., Liu, Y.C., Raza, M., Wang, Y.D., He, S.L.: Experimental demonstration of a multiphysics cloak: manipulating heat flux and electric current simultaneously. Phys. Rev. Lett. **113**, 205501 (2014)
18. Han, T.C., Yang, P., Li, Y., Lei, D.Y., Li, B.W., Hippalgaonkar, K., Qiu, C.-W.: Full-parameter omnidirectional thermal metadevices of anisotropic geometry. Adv. Mater. **30**, 1804019 (2018)
19. Li, J.X., Li, Y., Li, T.L., Wang, W.Y., Li, L.Q., Qiu, C.-W.: Doublet thermal metadevice. Phys. Rev. Appl. **11**, 044021 (2019)
20. Xu, L.J., Huang, J.P.: Metamaterials for manipulating thermal radiation: transparency, cloak, and expander. Phys. Rev. Appl. **12**, 044048 (2019)
21. Zhou, Z.Y., Shen, X.Y., Fang, C.C., Huang, J.P.: Programmable thermal metamaterials based on optomechanical systems. ES Energy Environ. **6**, 85 (2019)

22. Xu, L.J., Yang, S., Huang, J.P.: Effectively infinite thermal conductivity and zero-index thermal cloak. EPL **131**, 24002 (2020)
23. Nguyen, D.M., Xu, H.Y., Zhang, Y.M., Zhang, B.L.: Active thermal cloak. Appl. Phys. Lett. **107**, 121901 (2015)
24. Guo, J., Qu, Z.G.: Thermal cloak with adaptive heat source to proactively manipulate temperature field in heat conduction process. Int. J. Heat Mass Transf. **127**, 1212 (2018)
25. Xu, L.J., Yang, S., Huang, J.P.: Dipole-assisted thermotics: experimental demonstration of dipole-driven thermal invisibility. Phys. Rev. E **100**, 062108 (2019)

Chapter 6
Theory for Hele-Shaw Convective Cloaks: Bilayer Scheme

Abstract Thermal convection is one of the three basic heat transfer mechanisms, profoundly influencing the natural environment, social production, and daily life. However, the high complexity of governing equation, which describes the coupling of heat and mass transfer, makes it challenging to manipulate thermal convection at will in both theory and experiment. Here, we consider the heat transfer in Hele-Shaw cells, a widely-used model of Stokes flow between two parallel plates with a small gap, and apply the scattering-cancellation technology to construct convective thermal materials with bilayer structures and homogeneous isotropic materials. By tailoring thermal conductivity and viscosity, we demonstrate cloaking devices that can simultaneously hide obstacles from heat and fluid motion and verify their robustness under various thermal-convection environments by numerical simulations. Our results show that about 80% of the temperature and pressure disturbances in the background caused by obstacles can be eliminated by the cloak. The developed approach can be extended to control other convection-diffusion systems or multiphysics processes. The results pave a promising path for designing various metadevices such as concentrators or sensors.

Keywords Hele-Shaw flows · Convective cloaks · Bilayer scheme

6.1 Opening Remarks

Metamaterials [1] (and metadevices [2]), usually made of artificially-structured composites, have been a powerful tool to manipulate physical fields in many realms [1–4], and provide functions beyond naturally-occurring materials. One typical methodology to design metamaterials is the transformation optics [5] and its counterparts in other physical fields [3]. However, the requirement for inhomogeneous and anisotropic parameters makes it difficult to fabricate devices designed by transformation optics. To overcome this bottleneck, scattering-cancellation technology (SCT) has been developed and successfully used in electromagnetism [6] and other fields [7]. Generally speaking, SCT can realize a similar function to transformation optics, while it only needs bilayer or monolayer structures and homogeneous isotropic bulk mate-

© The Author(s) 2023   65
L.-J. Xu and J.-P. Huang, *Transformation Thermotics and Extended Theories*,
https://doi.org/10.1007/978-981-19-5908-0_6

rials. SCT is sometimes called 'solving the equation directly' sometimes. The main procedure of this method is inversely finding the coefficients (material parameters) of the equation according to the required solution with analytical techniques. One type of equation that SCT often deals with is the Laplacian-type equation, which is just a Laplace equation in a homogeneous isotropic medium. Laplacian-type equations can describe the magnetic scalar potential in static magnetic fields [8], the temperature or electrostatic potential in heat or electrical conduction [9–14], or the liquid pressure in a potential flow [15, 16]. Based on them, various bilayer or monolayer metamaterials have been realized in the mentioned scenarios and independent multiphysics [17–20].

However, although thermal convection is one of the primary modes of heat transfer and plays an essential role in nature and human society, effective techniques like SCT for manipulating it are still lacking. This dilemma may result from the complexity of its multiphysical governing equations. In detail, convective heat flux contains not only an advection term due to the movement of fluid medium but also a conductive term in the nonisothermal flow. Therefore, the governing equations consist of the conduction-advection heat equations, the law of continuity for fluid motion, and the Navier-Stokes equations, which make it challenging to apply transformation optics or SCT, especially the Navier-Stokes equation. As a result, although thermal metamaterials [21, 22] have been developed for more than a decade and show potential in practical applications such as thermal management of electronic devices, thermal camouflage imaging, and radiative cooling [23–25], the progress of metamaterials in thermal convection seems insufficient.

Current advances in convective thermal metamaterials mainly benefit from choosing an appropriate simplified model of the Navier-Stokes equations. For example, Darcy's law describes the creeping flow or Stokes flow (Reynolds number $Re \ll 1$) in porous media [26]. By engineering the permeability of porous media, some fluid-flow metamaterials have been designed [27, 28], and this technique has been combined with the tailoring method of thermal conductivity to design convective thermal metamaterials [29–33]. In theory, convective thermal metamaterials can control heat flux and flow field simultaneously [29, 30]. However, due to the limited practical means to tailor the permeability, reports on experimentally realizing such fluid-flow or convective metamaterials are still scarce. More recently, another hydrodynamic model has been used to control fluid motion, i.e., the Stokes flow inside two parallel plates, and a series of experimental works have been reported [34–37]. The gap between two plates is much smaller than the characteristic length of the other two spatial dimensions, so the model is also called the Hele-Shaw flow or Hele-Shaw cell [38]. As an extension of the Poiseuille flow [39], the Hele-Shaw flow is quite a fundamental model in many fields like viscous fingering [40], microfluidics [41], parametric resonance [42], and flow-induced choking [43]. The fluid pressure in the Hele-Shaw cell also satisfies a Laplacian-type equation [44], and SCT has been employed to construct a monolayer fluid-flow cloak in the cell [15].

Here, we develop SCT to control thermal convection in a Hele-Shaw cell and employ it to seek suitable thermal conductivity and viscosity of artificial structures. By surrounding an area with two layers of homogeneous isotropic material, we can get the desired temperature and pressure distribution (and thus heat flow and velocity distribution) inside and outside these two layers. Unlike previous works on SCT, which dealt with single or decoupled fields, we investigate how to apply SCT to a set of coupled equations and simultaneously regulate multiphysical fields. As an application, we design a bilayer convective (thermo-hydrodynamic) cloak that can prevent obstacles from simultaneously disturbing the external thermal and flow fields. We show how the respective cloaking conditions of heat conduction and fluid motion are combined and successfully work together in thermal convection. Our design is further verified by numerical simulation under various convective environments, showing tough robustness.

6.2 Governing Equation

Given that heat transfers within a Hele-Shaw cell, in which the fluid demonstrates a creeping flow. The governing equations of this model contain the heat transfer equation [45]

$$\nabla \cdot \left(-\kappa \nabla T + \rho C^P T \mathbf{v} \right) = 0, \tag{6.1}$$

the law of continuity for fluid motion

$$\nabla \cdot (\rho \mathbf{v}) = 0, \tag{6.2}$$

and the Hele-Shaw equation [44]

$$\mathbf{v} = -\frac{h^2}{12\mu} \nabla P, \tag{6.3}$$

which is a simplification of the Navier-Stokes equation. Here, T is the temperature and $\mathbf{v}$ is the velocity of fluid motion. κ, ρ, C^P and μ are the thermal conductivity, density, specific heat at constant pressure and the dynamic viscosity of the fluid, respectively. In addition, h is the depth of the Hele-Shaw cell and P is the fluid pressure. Strictly speaking, Eq. (6.3) gives the average velocity $\mathbf{v}(x, y)$ along the z-axis if the plates of cells are put on the x-y plane, so we can treat the three-dimensional (3D) model as a two-dimensional (2D) one. Applying the divergence operator on Eq. (6.3) and comparing it with Eq. (6.2), we can see

$$\nabla \cdot \left(\frac{\rho h^2}{12\mu} \nabla P \right) = 0. \tag{6.4}$$

In a region where ρ, μ and h are all constants (or the ratio $\rho h^2/\mu$ keeps the same), Eq. (6.4) is just a Laplace's equation, and thus the Hele-Shaw flow is a classical potential flow like the Darcy flow in the porous media. On the other hand, substituting

Eq. (6.3) into Eq. (6.1) to eliminate the velocity term, the heat transfer equation can be written as

$$\nabla \cdot \left(\kappa \nabla T + \frac{\rho C^P h^2 T}{12\mu} \nabla P \right) = 0. \tag{6.5}$$

When the velocity is zero everywhere, Eq. (6.5) is also a Laplacian-type equation known as the Fourier's law of heat conduction. As mentioned above, various bilayer metamaterials have been realized in pure heat conduction or potential flows. We aim to obtain the material parameters to realize similar functions as conduction under the convective environment.

6.3   Bilayer Scheme and Scattering-Cancellation Technology

Now we consider the case where both the thermal bias and the pressure bias are applied on the x direction (see the heat sources and pressures applied in Fig. 6.1; other boundary conditions will be discussed with simulation validation in Part C of the Supporting Information), and assume that

$$\nabla P = f(\mathbf{r}) \nabla T \tag{6.6}$$

in the whole space. Moreover, by doing a variable substitution $\varphi(\mathbf{r}) = f(\mathbf{r}) \frac{\rho C^P h^2}{12\mu\kappa}$, Eq. (6.5) can be rewritten as

$$\nabla \cdot (\kappa (\nabla T + \varphi(\mathbf{r}) T \nabla T)) = 0. \tag{6.7}$$

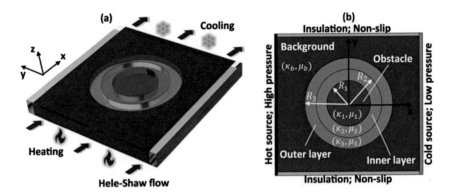

Fig. 6.1 Schematic design for a bilayer convective cloak in a Hele-Shaw cell. **a** The side view of the cell. **b** The top view of this quasi-two-dimensional model (in the x-y plane). Adapted from Ref. [46]

If φ is a constant in each domain, we can write the heat transfer equation with a compact form as

$$\nabla \cdot \left(\kappa \left(\nabla \left(T + \frac{1}{2}\varphi T^2 \right) \right) \right) = \nabla \cdot (\kappa \nabla \Phi) = 0, \qquad (6.8)$$

where $\Phi = T + \frac{1}{2}\varphi T^2$. This trick (similar to the Kirchhoff transformation in nonlinear heat conduction [47] which has been used in designing nonlinear thermal metamaterials [48]) makes the governing equation conform to the form of Laplace's equation if κ is also a constant. In the framework of bilayer metamaterials, we use subscript 1, 2, 3, and b to represent the central functional area inside the device, the inner layer, the outer layer, and the background outside the device, respectively (see Fig. 6.1b). The radius from inside to outside (corresponding to core, inner and outer layers in Fig. 6.1a) is R_1, R_2 and R_3.

SCT can be generalized as an inverse analytical calculation of possible material parameters in each region according to the desired physical field distribution in certain regions. However, whether we want to modulate temperature or pressure distribution in a thermal convection environment, these two variables (T and p) do not appear directly in the Laplace-like Eq. (6.8), but instead, another variable Φ that is a key difference between coupled multiphysics and single physics. Fortunately, under certain conditions (see the derivation details in Part A of the Supporting Information), bilayer devices such as invisibility cloaks can be realized by simultaneously modulating thermal conductivity and viscosity. This cloak work in both thermal and hydromechanical fields. Below we will deduce the designing parameters of such a cloak.

6.4 Convective Cloak Condition

As generally defined in metamaterials, cloaking can realize invisibility [5]. It means the scattering signals from an obstacle can be eliminated by a specific device surrounding it (named Criterion I which requires $T_b(r; r > R_3) = T_{\text{Ref}}(r; r > R_3)$ in heat transfer) and the flux cannot flow into the obstacle (named Criterion II which requires $c = 0$). For Laplacian-type governing equations with diffusion nature, scattering signals mean distortion of potential (such as temperature, fluid pressure and electrostatic potential) distribution in the background. Since Eq. (6.8) and the Fourier's law have the same form for Φ and T, we can expect that the thermal conductivity for a thermal cloak in convection is also similar to its counterpart in conduction. Of course, as mentioned before, we need to note that the independent variables and boundary conditions of these two equations are different. Therefore, the physical meanings of the corresponding conclusions are not exactly the same, unless there is no advection. In heat conduction, the condition for Criterion I using the general anisotropic monolayer structure has already been solved out [49], and its version for the isotropic bilayer scheme is

$$\kappa_1 = \frac{D_1 + D_2 + D_3 + D_4}{D_1 - D_2 - D_3 + D_4} \kappa_2, \tag{6.9}$$

where D_n ($n = 1, 2, 3, 4$) is given by

$$D_1 = -(\kappa_2 + \kappa_3)(\kappa_3 + \kappa_b) R_1^2, \tag{6.10a}$$

$$D_2 = -(-\kappa_2 + \kappa_3)(\kappa_3 + \kappa_b) R_2^2, \tag{6.10b}$$

$$D_3 = (\kappa_2 + \kappa_3)(\kappa_3 - \kappa_b) R_3^2, \tag{6.10c}$$

$$D_4 = (-\kappa_2 + \kappa_3)(\kappa_3 - \kappa_b) R_1^2 R_3^2 / R_2^2. \tag{6.10d}$$

In addition, previous works on preventing heat flux from entering into the cloak usually require that the inner layer is absolutely insulated, meaning that $\kappa_2 = 0$ [9–11]. In this case, κ_1 can take arbitrary real values, so the denominator in Eq. (6.9) must be zero, which can result in the familiar relationship [11]

$$\kappa_3 = \frac{R_3^2 + R_2^2}{R_3^2 - R_2^2} \kappa_b. \tag{6.11}$$

We must emphasize that, by now, Eq. (6.11) can only be seen as the condition for shielding scattering signals of Φ in thermal convection. Thermal conductivity engineering alone is not enough to achieve a cloak (for Φ) in convection, as our assumption on f or φ requires a certain viscosity distribution. From the condition that φ is a constant (see Part A of the Supporting Information), we should have

$$\mu_2 = \infty, \quad \mu_3^{-1} = \frac{R_3^2 + R_2^2}{R_3^2 - R_2^2} \mu_b^{-1}. \tag{6.12}$$

It is interesting that Eq. (6.12) is exactly the condition for a bilayer fluid-flow cloak [15, 16].

So far, we actually make a cloak for the potential Φ and pressure P. We should still verify the two criterions for T. First, from $\Phi_b = \Phi_{\mathrm{Ref}}$, $P_b = P_{\mathrm{Ref}}$ and the fact that the materials in reference and background are the same, we can obtain a differential equation for $T_b - T_{\mathrm{Ref}}$ as

$$\nabla (T_b - T_{\mathrm{Ref}}) = (T_{\mathrm{Ref}} - T_b) \frac{\rho_b C_b^P h_b^2}{12 \mu_b \kappa_b} \nabla P_b. \tag{6.13}$$

In a homogeneous medium, ∇P_{Ref} is a uniform field along the x axis, so there are two general solutions for Eq. (6.13), namely $T_b - T_{\mathrm{Ref}} \equiv 0$ and $T_b - T_{\mathrm{Ref}} \sim \exp\left(-\nabla_x P_b \frac{\rho_b C_b^P h_b^2}{12 \mu_b \kappa_b}\right)$. Since $T_b - T_{\mathrm{Ref}}$ must vanish on the left and right boundaries, the only possible solution is the trivial one, so Criterion I is met. Similarly, it can be deduced from $\nabla \Phi_1 = 0$ and $\nabla P_1 = 0$ that ∇T_1 equals zero. In conclusion, by tailoring thermal conductivity and viscosity based on Eqs. (6.11) and (6.12), the aim objects can be hidden in heat and fluid fields simultaneously within the artificial structures.

In our derivation, we do not give the assumption of whether the obstacle is a solid object or not. Under ideal conditions, the obstacle (and the inner layer) cannot move, so this concern does not matter. A perfect cloak ($\kappa_2 = 0$ and $\mu_2 = \infty$) does not care about the material inside since the inner layer isolates internal and external interactions. An interesting argument is that if the material inside the cloak has an extremely low conductivity ($\kappa_1 \to 0$) and high viscosity ($\mu_1 \to \infty$), Criterion II is met. Otherwise, an imperfect inner layer (with small and positive κ_2 and $1/\mu_2$) can be regarded as approximately insulated and immobile as long as

$$\frac{\mu_3}{\mu_2} \ll 1, \quad \frac{\mu_b}{\mu_2} \ll 1, \quad \frac{\kappa_2}{\kappa_3} \ll 1, \quad \frac{\kappa_2}{\kappa_b} \ll 1. \qquad (6.14)$$

In particular, when the inner layer and the central area are occupied by the same material, the bilayer cloak degenerates to a monolayer cloak.

6.5 Finite-Element Simulation

Now we verify our theoretical design by numerical simulation with the help of commercial finite-element software Comsol Multiphysics. As depicted in Fig. 6.1, the depth of 2D cell model is an extra parameter in the creeping flow module. In fact, besides the law of continuity, the governing equation of the Hele-Shaw flow or shallow channel approximation in Comsol Multiphysics is $\nabla p - \nabla \cdot \left(\mu \left(\nabla \mathbf{v} + \nabla \mathbf{v}^{\mathrm{T}} \right) \right) + \frac{12\mu}{h^2} \mathbf{v} = 0$. When $h \to 0$, this equation is reduced to Eq. (6.4). An alternative method is using the mathematics module to establish and solve Eq. (6.4). For the background material, we use water as a reference and set $\kappa_b = 0.6$ W m^{-1} K^{-1}, $\mu_b = 10^{-3}$ Pa s, $\rho_b = 1000$ kg m^{-3} and $C_b^P = 5000$ J kg^{-1} K^{-1}. The depth of the cell is set as $h = 2 \times 10^{-6}$ m, and the horizontal section (the x-y plane) is a square with side length equal to 2×10^{-4} m. The radius of the central region, inner layer and outer layer are respectively $R_1 = 0.25 \times 10^{-4}$ m, $R_2 = 0.4 \times 10^{-4}$ m, and $R_3 = 0.5 \times 10^{-4}$ m. In addition, the depth of the cell, the specific heat, and the density are set the same everywhere for the uniform product $\rho C^P h^2$. The applied temperature bias and pressure difference are 10 K and 500 Pa, respectively. The hot source (303.15 K) and the fluid inlet are set on the left boundary of the whole system, while the cold source (293.15 K) and the fluid outlet are both on the right side. We apply thermal insulation and non-slip conditions on the upper and bottom boundaries. If we do not consider the boundary layer, based on Eq. (6.3), the flow speed in the reference is 8.3×10^{-3} m/s. Using $2h_b$ as the characteristic linear dimension, the Reynolds number for the reference is 1.7×10^{-4}, which is consistent with the creeping flow hypothesis. In practice, we cannot achieve infinite thermal conductivity and zero viscosity for the cloak. Nevertheless, based on Eq. (6.14), we can set the inner layer

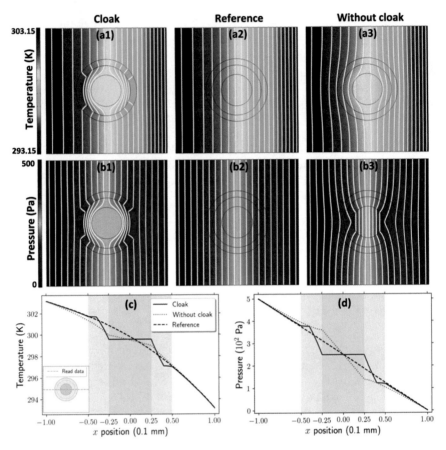

Fig. 6.2 Simulation results of a convective cloak and its contrast. **a1–a3** show the temperature distribution of the cloak, reference, and the case without a cloak, respectively, with twenty white isotherms. **b1–b3** illustrate the corresponding pressure distributions, also with twenty white isobars. **c** or **d** compares the temperature or pressure particularly in a chosen line segment $y = 0$ with data taken from **a1–b3**. The gray regions with different shades represent the bilayer structure (-0.5×10^{-4} m $< x < -0.25 \times 10^{-4}$ m and 0.25×10^{-4} m $< x < 0.5 \times 10^{-4}$ m) or the central obstacle (-0.25×10^{-4} m $< x < 0.25 \times 10^{-4}$ m) of the cloak. Adapted from Ref. [46]

approximately non-conductive and motionless with $\kappa_2 = 6 \times 10^{-4}$ W m^{-1} K^{-1} and $\mu_2 = 100$ Pa s. The parameters of the outer layer are $\kappa_3 = 2.73$ W m^{-1} K^{-1} and $\mu_3 = 2.20 \times 10^{-4}$ Pa s. Inside the convective cloak, we take $\kappa_1 = 2.4$ W m^{-1} K^{-1} and $\mu_1 = 0.01$ Pa s for the fluid material obstacle.

The simulation results of steady temperature and pressure performances are shown in Fig. 6.2a1–b3. Besides the cloak, 'Reference' represents the bare scenario with neither the cloak nor the obstacle, while 'Without cloak' scenario means putting only the obstacle into 'Reference.' According to isotherms and isobars, the temperature and pressure distributions of the convective cloak in the background are the same as the reference. The temperature and pressure inside the cloak are visually invariant in space. As a comparison, the obstacle distorts isotherms and isobars. In particular, for intuitive comparison, we detect the data from the line segment $y = 0$ (-10^{-4} m $< x < 10^{-4}$ m) in Fig. 6.2a1–b3 and plot them in **c** and **d** for the temperature and pressure, respectively. In both **c** and **d**, the blue dashed line ('Cloak') and black solid line ('Reference') match well in the background. Inside the cloak, the temperature and pressure patterns for the cloak show plateaus, which indicates the heat flux is almost zero, and the obstacle region approximately demonstrates no fluid flow. Further, we can define a ratio measuring the extent to which the disturbance of the external background by the obstacle is eliminated. The average absolute value of the temperature difference in the background area between the cloak and the reference is 0.016 K. Its counterpart between the reference and the case without a cloak is 0.095 K. So, 82% ($\approx 1 - 0.016/0.095$) of the temperature disturbance generated by the obstacle can be eliminated. The similar ratio for pressure can also be calculated as 77%. If we replace the fluid material obstacle with a solid one, the cloaking function for heat transfer and fluid flow still works well, although the fluid pressure inside the cloak is absent for a solid. Also, to make a perfect bilayer cloak with extreme parameters in numerical simulation, we can set the thermal insulation condition and the non-slip condition on the boundaries of the inner layer. However, we should notice the thermal and hydrodynamic fields in the central area (if fluids still occupy it) cannot be determined without giving extra information, although the background region would not be disturbed.

In Fig. 6.3, we show the heat flux density and velocity distributions for all the three convective devices and the reference. The advective heat flux density vector is defined as $\rho C^P (T - T_{\text{Amb}})\mathbf{v}$, where the ambient temperature T_{Amb} is 293.15 K. From a–d, it can be verified that the heat and mass fluxes are blocked in the inner layer and the obstacle, and the fluxes outside the bilayer structure are the same as the reference. In addition, we can find Reynolds number in the background for the cloak is Re $\approx 2.07 \times 10^{-3}$ if we use R_3 as the characteristic scale. It's known that the boundary layer can play an important role in thermal convection. Since the non-slip condition is applied on the upper and bottom boundaries (i.e., $y = \pm 10^{-4}$ m), the velocity should be zero, so we plot the flux data taken from the line segment $x = 0$ (-10^{-4} m $< y < 10^{-4}$ m) in Fig. 6.3e and f. We can see the velocity increases or decreases sharply near $x = -0.5 \times 10^{-4}$ m or $x = 0.5 \times 10^{-4}$ m in Fig. 6.3d, which corresponds to the laminar boundary layer. Also, the heat flux exhibits sharp changes near the boundary, but to a lesser extent than the change in velocity. This feature can be explained by the Péclet number Pe $=$ Re$\times$Pr (Pr is the Prandtl number). For the reference, Pe ≈ 0.7, so the conductive heat transport and the advective one are comparable in magnitude. Then, the advective heat flux demonstrating an obvious boundary layer plus the relatively uniform conductive heat flow results in the patterns in Fig. 6.3f. The Prandtl number is fixed for the same fluid material, so the Reynolds number affects the boundary layer. At the same time, according to our previous work [30], a larger Reynolds number also leads to a change in the temperature distribution pattern the isotherm area is not uniform. So we observe the cloaking performance

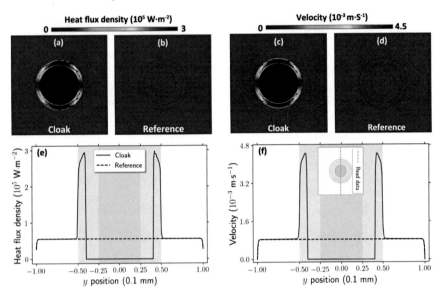

Fig. 6.3 Simulation results of heat flux density (**a** and **b**) and velocity distributions (**c** and **d**) for the cloak and the reference. The contour maps show the magnitude of heat flow or velocity vectors. The black arrows indicate the vector direction and the lengths of them also represent the vector size. **e** or **f** compares the heat flux density or speed particularly in a chosen line segment ($x = 0$) with data taken from **a**–**d**. Adapted from Ref. [46]

under different pressure differences in Part B of the Supporting Information and find our design works well.

We perform a three-dimensional (3D) simulation using the same material and structure parameters to mimic a more realistic working environment. Here, the shallow channel term $12\mu\mathbf{v}/h^2$ added in the 2D Navier-Stokes equations is not needed. The inertia term is also included in the governing equations so we actually use the full 3D incompressible Navier-Stokes equations $\nabla p - \nabla \cdot \left(\mu \left(\nabla \mathbf{v} + \nabla \mathbf{v}^T\right)\right) + \rho(\mathbf{v} \cdot \nabla)\mathbf{v} = 0$. The simulation results are shown in Fig. 6.4. Here, because the velocity on the surface $z = \pm h/2$ is zero, we should notice that the temperature or pressure distribution on the surface (mainly referring to the planes $z = \pm h/2$) is different from that on the central plane $z = 0$, and it is the latter that is of interest in 2D simulations. From Fig. 6.4, we can see that both the distributions on the surface and the cut plane show good cloaking effects in a low-speed flow environment.

6.6 Discussion

By now, the coupling between thermal field and fluid movement is unidirectional. In other words, only the velocity would influence the temperature distribution because we have not considered the thermal response of fluid properties like density and vis-

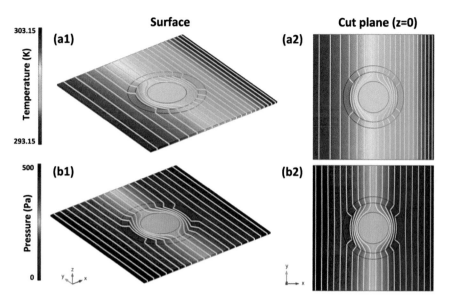

Fig. 6.4 Simulation results using 3D Navier-Stokes equations. **a1** and **b1** illustrate the temperature and pressure distributions of the surface, respectively. **b1** and **b2** show the distributions of the central x-y plane ($z = 0$.) Adapted from Ref. [46]

cosity, which might be important in a nonisothermal flow. Taking water as the working medium as we have done in his article, the density change is insignificant compared to viscosity under the applied thermal bias [50]. The dynamic viscosity of water can be expressed as a function of temperature with three parameters: $\mu = 10^{A + \frac{B}{T-C}}$ Pa s, where $A = -4.5318$, $B = 220.57$ K, and $C = 149.39$ K [51]. Taking $T = 20\,°C$, we can see $\mu \approx 1 \times 10^{-3}$ Pa s, which is just the value we have used for the background material in simulation. When $\Delta T = 10$ K, the thermal response of μ_b still has little influence on the temperature or pressure distributions no matter whether we assume μ_1, μ_2 and μ_3 change with temperature of the same magnitude or still let them temperature-independent. When the thermal bias increases, for example, to 50 K, and the four viscosities involved have the same temperature dependence, the functions like cloaking, sensing, and concentrating should not fail. Still, the pressure distribution will demonstrate uneven isobars. This variable viscosity (in fact, its reciprocal) behaves like a nonlinear thermal conductivity in bilayer conductive metadevices [48] (we can do a power series expansion to $1/\mu$ and get the polynomial form of temperature just like the nonlinear thermal conductivity often used in research). On the other hand, the viscous dissipation term can also be neglected in the framework of creeping flows, compared with the convective heat transfer. The discussion above might help to improve the feasibility of our design in potential practical applications.

Another important question is how to make κ and μ tunable in fluid materials. One idea is adding some inclusions or suspensions (like nanoparticles and even active matter) into the medium [52–54]. However, we must prevent the inclusions from moving from one domain to another and changing the spatial distribution of κ and μ. In some recent researches [34–36], solid pillars were put into the cell and fixed, which can enhance the effective viscosity of the solid-fluid structure. If viewed from another angle, this technique reduces the cell depth to zero without changing the viscosity of fluids. This technique also changes the thermal conductivity, specific heat, and density in the solid domain and thus influences the corresponding effective properties of the composites. So, the situation can be more complicated, involving tuning κ, μ, ρ, C^P, and even h. In this way, an effective medium theory considering heat transfer and fluid flow is needed to design suitable structures inversely. In Part D of the Supporting Information, we give a 3D cloak structure with only one fluid material by changing the depth of the outer layer and putting pillars in it. Although the parameter estimation is empirically given and relatively rough, our design does exhibit some invisibility effect.

6.7　Conclusion

In summary, through scattering-cancellation technology, we have established a framework to design bilayer convective metamaterials in a Hele-Shaw cell. We extend this approach to deal with coupled multiphysics. By engineering thermal conductivity and viscosity, we proposed a convective cloak that can realize thermal invisibility and hydrodynamic stealth at the same time. We also discuss the implications of the Reynolds number and directions of applied thermal bias and driving pressures, and the design shows robustness under different convective circumstances. Although we only consider circular layers surrounding a round-shape area, our design can be generalized to other geometries based on the existing and future research on Laplacian bilayer metamaterials, e.g., the elliptical structures [12, 48]. The material parameters needed in our design for each layer and the central area are homogeneous and isotropic, which could be achieved by sold-fluid composites. The related effective medium theory or inverse design technique remains developed. Our study might provide a promising method for feasible and flexible control of multiphysics processes.

6.8　Supporting Information

Part A: SCT Details for Thermal Convection
For the model shown in Fig. 6.1, the general solution of Φ_i ($i = 1, 2, 3$, and b) in the 2D scenario with the circular symmetry can be expressed as

$$\Phi_i = (A_{i0} + B_{i0} \ln r)(\alpha_{i0} + \beta_{i0}\theta) \tag{6.15}$$

$$+ \sum_{m=1}^{\infty} (A_{im}r^m + B_{im}r^{-m}) (\alpha_{im} \sin(m\theta) + \beta_{im} \cos(m\theta))$$

with polar coordinates (r, θ). By finding the right coefficients A_{im}, B_{im}, α_{im} and β_{im} based on the required manipulation function and certain boundary conditions, we can first obtain the inverse solution of thermal conductivities κ_i to realize the manipulation of Φ. Whether the required function can be realized for the temperature still needs some debates. Different from the familiar boundary conditions at infinity like $\nabla_x T (r = \infty) \sim \mathbf{e}_x$ ($\mathbf{e}_x$ is the unit vector along the x axis), here we should use $\nabla_x \Phi (r = \infty) \sim \mathbf{e}_x$ instead, and thus the temperature of a homogeneous medium could not vary linearly along the x axis. The degree of such non-uniformity depends on the value of φT, and the advection part behaves just like a nonlinear thermal conductivity. In addition, to guarantee φ is a constant in the whole system (thus T cannot be a multivalued function on the boundary of two domains), we must require

$$f_i(\mathbf{r}) \frac{\rho_i C_i^P h_i^2}{\mu_i \kappa_i} = f_j(\mathbf{r}) \frac{\rho_j C_j^P h_j^2}{\mu_j \kappa_j} \equiv C, \quad i \text{ or } j = 1, 2, 3, b. \tag{6.16}$$

Here C is a constant. Further, we can assume f, ρ, C^P and h are also constants in the whole space and obtain

$$\frac{1}{\mu_1 \kappa_1} = \frac{1}{\mu_2 \kappa_2} = \frac{1}{\mu_3 \kappa_3} = \frac{1}{\mu_b \kappa_b} \equiv C', \tag{6.17}$$

where C' is another constant. In other words, we should tailor μ to realize a convective metamaterial besides thermal conductivity engineering. It should be noted that Eq. (6.17) also gives the condition for a bilayer fluid-flow metamaterial, if we only consider Eqs. (6.2) and (6.3). We also need to neglect the difference between thermal insulation (the heat flow perpendicular to the boundary is zero) and non-slip (the velocity at the solid boundary is zero) boundary conditions. If the boundary layer is not significant, this neglect can be reasonable. For example, after calculating the thermal conductivity to avoid disturbing the distribution of Φ in the background, we can obtain the viscosity not to disturb P. Based on Eq. (6.4), tuning the ratio $\rho h^2/\mu$ is a more general strategy to make fluid-flow metamaterials. For simplicity, we only consider changing the conductivity and viscosity in the theory part. In this way, f is assumed to be a constant, and we must emphasize this is an approximation. When advection exits, f cannot be a strict constant, e.g., in a homogeneous medium. Nevertheless, we can keep this assumption and check how much the variable f will influence the performance of our design in the numerical simulation part.

In particular, for a convective cloak, we can find the assumption described by Eq. (6.16) can be relaxed by only requiring that it is valid in the background and outer layer. The zero conductivity and infinite viscosity in the inner layer have been enough to make Eq. (6.5) automatically satisfied in the inner layer and the obstacle.

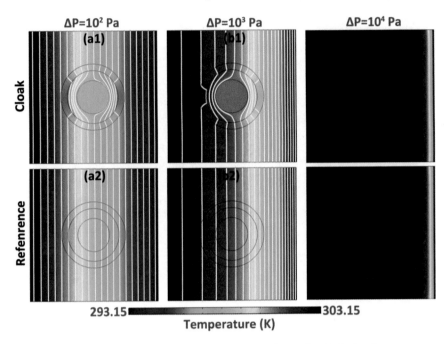

Fig. 6.5 Simulation results under different Reynolds numbers. **a1** and **a2** give the temperature distributions of the cloak and reference, respectively, when the pressure difference applied is 10^2 Pa. Similarly, **b1** and **b2** give those under a 10^3 Pa pressure difference, while **c1** and **c2** give those under a 10^4 Pa pressure difference. Adapted from Ref. [46]

As a result, we do not let $\mu_1/\mu_b = \kappa_b/\kappa_1$ or $\mu_2/\mu_b = \kappa_b/\kappa_2$ for the numerical simulations. What's more, since the boundary layers in Fig. 6.3 are quite thin, it will not undermine our assumption Eq. (6.16), i.e., the coefficient $f(\mathbf{r})$ in Eq. (6.6) should be a constant since the density, specific heat and cell depth are all kept invariant.

Part B: Performance with Different Reynolds Numbers
It is important to verify our conclusions under different Reynolds numbers (or Péclet numbers). Compared with pure conduction (Re = 0), a large Re can cause an obvious change in the temperature patterns, which means f is not strictly a constant even for the reference. Nevertheless, we can still test the performance of the cloak designed above. Here, we do three more simulations when ΔP takes 10^2 Pa, 10^3 Pa, and 10^4 Pa, respectively, and show the results of the cloak and reference in Fig. 6.5. Since the patterns of pressure distributions should not be changed under the Hele-Shaw regime, we only illustrate the temperature distributions here.

In Fig. 6.5a1 and a2, the pressure difference is 10^2 Pa, and the isotherms of the reference are almost evenly distributed. In (b1) and (b2), the pressure difference is 10^3 Pa, and the isotherms of the reference show a distinctly uneven distribution. The conductive flux can be neglected when the advective heat transfer is further enlarged in (c1) and (d1). Then the isotherms would be crowded on the side of the cold source (although we did not draw these overly dense isotherms), and the flow is isothermal almost everywhere. Thus the temperature gradient is close to zero except near the cold source. Anyway, we can see the cloaking effect is robust, and the assumption of creeping flow is still valid in the Hele-Shaw cell. Here, the temperature disturbance caused by the obstacle is eliminated by 95%, 81%, and 79%, respectively, for the three columns in Fig. 6.5. The cloaking ratio increases with the contribution of conductive heat transfer because our assumption that f is a constant is fulfilled when advection is absent.

Part C: Performance with Non-parallel Thermal and Pressure Biases

Now we turn to another aspect and consider the two applied biases are not in the same direction. Under this assumption, the relationship described by Eq. (6.6) should be revised. In general, we can decompose ∇P into two components which are parallel ($\nabla P_\parallel$) and perpendicular ($\nabla P_\perp$) to ∇T respectively, writing $\nabla P = P_\parallel + P_\perp$, and obtain

$$\nabla \cdot \left(\frac{\rho C^P h^2 T}{12\mu} \nabla P \right) = \frac{\rho C^P h^2 T}{12\mu} \nabla T \cdot \nabla P_\parallel. \tag{6.18}$$

$f(\mathbf{r})$ is now defined by $\nabla P_\parallel = f(\mathbf{r})\nabla T$ and equal to $(\nabla P \cdot \nabla T)/(\nabla T \cdot \nabla T)$. A special case is when the applied temperature difference and pressure difference are perpendicular to each other. In this case, we expect the patterns of temperature and pressure distributions approximately have a symmetry under rotating 90°. Then, f should be almost a constant (zero) [29] and we can apply SCT again and get the same parameters for the convective cloak.

To test our design, we let the pressure bias take 500 Pa along the y axis while the thermal bias is still kept at 10 K along the x axis. In Fig. 6.6a–d, we give the simulation results of the temperature and pressure distributions for both the cloak and the reference. The isotherms and isobars show the cloaking effect is achieved. More specifically, Fig. 6.6e and f are on a horizontal and a vertical line segments. We can also calculate the percentage of disturbances that are removed in this situation. The ratio for pressure deviation should be the same as its counterpart under parallel applied thermal and pressure biases. The ratio for temperature deviation is a little bit different, taking 79%.

Part D: Three-dimensional Convective Cloak

Here we propose a three-dimensional (3D) structure (see Fig. 6.7) to realize a convective cloak and use only one fluid material. The gray region in Fig. 6.7 illustrates the fluid domain. For simplicity, the inner layer of the cloak is provided as a solid thermal insulation material, so the structure is a monolayer one. Also, we can see that the outer layer has a larger depth than the background, which allows the fluid to have a smaller viscosity. Moreover, the larger depth of the fluid domain can enlarge the effective thermal conductivity due to a larger heat transfer cross-section. However, the increase in effective thermal conductivity caused by the depth alone does not exactly make the outer layer meet the conditions for a thermal cloak. More precisely, the thermal conductivity of the outer layer needs to be further improved. We

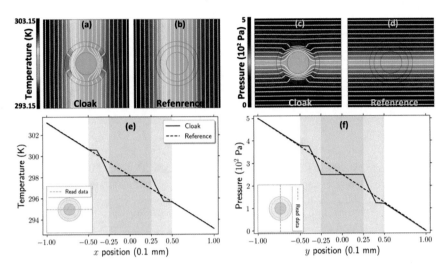

Fig. 6.6 **a** and **b** (or **c** and **d**) illustrate the temperature (or pressure) distribution of a convective cloak and the reference when the applied thermal bias and the pressure bias are vertical. In particular, **e** and **f** compare the temperature and pressure with data detected from the line $y = 0$ and line $x = 0$, respectively. Adapted from Ref. [46]

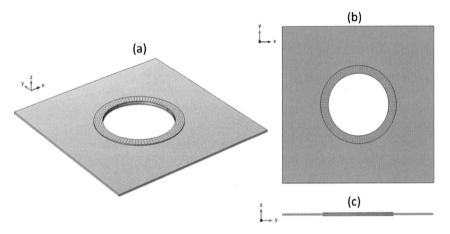

Fig. 6.7 Schematic diagram of the structure of the 3D convective cloak. The gray area represents the fluid domain and is symmetric concerning the central plane $z = 0$ (The center of the entire structure is the origin). **a** A 3D view. **b** A top view of x-y plane. We can see pillars (drawn as white small rounds in the enlarged image surrounded by a red box) placed in the deeper layer. The pillars form four radially equally spaced rings and exhibit a $4°$ rotational symmetry. **c** A view of y-z plane. Adapted from Ref. [46]

Fig. 6.8 Simulation results of the 3D convective cloak. **a** The temperature distribution of the surface. **b** The pressure distribution of the surface. Adapted from Ref. [46]

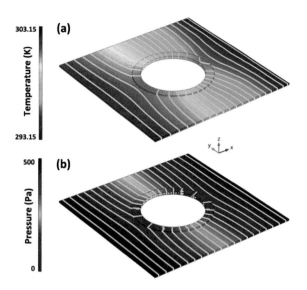

here combine the two methods used in the references [15, 34]. The outer layer is deeper than required for a normal fluid cloak in Ref. [15], and the too low viscosity is compensated by putting pillars [34] (see the black dots placed in the outer layer in Fig. 6.7a and b) to realize to fluid cloak again. Then, we can tune the thermal conductivity (and the heat capacity) of the pillars to meet the conditions for a convective cloak. The exact parameters of the pillars (volume fraction, thermal conductivity, and the heat capacity) are not easy to be solved through the existing effective medium theory.

As a rough estimation, we still use the same background material (water) and take the depth of the outer layer as $\sqrt{\dfrac{R_3^2 + R_2^2}{R_3^2 - R_2^2}} \approx 2.13$ times the background (the ratio is $\dfrac{R_3^2 + R_2^2}{R_3^2 - R_2^2}$ in Ref. [15] for a fluid cloak). In addition, four rings of cylindrical pillars are placed in the outer layer, and each ring consists of 90 pillars with a radius of $\sqrt{\dfrac{f_p(R_3^2 - R_2^2)}{N}}$. Here $N = 360$ is the total number of pillars, and f_p is the volume fraction of the pillars compared to the outer layer, which takes 0.32% in the following simulation. The material occupying the pillars can be air, soft matter, or solids. Here we set its thermal conductivity to be 40 W m^{-1} K^{-1}, its density to be 1000 kg m^{-3}, and its specific heat to be 5000 J kg^{-1} K^{-1}, which some mixture might achieve (e.g., copper and polydimethylsiloxane [55]). The simulation results are shown in Fig. 6.8. From the perspective of practical detection, here we give the temperature and pressure distribution of the surface. Although the cloaking effect is not perfect, compared with the case without a cloak (for example, see Fig. 6.2a3 and b3), we can see the bending of isotherms and isobars in the background region is alleviated to a certain extent. The parameters (including the volume fraction, geometry and thermal properties of the pillars, and the depth of the cloak layer) can be further optimized through analytical techniques and numerical methods.

Part E: Simulation Convergence Analysis

Since we adopt the finite element method to model thermal convection, a convergence analysis is necessary for reliability. We can get more accurate and convincing calculation results with the refinement of meshes. For example, we use five different sets of meshes, numbered 1 to 5, to execute independent simulations for the designed thermal cloak in Fig. 6.2. The size parameters of each set are shown in Table 6.1. Three groups of data at the positions ($x = \pm 0.5$ mm and $x = 0$) are extracted to compare the results produced by different grids. Bigger mesh numbers correspond to more elements, and 'Mesh 4' is the actual mesh used in Fig. 6.2.

We plot the temperature and pressure data read from $x = -0.5$ mm, $x = 0$ and $x = 0.5$ mm in Fig. 6.9. 'Mesh 4' is illustrated with solid lines, while its counterparts using other meshes are drawn in dashed lines with different colors. First, we observe Fig. 6.9a1, b1 and c1, demonstrating temperature comparisons. The latter two sets of mesh (numbered 4 and 5) produce smoother data lines than the first three sets (numbered 1, 2, and 3). In addition, the difference between 'Mesh 4' and 'Mesh 5' is very small, so the plots almost coincide, meaning that the simulation results of the temperature have converged to good accuracy. The same conclusions can be obtained for the pressure data in Fig. 6.9a2, b2 and c2. Therefore, the results using 'Mesh 4' in the previous simulations are credible.

Table 6.1 Mesh parameters. Five categories of meshes are adopted in modeling convective cloaks for convergence analysis. Adapted from Ref. [46]

	Max element size (m)	Min element size (m)	Total nodes	Nodes in $x = -0.5$ mm	Nodes in $x = 0$	Nodes in $x = 0.5$ mm
Mesh 1	7×10^{-6}	2×10^{-7}	7060	124	113	123
Mesh 2	5.6×10^{-6}	8×10^{-8}	8305	138	126	138
Mesh 3	2.6×10^{-6}	3×10^{-8}	13155	213	202	210
Mesh 4	1.34×10^{-6}	4×10^{-9}	31754	337	351	342
Mesh 5	6.34×10^{-7}	2×10^{-9}	138141	672	710	674

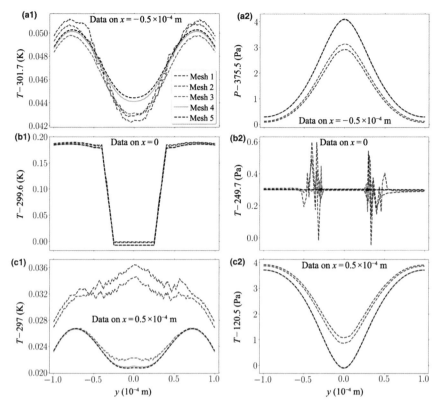

Fig. 6.9 Simulation results of the convective cloak using different meshes. **a1** and **a2** show the temperature and pressure data on $x = -5 \times 10^{-4}$ m. **b1** and **b2** show the corresponding data on $x = 0$, while **c1** and **c2** show those on $x = 5 \times 10^{-4}$ m. Adapted from Ref. [46]

6.9 Exercise and Solution

Exercise

1. Derive Eq. (6.9) in detail.

Solution

1. See Ref. [49].

References

1. Chen, H., Chan, C.T., Sheng, P.: Transformation optics and metamaterials. Nature Mater. **9**, 387 (2010)
2. Zhuledev, N.I., Kivshar, Y.S.: From metamaterials to metadevices. Nature Mater. **11**, 917 (2012)
3. Kadic, M., Bückmann, T., Schittny, R., Wegener, M.: Metamaterials beyond electromagnetism. Rep. Prog. Phys. **76**, 126501 (2013)
4. Fu, X., Cui, T.J.: Recent progress on metamaterials: From effective medium model to real-time information processing system. Prog. Quant. Electron. **67**, 100223 (2019)
5. Pendry, J.B., Schurig, D., Smith, D.R.: Controlling electromagnetic fields. Science **312**, 1780 (2006)
6. Alù, A., Engheta, N.: Achieving transparency with plasmonic and metamaterial coatings. Phys. Rev. E **72**, 016623 (2005)
7. P.-Y. Chen, J. Soric, and A. Alù, Invisibility and cloaking based on scattering cancellation, Adv. Mater. **24**, OP281COP304 (2012)
8. Gömöry, F., Solovyov, M., Šouc, J., Navau, C., Prat-Camps, J., Sanchez, A.: Experimental realization of a magnetic cloak. Science **335**, 1466 (2012)
9. Xu, H., Shi, X., Gao, F., Sun, H., Zhang, B.: Ultrathin three-dimensional thermal cloak. Phys. Rev. Lett. **112**, 054301 (2014)
10. Han, T., Bai, X., Gao, D., Thong, J.T.L., Li, B., Qiu, C.-W.: Experimental demonstration of a bilayer thermal cloak. Phys. Rev. Lett. **112**, 054302 (2014)
11. Han, T., Bai, X., Thong, J.T.L., Li, B., Qiu, C.-W.: Full control and manipulation of heat signatures: cloaking, camouflage and thermal metamaterials. Adv. Mater. **26**, 1731 (2014)
12. Han, T., Yang, P., Li, Y., Lei, D., Li, B., Hippalgaonkar, K., Qiu, C.-W.: Full-parameter omni-directional thermal metadevices of anisotropic geometry. Adv. Mater. **30**, 1804019 (2018)
13. Li, Y., Liu, C., Li, P., Lu, T., Chen, C., Guo, Z., Su, Y., Qiao, L., Zhou, J., Bai, Y.: Scattering cancellation by a monolayer cloak in oxide dispersion-strengthened alloys. Adv. Funct. Mater. **30**, 2003270 (2020)
14. Xu, G., Zhou, X., Zhang, J.: Bilayer thermal harvesters for concentrating temperature distribution. Int. J. Heat Mass Transfer **142**, 118434 (2019)
15. Tay, F.Y., Zhang, Y.M., Xu, H.Y., Goh, H.H., Luo, Y., Zhang, B.L.: A metamaterial-free fluid flow cloak. Natl. Sci. Rev. nwab205 (2021)
16. Wang, B., Shih, T.-M., Xu, L., Dai, G., Huang, J.: Intangible hydrodynamic cloaks for convective flows. Phys. Rev. Appl. **15**, 034014 (2021)
17. Ma, Y., Liu, Y., Raza, M., Wang, Y., He, S.: Experimental demonstration of a multiphysics cloak: manipulating heat flux and electric current simultaneously. Phys. Rev. Lett. **113**, 205501 (2014)
18. Lan, C., Bi, K., Gao, Z., Li, B., Zhou, J.: Achieving bifunctional cloak via combination of passive and active schemes. Appl. Phys. Lett. **109**, 201903 (2016)
19. Zhang, X., He, X., Wu, L.: A bilayer thermal-electric camouflage device suitable for a wide range of natural materials. Compos. Struct. **261**, 113319 (2021)
20. Yang, T., Bai, X., Gao, D., Wu, L., Li, B., Thong, J.T.L., Qiu, C.-W.: Invisible sensor: simultaneous sensing and camouflaging in multiphysical fields. Adv. Mater. **27**, 7752 (2015)
21. Fan, C.Z., Gao, Y., Huang, J.P.: Shaped graded materials with an apparent negative thermal conductivity. Appl. Phys. Lett. **92**, 251907 (2008)
22. Narayana, S., Sato, Y.: Heat flux manipulation with engineered thermal materials. Phys. Rev. Lett. **108**, 214303 (2012)
23. Yang, S., Wang, J., Dai, G., Yang, F., Huang, J.: Controlling macroscopic heat transfer with thermal metamaterials: theory, experiment and application. Phys. Rep. **908**, 1 (2021)
24. Li, Y., Li, W., Han, T., Zheng, X., Li, J., Li, B., Fan, S., Qiu, C.-W.: Transforming heat transfer with thermal metamaterials and devices. Nat. Rev. Mater. **6**, 488 (2021)
25. Hu, R., Xi, W., Liu, Y., Tang, K., Song, J., Luo, X., Wu, J., Qiu, C.-W.: Thermal camouflaging metamaterials. Mater. Today **45**, 120 (2021)

26. Darcy, H.: Les Fontaines Publiques de la Ville de Dijon. Dalmont, Paris (1856)
27. Urzhumov, Y.A., Smith, D.R.: Fluid flow control with transformation media. Phys. Rev. Lett. **107**, 074501 (2011)
28. Bowen, P.T., Urzhumov, Y.A., Smith, D.R.: Wake control with permeable multilayer structures: the spherical symmetry case. Phys. Rev. E **92**, 063030 (2015)
29. Dai, G., Shang, J., Huang, J.: Theory of transformation thermal convection for creeping flow in porous media: cloaking, concentrating, and camouflage. Phys. Rev. E **97**, 022129 (2018)
30. Dai, G., Huang, J.: A transient regime for transforming thermal convection: cloaking, concentrating and rotating creeping flow and heat flux. J. Appl. Phys. **124**, 235103 (2018)
31. Yeung, W.-S., Mai, V.-P., Yang, R.-J.: Cloaking: Controlling thermal and hydrodynamic fields simultaneously. Phys. Rev. Appl. **13**, 064030 (2020)
32. Yang, F., Xu, L., Huang, J.: Thermal illusion of porous media with convection-diffusion process: transparency, concentrating, and cloaking. ES Energy Environ. **6**, 45 (2019)
33. Xu, L., Huang, J.: Chameleonlike metashells in microfluidics: a passive approach to adaptive responses. Sci. China-Phys. Mech. Astron. **63**, 228711 (2020)
34. Park, J., Youn, J.R., Song, Y.S.: Hydrodynamic metamaterial cloak for drag-free flow. Phys. Rev. Lett. **123**, 074502 (2019)
35. Park, J., Youn, J.R., Song, Y.S.: Fluid-flow rotator based on hydrodynamic metamaterial. Phys. Rev. Appl. **12**, 061002 (2019)
36. Park, J., Youn, J.R., Song, Y.S.: Metamaterial hydrodynamic flow concentrator. Extreme Mech. Lett. **42**, 101061 (2021)
37. Boyko, E., Bacheva, V., Eigenbrod, M., Paratore, F., Gat, A.D., Hardt, S., Bercovici, M.: Microscale hydrodynamic cloaking and shielding via electro-osmosis. Phys. Rev. Lett. **126**, 184502 (2021)
38. Hele-Shaw, H.S.: The flow of water. Nature **58**, 34 (1898)
39. Sutera, S.P., Skalak, R.: The history of Poiseuille's law. Annu. Rev. Fluid Mech. **25**, 1 (1993)
40. Saffman, P.G.: Viscous fingering in Hele-Shaw cells. J. Fluid Mech. **173**, 73 (1986)
41. Stone, H.A., Stroock, A.D., Ajdari, A.: Engineering flows in small devices: microfluidics toward a lab-on-a-chip. Annu. Rev. Fluid Mech. **25**, 381 (2004)
42. Rajchenbach, J., Leroux, A., Clamond, D.: New standing solitary waves in water. Phys. Rev. Lett. **107**, 024502 (2011)
43. Box, F., Peng, G.G., Pihler-Puzović, D., Juel, A.: Flow-induced choking of a compliant Hele-Shaw cell. Proc. Natl. Acad. Sci. U.S.A. **117**, 30228 (2020)
44. Panton, R.L.: Incompressible Flow, 4th edn. Wiley, Hoboken (2013)
45. Bejan, A.: Convection Heat Transfer, 4th edn. Wiley, Hoboken (2013)
46. Dai, G.L., Zhou, Y.H., Wang, J., Yang, F.B., Qu, T., Huang, J.P.: Convective cloak in Hele-Shaw cells with bilayer structures: Hiding objects from heat and fluid motion simultaneously. Phys. Rev. Appl. **17**, 044006 (2022)
47. Kirchhoff, G.: Vorlesungen über die Theorie der Wärme. Teubner, Leipzig (1894)
48. Xu, L., Huang, J.: Metamaterials for manipulating thermal radiation: transparency, cloak, and expander. Phys. Rev. Appl. **12**, 044048 (2019)
49. Xu, L., Yang, S., Huang, J.: Passive metashells with adaptive thermal conductivities: chameleonlike behavior and its origin. Phys. Rev. Appl. **11**, 054071 (2019)
50. Haynes, W.M.: CRC Handbook of Chemistry and Physics, 97th edn. CRC Press, Boca Raton (2017)
51. Viswanath, D.S., Ghosh, T., Prasad, D.H.L., Dutt, N.V.K., Rani, K.Y.: Viscosity of Liquids: Theory, Estimation, Experiment, and Data. Springer, Dordrecht (2007)
52. Liu, M.-S., Lin, M.C.-C., Tsai, C.Y., Wang, C.-C.: Enhancement of thermal conductivity with Cu for nanofluids using chemical reduction method. Int. J. Heat Mass Transf. **49**, 3028 (2006)
53. Tuteja, A., Duxbury, P.M., Mackay, M.E.: Multifunctional nanocomposites with reduced viscosity. Macromolecules **40**, 9427 (2007)
54. Rafaï, S., Jibuti, L., Peyla, P.: Effective viscosity of microswimmer suspensions. Phys. Rev. Lett. **104**, 098102 (2010)
55. Shang, J., Tian, B.Y., Jiang, C.R., Huang, J.P.: Digital thermal metasurface with arbitrary infrared thermogram. Appl. Phys. Lett. **113**, 261902 (2018)

Chapter 7
Theory for Coupled Thermoelectric Metamaterials: Bilayer Scheme

Abstract In this chapter, we theoretically design bilayer thermoelectric metamaterials based on the generalized scattering-cancellation method. By solving the governing equations directly, we formulate the specific parameter requirements for desired functionalities beyond existing single-field or decoupled multi-field Laplacian metamaterials. Unlike the recently reported transformation thermoelectric flows, bilayer schemes do not require inhomogeneity and anisotropy in constitutive materials. Finite-element simulations confirm the analytical results and show robustness under various exterior conditions. Feasible experimental design with naturally occurring materials is also proposed for further proof-of-principle verification. Our theoretical method may be extended to other coupled multiphysical systems such as thermo-optics, thermomagnetics, and optomechanics.

Keywords Thermoelectric coupling · Bilayer scheme · Multiphysical field

7.1 Opening Remarks

Metamaterials have shown superior control ability beyond naturally occurring materials in both wave [1–9] and diffusion [10–18] systems. The transformation theory [1–4, 10, 11] and scattering-cancellation method [8, 9, 12–14], as two common approaches for manipulating physical fields, have achieved great success in artificial structure design. In particular, the latter is based on solving steady-state governing equations directly under given boundary conditions, leading to isotropic and homogeneous design parameters. However, if multiple fields act on an individual system, for example, there exist heat and electric fluxes simultaneously [19–21], the governing equations are hard to handle because of the newly-introduced coupling terms induced by thermoelectric (TE) effects. Appropriate theoretical methods need to be developed for designing such multiphysical metamaterials.

   Early research on tailoring TE fields focused on the decoupled cases, which means that heat and current flows transfer independently without interaction [22–26]. This simplified hypothesis facilitates the generalization of transformation theory or scattering-cancellation method from extensively-studied single physics to multi-

© The Author(s) 2023   87
L.-J. Xu and J.-P. Huang, *Transformation Thermotics and Extended Theories*,
https://doi.org/10.1007/978-981-19-5908-0_7

physics. Nevertheless, it usually deviates from actual situations because the coupling terms are omitted. Recently, transformed TE metamaterials were reported [27, 28], which extended the transformation theory from controlling a single field to coupled TE field. The form invariance of TE governing equations under coordinate transformation remains valid, and corresponding transformation rules on physical parameters are deduced. However, inhomogeneous and anisotropic TE materials are still required, just as their counterparts in single physics. Although some laminar-structure schemes with natural TE materials are proposed for mimicking the predicated TE parameters [27–30], experimental realization remains lacking due to the complexity of manufacture and availability of materials. Considering the challenges mentioned above, the scattering-cancellation method, which facilitates manufacture with simplified structures and homogeneous isotropic materials, could be a feasible route to practical implementation in TE control.

We propose a bilayer scheme based on the scattering-cancellation method for manipulating TE fields with naturally occurring TE materials. By introducing a generalized auxiliary potential, we construct Laplacian-form governing equations. We then derive the required thermal conductivity, electrical conductivity, and the Seeback coefficient for achieving cloaking, concentrating, and sensing functionalities. Finite-element simulations confirm our theoretical design and show the robustness of the proposed bilayer design under various conditions. Compared with the transformation TE theory, anisotropy and inhomogeneity are no longer necessities, making the manufacturing more convenient. The theoretical results and device behaviors can be naturally extended to other coupled multiphysics.

7.2 Theoretical Foundation

Let us consider a steady TE transport process where physical parameters are scalar at each local position. That is, the isotropy of TE materials is stipulated. In such an isotropic system, the governing equations can be described by [21]

$$j = -\sigma \nabla \mu - \sigma S \nabla T, \tag{7.1}$$

$$\nabla \cdot j = 0, \tag{7.2}$$

$$q = -\kappa \nabla T + T S j, \tag{7.3}$$

$$\nabla \cdot q = -\nabla \mu \cdot j, \tag{7.4}$$

where q and j are thermal and electric flows respectively, T and μ refer to temperature and electric potentials, and κ and σ denote to scalar thermal and electrical conductivities. S is Seeback coefficient for coupling heat and current flows. We define U as an auxiliary generalized potential, which is expressed as

$$U = \mu + TS. \tag{7.5}$$

Combining Eqs. (7.1)–(7.5), two identical relations about U can be obtained as

$$\sigma \nabla^2 U = 0 \tag{7.6}$$

and

$$\kappa \nabla^2 T = \sigma \nabla U \cdot \nabla U. \tag{7.7}$$

Note that Eq. (7.6) has a Laplacian form, so it is possible to map the field distribution of U by tailoring σ in a bilayer structure with a similar method employed in single-physics cases [12, 13]. Then we resort to remolding Eq. (7.7) for detecting the direction relation between ∇T and ∇U. The Poisson equation Eq. (7.7) has the solution consisting of two parts. One is the general solution of its corresponding Laplace equation

$$\kappa \nabla^2 T = 0. \tag{7.8}$$

The other is the particular solution. We can see the identical relation

$$\kappa \nabla T = \sigma U \nabla U \tag{7.9}$$

should always be valid to make Eq. (7.7) be satisfied. This can be deduced by taking the divergence of Eq. (7.9) in both sides as

$$\kappa \nabla^2 T = \sigma \nabla (U \nabla U) = \sigma (\nabla U \cdot \nabla U + U \nabla^2 U) = \sigma \nabla U \cdot \nabla U. \tag{7.10}$$

Then we can conclude that ∇T is always parallel to ∇U in its particular solution. Now we are in the position to discuss the relation between ∇T and ∇U in the general solution. T will thus be manipulated like U. Combining Eqs. (7.6) and (7.8), which are both Laplace equations, we can get the following conditions to make ∇T parallel to ∇U. Condition I is

$$S = S_0, \tag{7.11}$$

indicating that S keeps invariant in the whole space. Condition II is

$$\sigma = C\kappa, \tag{7.12}$$

where C is a constant for keeping σ and κ proportional in the whole space. Condition III relies on boundary condition settings. It means that external thermal and electrical fields should be parallel for ensuring homodromous ∇U and ∇T at each point. These three conditions enable us to map T distribution by tailoring U, which is described by a Laplacian-form governing equation. Then, we can define $f(\mathbf{r})$, a coordinate-dependent scalar function, to denote the relationship between ∇U and ∇T as

$$\nabla U = f(\mathbf{r}) \nabla T. \tag{7.13}$$

Next, we will handle the electrical potential μ. Note that S is constant, by combining Eqs. (7.5) and (7.13) together, we can obtain

$$\nabla \mu = (f(r) - S)\nabla T. \tag{7.14}$$

Evidently, $\nabla \mu$ is also parallel to ∇T and ∇U. So once Eqs. (7.11) and (7.12) are satisfied simultaneously, and the boundary temperature and potential fields are parallel, we can manipulate TE flows. Since bilayer is the most simplified structure for realizing specific functionalities such as cloaking, concentrating, and sensing with isotropic materials in a single field [12, 32, 33], we design TE cloaking, invisible sensing, and concentrating devices with bilayer configurations for verification. More layers will achieve the same effects but cannot improve the behaviors, which has been discussed sufficiently in many single-field metamaterial research.

We design three different functionalities in a size-fixed bilayer structure with background thermal conductivity κ_b and electrical conductivity σ_b, as shown in Fig. 7.1. For simplification without loss of generality, we only consider two-dimensional cases, which can readily be transferred to three-dimensional systems. According to the deductions above, the parameter requirements, i.e., Eqs. (7.11) and (7.12), should be satisfied simultaneously. And some additional conditions for realizing different functions are required. We set σ_0, σ_1, σ_2 as respective electrical conductivities from the center to the outer layer. Same definitions are employed for κ_0, κ_1, κ_2. Detailed parameter settings are as follows.

For cloaking [12], which prevents TE flows from running into the center without distorting the ambient temperature and potential distributions outside, as shown in Fig. 7.1b, the additional conditions for the inner layer should be

$$\sigma_1 \approx 0, \tag{7.15}$$

which make the inner layer a nearly-perfect thermal/electrical insulation material. And the outer layer should be

$$\sigma_2 = \sigma_b(r_2{}^2 + r_1{}^2)/(r_2{}^2 - r_1{}^2), \tag{7.16a}$$

guarantying no distortion of ambient temperature and potential outside.

For invisible sensing [32], which maintains the original temperature and potential in both center and background regions for obtaining accurate sensor effects, as shown in Fig. 7.1c, the additional conditions are found as

$$\sigma_1 = \left[\sigma_0 A_3 - \sigma_b A_1 + \sqrt{(\sigma_0 - \sigma_b)\left(\sigma_0 A_2^2 - \sigma_b A_1^2\right)} \right] / A_5, \tag{7.17a}$$

$$\sigma_2 = \left[\sigma_0 A_2 - \sigma_b A_4 - \sqrt{(\sigma_0 - \sigma_b)\left(\sigma_0 A_2^2 - \sigma_b A_1^2\right)} \right] / A_6, \tag{7.17b}$$

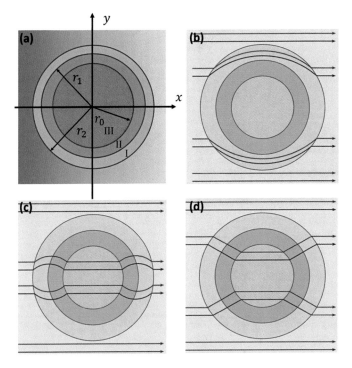

Fig. 7.1 **a** Schematic diagram of bilayer TE metamaterials. The core is marked as region III with electrical conductivity σ_0 and thermal conductivity κ_0. The inner layer is marked as region II with σ_1 and κ_1. The outer layer is marked as region I with σ_2 and κ_2. The background in gray has σ_b and κ_b. The electrical conductivity, thermal conductivity, and Seeback coefficient S are homogeneous isotropic scalars in each region. The Cartesian coordinate (x-y) is built on designed metamaterials with the overlapping origin and center point. **b** Illustration of a TE cloak. **c** Illustration of a TE invisible sensor. **d** Illustration of a TE concentrator. Red and blue lines represent heat and electrical fluxes, respectively, in **b**–**d**. Adapted from Ref. [31]

where

$$A_1 = r_0^2(r_1^2 + r_2^2) + r_1^2(r_1^2 - 3r_2^2), \tag{7.18a}$$

$$A_2 = r_0^2(3r_1^2 - r_2^2) - r_1^2(r_1^2 + r_2^2), \tag{7.18b}$$

$$A_3 = r_0^2(2r_0^2 - r_1^2 - r_2^2) + r_1^2(r_1^2 - r_2^2), \tag{7.18c}$$

$$A_4 = r_1^2(r_0^2 - r_1^2) + r_2^2(r_0^2 + r_1^2 - 2r_2^2), \tag{7.18d}$$

$$A_5 = 2(r_0^2 - r_1^2)(r_0^2 - r_2^2), \tag{7.18e}$$

$$A_6 = 2(r_0^2 - r_2^2)(r_1^2 - r_2^2). \tag{7.18f}$$

κ_1 and κ_2 follow the formally-similar parameter requirements as σ_1 and σ_2. It is noted that two sets of parameters are available in sensing design within a fixed geometry structure. We arbitrarily adopt one of them here.

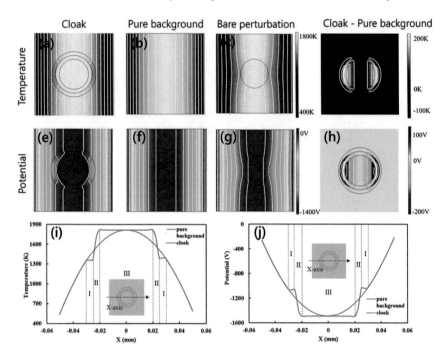

Fig. 7.2 Simulation results of the TE cloak under parallel external thermal and electrical fields. Isothermal or isopotential lines are marked in white. **a** Temperature distribution of the matrix plus cloak. **b** Temperature distribution of the pure matrix. **c** Temperature distribution of the bare perturbation. **d** Temperature difference distribution between **a** and **b**. **e** Potential distribution of the matrix with a cloak. **f** Potential distribution of the pure matrix as a reference. **g** Potential distribution of the bare perturbation. **h** Potential difference distribution between **d** and **e**. **i** Quantitative temperature comparison between **a** and **b** at the chosen line, which crosses the origin along the x axis. **j** Quantitative potential comparison between **d** and **e** at the chosen line, which crosses the origin along the x axis. Different regions (I, II, and III) are indicated in **i** and **j**, corresponding to the model in Fig. 7.1a. Backgrounds are outside region I. Adapted from Ref. [31]

For concentrating [33], which can enhance the gradients of temperature and potential in the center without distorting the ambient ones, as shown in Fig. 7.1d, the additional condition for σ_0, σ_1, and σ_2 can be written as

$$
\begin{aligned}
\sigma_0 = &[r_2^2 r_0^2 (\sigma_2 - \sigma_1) (\sigma_2 - \sigma_b) / r_1^2 - r_0^2 (\sigma_1 + \sigma_2) (\sigma_b + \sigma_2) \\
&+ r_1^2 (\sigma_1 - \sigma_2) (\sigma_b + \sigma_2) + r_2^2 (\sigma_1 + \sigma_2) (\sigma_2 - \sigma_b)]\sigma_1 / \\
&[r_2^2 r_0^2 (\sigma_2 - \sigma_1) (\sigma_2 - \sigma_b) / r_1^2 - r_0^2 (\sigma_1 + \sigma_2) (\sigma_b + \sigma_2) \\
&+ r_1^2 (\sigma_2 - \sigma_1) (\sigma_b + \sigma_2) - r_2^2 (\sigma_1 + \sigma_2) (\sigma_2 - \sigma_b)],
\end{aligned}
\tag{7.19}
$$

which is obtained by solving the Laplacian equation and then set the coefficient of the nonlinear term of the ambient potential zero. Similar forms of the relation between κ_0, κ_1, and κ_2 are also requested. Given that all the required conditions are met, the

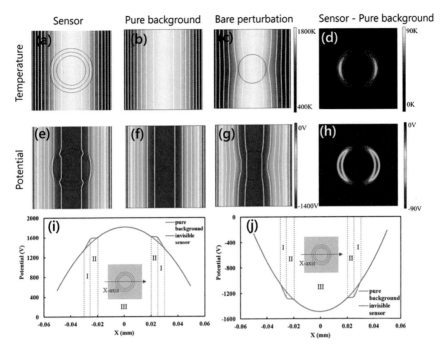

Fig. 7.3 Simulation results of the TE invisible sensor under parallel thermal and electrical boundary conditions. All figure arrangements are the same as Fig. 7.2 except the functionality of the central device. Adapted from Ref. [31]

ratio of the temperature/potential gradient in the center to the temperature/potential gradient in the background, which can describe the efficiency of concentrating, can be tailored by changing the dimensions and conductivities of the layers. So far, we have listed three sets of parameters for achieving three functionalities in TE transport. It is noted that the rationality of generalization from single physics to coupled multiphysics is established on the basis that Eqs. (7.11) and (7.12) should be satisfied simultaneously.

7.3 Finite-Element Simulation

We perform finite-element simulations with the commercial software COMSOL Multiphysics to confirm the proposed theoretical models. A two-dimensional bilayer structure of $r_0 = 0.02$ m, $r_1 = 0.025$ m, and $r_2 = 0.03$ m is employed. The bilayer structure is embedded at the center of a matrix, whose length is 0.11 m, as shown in Fig. 7.1a. To demonstrate the functionalities of the cloak, invisible sensor, and concentrator, we obtain three sets of thermal conductivity, electrical conductivity, and Seebeck coefficient for each case, as listed in Table 7.1. For boundary conditions, tem-

Table 7.1 Simulation parameter settings of TE cloaking, invisible sensing and concentrating. For simplicity, Seebeck coefficient is set as 1 in all regions. (This value is much larger than common materials, which will induce stong coupling effects between heat and electricity.) Adapted from Ref. [31]

	Cloak	Invisible sensor	Concentrator
Thermal conductivity (W/(m·K))			
κ_0	50	50	0.21
κ_1	0.01	378.5	10
κ_2	554.54	58.5	277
κ_b	100	100	50
Electrical conductivity (S/m)			
σ_0	50	50	0.21
σ_1	0.01	378.5	10
σ_2	554.54	58.5	277
σ_b	100	100	50
Seebeck coefficient (V/m)			
S	1	1	1

perature and potential gradients should be parallel. So we set boundary conditions as follows. The temperatures of the left and right boundaries are 273.15 K and 333.15 K. The potentials of the left and right boundaries are 0 V and 50 V. Upper and lower boundaries are thermally and electrically insulated. To show the effectiveness and accuracy of these three metamaterials, we also compare them with bare-perturbation and pure-background results. We perform simulations of these references under the same boundary conditions and plot the temperature and potential distribution of metamaterials and references. Differences in temperature and potential distribution illustrate the changes in temperature and potential between the metamaterials and pure backgrounds. These simulation results of cloak, concentrator and invisible sensor are demonstrated in Figs. 7.2, 7.3 and 7.4.

As shown in Figs. 7.2d and h, 7.3d and h, 7.4d and h, both the temperature and potential differences in backgrounds are nearly zero, which means none of these three metamaterials have distorted the ambient temperatures or potentials. This is also confirmed by the overlapping parts of the curves in Figs. 7.2i and j, 7.3i and j, 7.4i and j. As contrast, in Fig. 7.2c and g, 7.3c and g, 7.4c and g, the ambient temperatures and potentials are manifestly distorted by the bare perturbations. For the cloak, we can see in Fig. 7.2a and e or i and j, the temperature and potential gradients at the center are nearly zero, which means that thermal and electric flows are prevented from running into the center. For the sensor, which refers to the core region coated by the bilayer structure in Fig. 7.3a and e, it can be intuitively seen that the core temperature and potential are consistent before and after the sensor is embedded. In Fig. 7.3i and j, the curves of metamaterials and references fit well at the core and ambient regions. Therefore, we may safely say that the sensor can measure

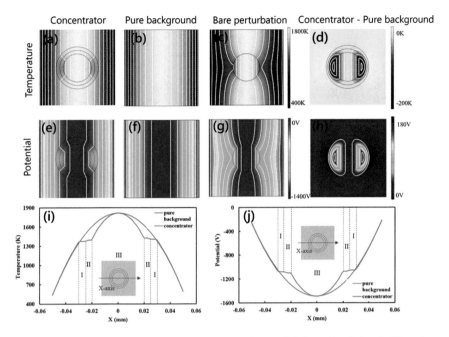

Fig. 7.4 Simulation results of the TE concentrator under parallel thermal and electrical boundary conditions. All figure arrangements are the same as Fig. 7.2 except the functionality of the central device. Adapted from Ref. [31]

the ambient temperature and potential without introducing any distortion. For the concentrator, Fig. 7.4a and e show that both the temperature and potential gradients in the core are greater than the ambient. From Fig. 7.4i and j, we can see more clearly that along the x-axis, the temperature and potential gradients are enhanced at the center.

To verify that only under the condition ∇T is parallel to $\nabla \mu$ can our design be exactly effective, we perform two simulations for the cloak when ∇T is not parallel to $\nabla \mu$, see Fig. 7.5. We set the upper and lower boundary temperatures in the upper two panels as 273.15 K and 333.15 K and potentials as 0 V and 50 V, respectively. In the lower two panels, a linear point heat source with the power of 6×10^6 W m^{-3} K^{-1} is applied at the left-bottom corner of the matrix, whose position is $(-0.049, -0.049)$ cm. The neighbor sides of the source are insulated, and the temperature of the remaining two sides is kept at 300 K. The results are shown in Fig. 7.5. Along the x axis, the difference between the ambient temperature (potential) of the pure matrix and the matrix with a cloak has some minor gaps. The designed schemes are not strictly accurate under nonparallel external fields. But it can still be regarded as a well approximated result based on the curves in Fig. 7.5c, f, i, and l, showing great accordance at background regions. The robustness of our design makes it adaptive under multiple complicated conditions.

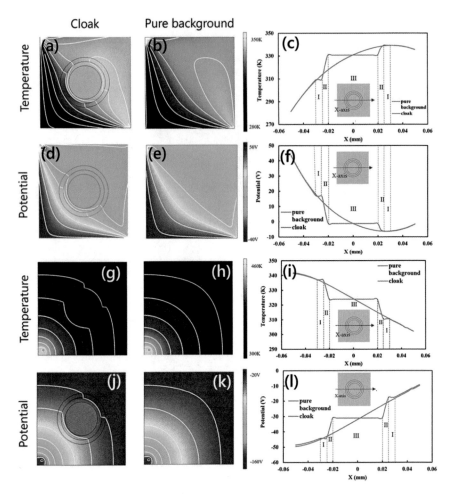

Fig. 7.5 a–f Simulation results of the TE cloak under the perpendicular boundary temperature and potential fields. Isothermal or isopotential lines are marked in white. **a** Temperature distribution of the matrix plus cloak. **b** Temperature distribution of the pure matrix. **c** Quantitative temperature comparison between **a** (cloak) and **b** (reference) at the chosen line, which crosses the origin along the *x* axis. **d** Potential distribution of the matrix with a cloak. **e** Potential distribution of the pure matrix. **f** Quantitative potential comparison between **d** (cloak) and **e** (reference) at the chosen line, which crosses the origin along the *x* axis. Different regions (I, II, and III) are indicated in **g** and **h**, corresponding to the model in Fig. 7.1a. **g–l** Simulation results of the TE cloak under the *y*-direction external potential fields and point heat sources at the left-bottom corner. Adapted from Ref. [31]

7.4 Discussion

Although actual materials may not perfectly meet the requirements put forward in our theory, we further verify that it is possible that practical realization to an approximate extent can be achieved. Many TE materials, such as ionic-conducting materials, can yield a large variety of TE characteristics due to various mechanisms and tuning methods such as changing the doping ratio [34] or humidity [35]. Therefore, this provides the physical possibility for searching for available materials. Compared with transformation optics requiring extremely anisotropic and inhomogeneous properties, though the proposed scattering cancellation methodology cannot achieve some effects such as rotating, our scheme will yield isotropic and homogeneous parameters to achieve the same effects of cloaking, concentrating, and sensing. Once we have suitable TE materials, the bilayer design will make it easier to manufacture corresponding metamaterials. Another issue is that the role of contact resistance, especially the thermal contact resistance (TCR), may affect the practical results [36]. TCR arises due to limited contact areas at the interface and lattice mismatch at the boundaries of different materials. According to the acoustic mismatch or diffusive mismatch model, the latter is usually too slight to be considered at the macroscale. In most reported macroscale experiments, the former is usually eliminated by "solid plus soft matter" structures. Even without such structures, the experimental results of a decoupled TE sensor, based on common metals, are in accord with the theory, ignoring the contact resistance [25].

7.5 Conclusion

In conclusion, we have built a scattering-cancellation method for manipulating coupled TE transport and designed three representative devices with bilayer schemes. Considering that TE governing equations are no longer Laplacian forms, additional constraint conditions are required beyond single-field cases. Our deduced requirements of constant Seebeck coefficient and proportional thermal/electrical conductivities echo with the results of the transformation TE method [27, 28] under homogeneous isotropic background conditions. And we further point out that the external TE distribution will not be affected only by applying parallel external thermal and electrical fields on the devices. However, simulation results also verify the robustness of our design under other boundary conditions, which can broaden the practical application range. Our work may provide hints for manipulating coupled multiphysical fields beyond single-physics Laplacian transport, which doubtlessly simplifies the requirements on materials and structures of existing transformation metamaterials. Moreover, since TE effects are widely utilized in practical applications, ranging from generating electric power from waste heat to solid-state-based cooling down, our work may help facilitate device preparation and raise energy conversion efficiency.

7.6 Exercise and Solution

Exercise

1. Derive the clear relations about U, including boundary conditions and parameter requirements.

Solution

1. The introduction of auxiliary generalized potential U and the analyses on corresponding boundary condition settings are provided. First, we consider U in a certain domain. Combing Eqs. (7.1) and (7.2), we can obtain

$$\nabla \cdot \sigma \nabla (\mu + ST) = 0. \tag{7.20}$$

Considering Eq. (7.5), we can write

$$\nabla \cdot \sigma \nabla U = 0. \tag{7.21}$$

Substituting Eq. (7.4) into Eq. (7.3), we have

$$-\kappa \nabla^2 T + S\nabla T \cdot \boldsymbol{j} + ST\nabla \cdot \boldsymbol{j} = -\nabla \mu \cdot \boldsymbol{j}. \tag{7.22}$$

According to Eq. (7.1), that is $\nabla \cdot \boldsymbol{j} = 0$, Eq. (7.22) can be simplified as

$$\kappa \nabla^2 T = (\nabla \mu + S\nabla T) \cdot \boldsymbol{j}. \tag{7.23}$$

Substituting Eqs. (7.2) and (7.5) into Eq. (7.23), we can thus obtain another equation about U as

$$\kappa \nabla^2 T = \sigma \nabla U \cdot \nabla U. \tag{7.24}$$

Now let us discuss the boundary condition settings of U. Apparently, U is a combination of T and μ. For T and μ, the boundary behaviors are already known as

$$T_i = T_{i+1}, \tag{7.25a}$$

$$\kappa_i \frac{\partial T_i}{\partial r} = \kappa_{i+1} \frac{\partial T_{i+1}}{\partial r}, \tag{7.25b}$$

$$\mu_i = \mu_{i+1}, \tag{7.25c}$$

$$\sigma_i \frac{\partial \mu_i}{\partial r} = \sigma_{i+1} \frac{\partial \mu_{i+1}}{\partial r}, \tag{7.25d}$$

where i and $i + 1$ denote two adjacent domains. Because U satisfies Laplace equation Eq. (7.6), to make U be manipulated by tailoring σ in a way similar to that proposed by Ref. [12], similar boundary behaviors will also be required

$$U_i = U_{i+1},$$ (7.26a)

$$\sigma_i \frac{\partial U_i}{\partial r} = \sigma_{i+1} \frac{\partial U_{i+1}}{\partial r}.$$ (7.26b)

According to Eq. (7.5), we can rewrite Eq. (7.26) as

$$\mu_i + S_i T_i = \mu_{i+1} + S_{i+1} T_{i+1},$$ (7.27a)

$$\sigma_i \frac{\partial \mu_i}{\partial r} + \sigma_i S_i \frac{\partial T_i}{\partial r} = \sigma_{i+1} \frac{\partial \mu_{i+1}}{\partial r} + \sigma_{i+1} S_{i+1} \frac{\partial T_{i+1}}{\partial r}.$$ (7.27b)

Substituting Eqs. (7.25a), (7.25c) into (7.27a), we have

$$S_i = S_{i+1},$$ (7.28)

from which the conclusion that S should keep invariant in all domains can be easily deduced. Meanwhile, Eq. (7.27b) can be rewritten as

$$\sigma_i \frac{\partial \mu_i}{\partial r} + \frac{\sigma_i S_i}{\kappa_i} \frac{\kappa_i \partial T_i}{\partial r} = \sigma_{i+1} \frac{\partial \mu_{i+1}}{\partial r} + \frac{\sigma_{i+1} S_{i+1}}{\kappa_{i+1}} \frac{\kappa_{i+1} \partial T_{i+1}}{\partial r}.$$ (7.29)

Hence substituting Eqs. (7.25b), (7.25d) into (7.27b), we have

$$\frac{\sigma_i S_i}{\kappa_i} = \frac{\sigma_{i+1} S_{i+1}}{\kappa_{i+1}}.$$ (7.30)

Making use of the Eq. (7.28), we can eventually obtain

$$\frac{\sigma_i}{\kappa_i} = \frac{\sigma_{i+1}}{\kappa_{i+1}},$$ (7.31)

from which a generalized conclusion that σ and κ are proportional between different domains, i.e., condition II or Eq. (7.13), can be easily deduced. For condition III, since ∇T and $\nabla \mu$ should be parallel where there are sources or boundary temperatures/potentials, it is obvious that the sources or boundary temperatures/potentials should appear in pairs.

References

1. Pendry, J.B., Schurig, D., Smith, D.R.: Controlling electromagnetic fields. Science **312**, 1780 (2006)
2. Chen, H.Y., Chan, C.T., Sheng, P.: Transformation optics and metamaterials. Nat. Mater. **9**, 387 (2010)
3. Pendry, J.B., Aubry, A., Smith, D.R., Maier, S.A.: Transformation optics and subwavelength control of light. Science **337**, 549 (2012)

4. Xu, L., Chen, H.Y.: Conformal transformation optics. Nat. Photonics **9**, 15 (2015)
5. Yang, Z., Mei, J., Yang, M., Chan, N.H., Sheng, P.: Membrane-type acoustic metamaterial with negative dynamic mass. Phys. Rev. Lett. **101**, 204301 (2008)
6. Zigoneanu, L., Popa, B.I., Cummer, S.A.: Three-dimensional broadband omnidirectional acoustic ground cloak. Nat. Mater. **13**, 352 (2014)
7. Cummer, S.A., Christensen, J., Alù, A.: Controlling sound with acoustic metamaterials. Nat. Rev. Mater. **1**, 16001 (2016)
8. Alù, A., Engheta, N.: Achieving transparency with plasmonic and metamaterial coatings. Phys. Rev. E **72**, 016623 (2005)
9. Gomory, F., Solovyov, M., Souc, J., Navau, C., Prat-Camps, J., Sanchez, A.: Experimental realization of a magnetic cloak. Science **335**, 1466 (2012)
10. Fan, C.Z., Gao, Y., Huang, J.P.: Shaped graded materials with an apparent negative thermal conductivity. Appl. Phys. Lett. **92**, 251907 (2008)
11. Chen, T.Y., Weng, C.-N., Chen, J.-S.: Cloak for curvilinearly anisotropic media in conduction. Appl. Phys. Lett. **93**, 114103 (2008)
12. Han, T.C., Bai, X., Gao, D.L., Thong, J.T.L., Li, B.W., Qiu, C.-W.: Experimental demonstration of a bilayer thermal cloak. Phys. Rev. Lett. **112**, 054302 (2014)
13. Xu, H.Y., Shi, X.H., Gao, F., Sun, H.D., Zhang, B.L.: Ultrathin three-dimensional thermal cloak. Phys. Rev. Lett. **112**, 054301 (2014)
14. Su, C., Xu, L.J., Huang, J.P.: Nonlinear thermal conductivities of core-shell metamaterials: rigorous theory and intelligent application. EPL **130**, 34001 (2020)
15. Maldovan, M.: Sound and heat revolutions in phononics. Nature **503**, 209 (2013)
16. Li, Y., Li, W., Han, T.C., Zheng, X., Li, J.X., Li, B.W., Fan, S.H., Qiu, C.-W.: Transforming heat transfer with thermal metamaterials and devices. Nat. Rev. Mater. **6**, 488 (2021)
17. J. Wang, G. L. Dai, and J. P. Huang, Thermal metamaterial: fundamental, application, and outlook. Isience **23**, 101637 (2020)
18. Yang, S., Wang, J., Dai, G.L., Yang, F.B., Huang, J.P.: Controlling macroscopic heat transfer with thermal metamaterials: theory, experiment and application. Phys. Rep. **908**, 1 (2021)
19. Bell, L.E.: Cooling, heating, generating power, and recovering waste heat with thermoelectric systems. Science **321**, 1457 (2008)
20. Domenicali, C.A.: Irreversible thermodynamics of thermoelectricity. Rev. Mod. Phys. **26**, 1103 (1954)
21. Biswas, K., He, J.Q., Blum, I.D., Wu, C.-I., Hogan, T.P., Seidman, D.N., Dravid, V.P., Kanatzidis, M.G.: High-performance bulk thermoelectrics with all-scale hierarchical architectures. Nature **489**, 11439 (2012)
22. Li, J.Y., Gao, Y., Huang, J.P.: A bifunctional cloak using transformation media. J. Appl. Phys. **108**, 074504 (2010)
23. Moccia, M., Castaldi, G., Savo, S., Sato, Y., Galdi, V.: Independent manipulation of heat and electrical current via bifunctional metamaterials. Phys. Rev. X **4**, 021025 (2014)
24. Ma, Y.G., Liu, Y.C., Raza, M., Wang, Y.D., He, S.L.: Experimental demonstration of a multiphysics cloak: manipulating heat flux and electric current simultaneously. Phys. Rev. Lett. **113**, 205501 (2014)
25. Yang, T.Z., Bai, X., Gao, D.L., Wu, L.Z., Li, B.W., Thong, J.T.L., Qiu, C.-W.: Invisible sensors: simultaneous sensing and camouflaging in multiphysical fields. Adv. Mater. **27**, 7752 (2015)
26. Lan, C.W., Bi, K., Fu, X.J., Li, B., Zhou, J.: Bifunctional metamaterials with simultaneous and independent manipulation of thermal and electric fields. Opt. Express **24**, 23072 (2016)
27. Stedman, T., Woods, L.M.: Cloaking of thermoelectric transport. Sci. Rep. **7**, 6988 (2017)
28. Shi, W., Stedman, T., Woods, L.M.: Transformation optics for thermoelectric flow. J. Phys-Energy **1**, 025002 (2019)
29. Wang, J., Shang, J., Huang, J.P.: Negative energy consumption of thermostats at ambient temperature: electricity generation with zero energy maintenance. Phys. Rev. Appl. **11**, 024053 (2019)
30. Shi, W., Stedman, T., Woods, L.M.: Thermoelectric transport control with metamaterial composites. J. Appl. Phys. **128**, 025104 (2020)

31. Qu, T., Wang, J., Huang, J.P.: Manipulating thermoelectric fields with bilayer schemes beyond Laplacian metamaterials. EPL **135**, 54004 (2021)
32. Xu, L.J., Yang, S., Huang, J.P.: Effectively infinite thermal conductivity and zero-index thermal cloak. EPL **131**, 24002 (2020)
33. Xu, G.Q., Zhou, X., Zhang, J.Y.: Bilayer thermal harvesters for concentrating temperature distribution. Int. J. Heat Mass Transf. **142**, 118434 (2019)
34. He, X., Cheng, H., Yue, S., Ouyang, J.: Quasi-solid state nanoparticle/(ionic liquid) gels with significantly high ionic thermoelectric properties. J. Mater. Chem. A **8**, 10813 (2020)
35. Kim, S.L., Lin, H.T., Yu, C.: Thermally chargeable solid-state supercapacitor. Adv. Energy Mater. **6**, 1600546 (2016)
36. Zheng, X., Li, B.W.: Effect of interfacial thermal resistance in a thermal cloak. Phys. Rev. Appl. **13**, 024071 (2020)

Chapter 8
Theory for Enhanced Thermal Concentrators: Thermal Conductivity Coupling

Abstract In this chapter, we propose the theory of conductivity coupling to solve the problem that the concentrating efficiency of a thermal concentrator is restricted by its geometric configuration. We first discuss a monolayer scheme with an isotropic thermal conductivity, which can break the upper limit but is still restricted by the geometric structure. We further explore another degree of freedom by considering the monolayer scheme with an anisotropic thermal conductivity or adding the second shell with an isotropic thermal conductivity, thereby freeing the concentrating efficiency from the geometric configuration. Finite-element simulations are performed to confirm the theoretical predictions, and experimental suggestions are also provided to improve feasibility. These results may have potential applications for thermal camouflage and provide insights into other diffusive systems such as static magnetic fields and DC fields for achieving similar behaviors.

Keywords Thermal concentrating efficiency · Thermal conductivity coupling · Apparently negative thermal conductivity

8.1 Opening Remarks

The theory of transformation thermotics [1, 2] has promoted an advanced control of heat transfer based on thermal metamaterials [3, 4]. As a representative example, a thermal concentrator [5–24] can increase its interior temperature gradient without distorting its exterior one. So far, many schemes have been proposed to design thermal concentrators. The initial explorations are based on the theory of transformation thermotics [5–14] which is a bridge linking space transformations and material transformations. Therefore, the effect of thermal concentrating can be achieved by coating a region (i.e., the core) with a designed shell (i.e., the thermal concentrator). This scheme has three features: (I) the thermal conductivities inside and outside the shell are identical; (II) the shell has an anisotropic thermal conductivity that is commonly realized by a layered structure [15–19]; and (III) both temperature gradient and heat flux are enhanced in the core. An alternative scheme is based on the effective medium theory [20–22] with also three features: (I) the thermal conductivity inside the shell

L.-J. Xu and J.-P. Huang, *Transformation Thermotics and Extended Theories*,
https://doi.org/10.1007/978-981-19-5908-0_8

should be smaller than that outside the shell; (II) the shell requires only a homogeneous and isotropic thermal conductivity; and (III) temperature gradient increases but heat flux decreases in the core. Recently, topology optimization has also become a powerful tool for designing thermal concentrators [23, 24], which largely reduces the requirements for materials and structures [25–28].

Despite varieties of schemes, the concentrating efficiency of a thermal concentrator, commonly reflected in the ratio of its interior to exterior temperature gradients, has an upper limit. Specifically, when a circular concentrator with inner radius r_c and outer radius r_s is designed, the upper limit for the concentrating efficiency is $\eta = r_s/r_c$ [5–24], indicating that the isotherms in the shell are completely compressed to the core. To reach the upper limit, the theory of transformation thermotics requires a shell with an extremely anisotropic thermal conductivity [29–32], and the effective medium theory needs to fabricate a core with a near-zero thermal conductivity [20–22]. However, breaking the upper limit for concentration efficiency is still challenging.

To solve the problem, we investigate a monolayer scheme and two extended schemes with the coupling of thermal conductivities. These three schemes feature the simultaneous concentrating of heat flux and temperature gradient with only homogeneous materials. More importantly, they contribute to much higher efficiency than existing schemes. Nevertheless, apparent negative thermal conductivities are required, which can be effectively realized with external heat energy and have been applied to design thermal metamaterials [33–36].

8.2 Monolayer Scheme with Isotropic Thermal Conductivity

We first discuss a monolayer scheme in the Cartesian coordinate system x_i ($i = 1, 2, 3$ for three dimensions and $i = 1, 2$ for two dimensions). A confocal core-shell structure is embedded in a background (Fig. 8.1a). The semi-axis of the core (or shell) along the x_i axis is denoted as r_{ci} (or r_{si}). The thermal conductivities of the core, shell, and background are denoted as κ_c, κ_s, and κ_b, respectively. The conversion between the Cartesian coordinates x_i and elliptical (or ellipsoidal) coordinates ρ_j is

$$\sum_i \frac{x_i^2}{\rho_j + r_{ci}^2} = 1, \tag{8.1}$$

with parameters of $j = 1, 2, 3$ for three dimensions and $j = 1, 2$ for two dimensions. The coordinate ρ_1 ($> -r_{ci}^2$) denotes an elliptical (or ellipsoidal) boundary. For example, the inner and outer boundaries of the shell can be denoted as $\rho_1 = \rho_c$ ($= 0$) and $\rho_1 = \rho_s$, respectively. In the presence of a uniform thermal field along the x_i axis, thermal conduction equation can be expressed in the elliptical (or ellipsoidal) coordinate system as [38]

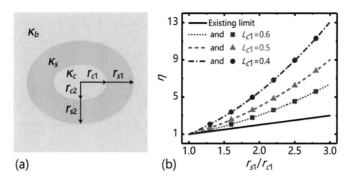

(a)   (b)

Fig. 8.1 **a** Monolayer scheme with an isotropic thermal conductivity. **b** Concentrating efficiency η as a function of geometric configuration r_{s1}/r_{c1}. Lines and points denote theoretical results and simulation results, respectively. Adapted from Ref. [37]

$$\frac{\partial}{\partial \rho_1}\left[g\left(\rho_1\right)\frac{\partial T}{\partial \rho_1}\right] + \frac{g\left(\rho_1\right)}{\rho_1 + r_{ci}^2}\frac{\partial T}{\partial \rho_1} = 0, \tag{8.2}$$

with a definition of $g\left(\rho_1\right) = \prod_i \left(\rho_1 + r_{ci}^2\right)^{1/2}$. For three dimensions, $4\pi g\left(\rho_1\right)/3$ represents the volume of an ellipsoid. For two dimensions, $\pi g\left(\rho_1\right)$ denotes the area of an ellipse. The temperature distributions along the x_i axis in the core T_{ci}, shell T_{si}, and background T_{bi} can be expressed as [38]

$$\begin{cases} T_{ci} = A_{ci}x_i, \\ T_{si} = [A_{si} + B_{si}\phi_i\left(\rho_1\right)]x_i, \\ T_{bi} = [A_{bi} + B_{bi}\phi_i\left(\rho_1\right)]x_i, \end{cases} \tag{8.3}$$

with a definition of $\phi_i\left(\rho_1\right) = \int_{\rho_c}^{\rho_1}\left[\left(\rho_1 + r_{ci}^2\right)g\left(\rho_1\right)\right]^{-1}d\rho_1$. A_{ci}, A_{si}, B_{si}, and B_{bi} can be determined by the continuities of temperature and normal heat flux. Since the temperature distribution in the background should be undistorted, we take $B_{bi} = 0$ and then obtain

$$\kappa_b = \frac{L_{ci}\kappa_c + (1 - L_{ci})\kappa_s + (1 - L_{si})\left(\kappa_c - \kappa_s\right)f}{L_{ci}\kappa_c + (1 - L_{ci})\kappa_s - L_{si}\left(\kappa_c - \kappa_s\right)f}\kappa_s, \tag{8.4}$$

with a definition of $f = g\left(\rho_c\right)/g\left(\rho_s\right) = \prod_i r_{ci}/r_{si}$. The shape factor L_{c1} (or L_{s1}) reflects the flattening degree of the ellipse, and the larger the shape factor is, the more flattening the ellipse is.

Then, the concentrating efficiency of a thermal concentrator can then be defined as the ratio of its interior and exterior temperature gradients (taking $B_{bi} = 0$),

$$\eta = \frac{\nabla T_{ci}}{\nabla T_{bi}} = \frac{A_{ci}}{A_{bi}} = \frac{\kappa_s}{L_{ci}\kappa_c + (1 - L_{ci})\kappa_s - L_{si}\left(\kappa_c - \kappa_s\right)f}. \tag{8.5}$$

For a two-dimensional circular case with $L_{ci} = L_{si} = 1/2$, Eq. (8.5) can be reduced to

$$\eta = \frac{2\kappa_s}{\kappa_c + \kappa_s - (\kappa_c - \kappa_s) f}. \tag{8.6}$$

For a three-dimensional spherical case with $L_{ci} = L_{si} = 1/3$, Eq. (8.5) can be reduced to

$$\eta = \frac{3\kappa_s}{\kappa_c + 2\kappa_s - (\kappa_c - \kappa_s) f}. \tag{8.7}$$

We also consider the same thermal conductivities inside and outside the shell and then obtain two coupling conditions to satisfy $\kappa_c = \kappa_b$,

$$\kappa_s = \kappa_c, \tag{8.8}$$

$$-\frac{1 - L_{ci} - (1 - L_{si}) f}{L_{ci} - L_{si} f} \kappa_s = \kappa_c. \tag{8.9}$$

Equation (8.8) leads to a trivial case with $\kappa_c = \kappa_s = \kappa_b$ and $\eta = 1$. However, if we apply the coupling condition described by Eq. (8.9), the concentrating efficiency largely increases,

$$\eta = f^{-1} = \prod_i r_{si}/r_{ci}. \tag{8.10}$$

Clearly, the concentrating efficiency exceeds the upper limit for existing thermal concentrators $\eta = r_{s1}/r_{c1}$, and a smaller L_{c1} yields a larger η (Fig. 8.1b). However, the geometric configuration still restricts the concentrating efficiency, so we further consider the following two schemes by adding another degree of freedom.

8.3 Monolayer Scheme with Anisotropic Thermal Conductivity

We further consider a shell with an anisotropic thermal conductivity. Since it is not convenient to unify two and three dimensions, we discuss them independently. Nevertheless, the conclusion of three dimensions is similar to that of two dimensions. We first discuss a two-dimensional circular shell with inner and outer radii of r_c and r_s, respectively (Fig. 8.2a). Thermal conduction equation can be written in the cylindrical coordinate system (r, θ) as [29]

$$\frac{1}{r}\frac{\partial}{\partial r}\left(r\kappa_{srr}\frac{\partial T}{\partial r}\right) + \frac{1}{r}\frac{\partial}{\partial \theta}\left(\kappa_{s\theta\theta}\frac{\partial T}{r\partial \theta}\right) = 0, \tag{8.11}$$

where κ_{srr} and $\kappa_{s\theta\theta}$ are the radial and tangential thermal conductivities of the shell, respectively. The temperature distributions of the core T_c, shell T_s, and background

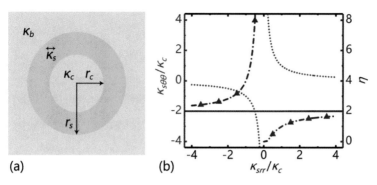

Fig. 8.2 a Monolayer scheme with an anisotropic thermal conductivity. **b** $\kappa_{s\theta\theta}/\kappa_c$ and η as a function of κ_{srr}/κ_c when $r_s/r_c = 2$. Lines and points denote theoretical results and simulation results, respectively. Adapted from Ref. [37]

T_b can be written as [29]

$$\begin{cases} T_c = A_c r \cos\theta, \\ T_s = \left(A_s r^{d_{s1}} + B_s r^{d_{s2}}\right)\cos\theta, \\ T_b = \left(A_b r + B_b r^{-1}\right)\cos\theta, \end{cases} \tag{8.12}$$

with definitions of $d_{s1} = \sqrt{\kappa_{s\theta\theta}/\kappa_{srr}}$ and $d_{s2} = -\sqrt{\kappa_{s\theta\theta}/\kappa_{srr}}$. A_c, A_s, B_s, and B_b are four constants to be determined by the boundary conditions. By taking $B_b = 0$, we can further derive

$$\kappa_b = \frac{d_{s1}\left(\kappa_c - d_{s2}\kappa_{srr}\right) - d_{s2}\left(\kappa_c - d_{s1}\kappa_{srr}\right) f^{(d_{s1}-d_{s2})/2}}{\kappa_c - d_{s2}\kappa_{srr} - \left(\kappa_c - d_{s1}\kappa_{srr}\right) f^{(d_{s1}-d_{s2})/2}}\kappa_{srr}, \tag{8.13}$$

with a definition of $f = r_c^2/r_s^2$. We also define the concentrating efficiency as

$$\eta = \frac{A_c}{A_b} = \frac{(d_{s1} - d_{s2})\,\kappa_{srr}\,f^{(d_{s1}-1)/2}}{\kappa_c - d_{s2}\kappa_{srr} - \left(\kappa_c - d_{s1}\kappa_{srr}\right) f^{(d_{s1}-d_{s2})/2}}. \tag{8.14}$$

For an isotropic case with $d_{s1} = -d_{s2} = 1$, Eq. (8.14) can be simplified as

$$\eta = \frac{2\kappa_{srr}}{\kappa_c + \kappa_{srr} - (\kappa_c - \kappa_{srr})\,f}, \tag{8.15}$$

which has the same form as Eq. (8.6) in Sect. 8.2.

We also obtain two coupling conditions for $\kappa_c = \kappa_b$,

$$d_{s1}\kappa_{srr} = \kappa_c, \tag{8.16}$$

$$d_{s2}\kappa_{srr} = \kappa_c. \tag{8.17}$$

Equations (8.16) and (8.17) can be unified as

$$\kappa_{srr}\kappa_{s\theta\theta} = \kappa_c^2,\tag{8.18}$$

which is plotted with the dotted line in Fig. 8.2b.

When the thermal conductivities of the core and shell satisfy Eq. (8.18), Eq. (8.14) becomes

$$\eta = f^{-(1-\kappa_c/\kappa_{srr})/2} = (r_s/r_c)^{1-\kappa_c/\kappa_{srr}},\tag{8.19}$$

which is plotted with the dashed-dotted line in Fig. 8.2b. Obviously, the minimum value $\eta \to 0$ appears when $\kappa_{srr}/\kappa_c \to 0^+$, and the maximum value $\eta \to \infty$ appears when $\kappa_{srr}/\kappa_c \to 0^-$. Moreover, we can observe $\eta \to r_s/r_c$ when $\kappa_{srr}/\kappa_c \to \pm\infty$, which is just the upper limit for existing concentrating efficiency (see the solid line in Fig. 8.2b). If the thermal conductivity of the shell is isotropic and nontrivial $\kappa_{srr}/\kappa_c = 1/d_{s2} = -1$, the concentrating efficiency also exceeds the upper limit and becomes $\eta = r_s^2/r_c^2$, which is in accordance with the two-dimensional conclusion in Sect. 8.2. Therefore, the concentrating efficiency can exceed the upper limit and even approach infinity when $\kappa_{srr}/\kappa_c \to 0^-$.

8.4 Bilayer Scheme with Isotropic Thermal Conductivity

We then consider the second shell whose isotropic thermal conductivity and semi-axis along the x_i axis are denoted as κ_t and r_{ti}, respectively (Fig. 8.3a). With the conclusion of the monolayer scheme (Eq. (8.4)), the effective thermal conductivity of the core and the first shell κ_{cs} can be calculated by

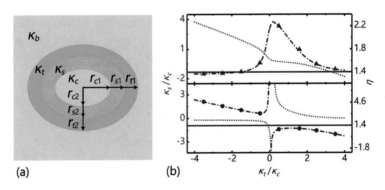

(a)   (b)

Fig. 8.3 a Bilayer scheme with isotropic thermal conductivities. **b** κ_s/κ_c and η as a function of κ_t/κ_c when $r_{s1}/r_{c1} = 1.2, r_{t1}/r_{c1} = 1.4$, and $L_{c1} = 1/3$. Lines and points denote theoretical results and simulation results, respectively. Adapted from Ref. [37]

$$\kappa_{cs} = \frac{L_{ci}\kappa_c + (1 - L_{ci})\,\kappa_s + (1 - L_{si})\,(\kappa_c - \kappa_s)\,f}{L_{ci}\kappa_c + (1 - L_{ci})\,\kappa_s - L_{si}\,(\kappa_c - \kappa_s)\,f}\kappa_s. \tag{8.20}$$

We then treat the core and the first shell as an effective core with an effective thermal conductivity of κ_{cs}, so we can further derive

$$\kappa_b = \frac{L_{si}\kappa_{cs} + (1 - L_{si})\,\kappa_t + (1 - L_{ti})\,(\kappa_{cs} - \kappa_t)\,p}{L_{si}\kappa_{cs} + (1 - L_{si})\,\kappa_t - L_{ti}\,(\kappa_{cs} - \kappa_t)\,p}\kappa_t, \tag{8.21}$$

with a definition of $p = g\,(\rho_s)\,/g\,(\rho_t) = \prod_i r_{si}/r_{ti}$. ρ_t denotes the outer boundary of the second shell. L_{ti} is the shape factor of the second shell along the x_i axis,

$$L_{ti} = \frac{g\,(\rho_t)}{2} \int_{\rho_t}^{\infty} \left[(\rho_1 + r_{ci}^2)\,g\,(\rho_1)\right]^{-1} d\rho_1. \tag{8.22}$$

We can also express the concentrating efficiency as

$$\eta = \frac{A_{ci}}{A_{bi}} = \frac{\kappa_s\kappa_t}{\lambda_1 + \lambda_2 + \lambda_3}, \tag{8.23}$$

where λ_1, λ_2, and λ_3 take the form of

$$\begin{cases} \lambda_1 = [L_{ci}\kappa_c + (1 - L_{ci})\,\kappa_s]\,[L_{si}\kappa_s + (1 - L_{si})\,\kappa_t - L_{ti}\,(\kappa_s - \kappa_t)\,p], \\ \lambda_2 = -L_{ti}\,(\kappa_c - \kappa_s)\,[(1 - L_{si})\,\kappa_s + L_{si}\kappa_t]\,fp, \\ \lambda_3 = L_{si}\,(1 - L_{si})\,(\kappa_c - \kappa_s)\,(\kappa_s - \kappa_t)\,f. \end{cases} \tag{8.24}$$

As a more general model, the bilayer scheme can also be reduced to the monolayer scheme in Sect. 8.2 at two certain conditions. When $\kappa_c = \kappa_s$, Eq. (8.23) can be reduced to

$$\eta = \frac{\kappa_t}{L_{si}\kappa_s + (1 - L_{si})\,\kappa_t - L_{ti}\,(\kappa_s - \kappa_t)\,p}. \tag{8.25}$$

When $\kappa_s = \kappa_t$, Eq. (8.23) becomes

$$\eta = \frac{\kappa_t}{L_{ci}\kappa_c + (1 - L_{ci})\,\kappa_t - L_{ti}\,(\kappa_c - \kappa_t)\,fp}. \tag{8.26}$$

Obviously, Eqs. (8.25) and (8.26) have similar forms as Eq. (8.5) in Sect. 8.2. We can also derive two coupling conditions for $\kappa_c = \kappa_b$,

$$M\,(\kappa_s,\,\kappa_t) = \kappa_c, \tag{8.27}$$

$$N\,(\kappa_s,\,\kappa_t) = \kappa_c. \tag{8.28}$$

M and N are two analytical functions. Therefore, one κ_t corresponds to two κ_s for satisfying $\kappa_c = \kappa_b$, i.e., $\kappa_s = m(\kappa_t)$ being a continuous function (see the dotted line in the upper inset of Fig. 8.3b) and $\kappa_s = n(\kappa_t)$ being a quasi-hyperbolic function (see the dotted line in the lower inset of Fig. 8.3b). We do not express the concrete forms of m and n because they are too complicated.

When Eq. (8.27) is satisfied, the upper limit of $\eta = r_{t1}/r_{c1}$ can be broken, but the concentrating efficiency can still not tend to infinity (see the dashed-dotted line in the upper inset of Fig. 8.3b). Moreover, Eq. (8.27) contains two special cases that can be reduced to the conclusion in Sect. 8.2. One features a concentrating efficiency of $\eta = f^{-1}$ with the same thermal conductivities of the second shell and core,

$$-\frac{1 - L_{ci} - (1 - L_{si})f}{L_{ci} - L_{si}f}\kappa_s = \kappa_t = \kappa_c. \tag{8.29}$$

The other features a concentrating efficiency of $\eta = p^{-1}$ with the same thermal conductivities of the first shell and core,

$$\kappa_s = -\frac{1 - L_{si} - (1 - L_{ti})p}{L_{si} - L_{ti}p}\kappa_t = \kappa_c. \tag{8.30}$$

Fortunately, Eq. (8.28) can lead to an infinite efficiency. $\kappa_t/\kappa_c \to 0^-$ and $\kappa_t/\kappa_c \to 0^+$, respectively, yield $\eta \to \infty$ and $\eta \to -\infty$, and the thermal conductivity of the first shell satisfies

$$-\frac{1 - L_{ci} - (1 - L_{si})f}{L_{ci} + (1 - L_{si})f}\kappa_s \approx \kappa_c. \tag{8.31}$$

Meanwhile, $\kappa_t/\kappa_c \to \mp\infty$ can also lead to $\eta \to \pm\infty$, and the thermal conductivity of the first shell satisfies

$$-\frac{1 - L_{ci} + L_{si}f}{L_{ci} - L_{si}f}\kappa_s \approx \kappa_c. \tag{8.32}$$

Moreover, Eq. (8.28) also contains a special case that can be reduced to the conclusion in Sect. 8.2. That is, the concentrating efficiency of $\eta = (fp)^{-1}$ occurs when the two shells have the same thermal conductivities,

$$\kappa_c = -\frac{1 - L_{ci} - (1 - L_{ti})fp}{L_{ci} - L_{ti}fp}\kappa_s = -\frac{1 - L_{ci} - (1 - L_{ti})fp}{L_{ci} - L_{ti}fp}\kappa_t. \tag{8.33}$$

There is another case for $\eta = (fp)^{-1}$ if the thermal conductivities of the two shells satisfy

$$\kappa_c = -\frac{1 - L_{ci} - (1 - L_{si})f}{L_{ci} - L_{si}f}\kappa_s = -\frac{1 - L_{si} - (1 - L_{ti})p}{L_{si} - L_{ti}p}\kappa_t. \tag{8.34}$$

Conductivity coupling occurs layer by layer in this case. The core is coupled with the first shell described by Eq. (8.9). Then, they are treated as an effective core with an effective thermal conductivity of κ_c. The effective core is then coupled with the second shell described by the similar form of Eq. (8.9).

Another unique feature of Eq. (8.28) is the concentrating efficiency of $\eta < 0$ when the thermal conductivity of the second shell satisfies

$$\kappa_t > \frac{1 - L_{si} + L_{ti} p}{1 - L_{si} - (1 - L_{ti}) p} \kappa_c, \tag{8.35}$$

or

$$0 < \kappa_t < \frac{L_{si} - L_{ti} p}{L_{si} + (1 - L_{ti}) p} \kappa_c, \tag{8.36}$$

indicating that the temperature gradient in the core changes its direction.

Then we can draw a brief conclusion for these three schemes. The monolayer scheme with an isotropic thermal conductivity can break the upper limit but is still restricted by its geometric configuration. To be free from geometric configurations, we further consider the monolayer scheme with an anisotropic thermal conductivity and the bilayer scheme with isotropic thermal conductivities. For the former, the efficiency can tend to infinity with $\kappa_{srr}/\kappa_c \to 0^-$. For the latter, the efficiency can also reach infinity when $\kappa_t/\kappa_c \to 0^-$ or $\kappa_t/\kappa_c \to -\infty$. Moreover, the latter features $\eta < 0$ if the coupling condition is appropriately chosen.

8.5 Finite-Element Simulation

We also perform finite-element simulations to confirm the theories with COMSOL Multiphysics. From a practical perspective, although interfacial thermal resistance exists widely, its effect at the macroscopic scale is not dominant, so it is reasonable to ignore it. Without loss of generality, we consider a two-dimensional case with size $10 \times 10 \, \text{cm}^2$ and set the core and background thermal conductivities as $1 \, \text{W m}^{-1} \, \text{K}^{-1}$. The left boundaries are set at $313 \, \text{K}$, the right boundaries are set at $273 \, \text{K}$, and the upper and lower boundaries are adiabatic. To compare the concentrating efficiency of different thermal concentrators, we introduce a dimensionless temperature of $T^* = 100(T - T_0)/T_0$ and a dimensionless position of $x^* = x/w$, where T_0 and w denote the central temperature and half-length of the system, respectively.

Thermal concentrating aims to increase the temperature gradient in the core without distorting that in the background. In order to confirm Eqs. (8.5), (8.8), and (8.9) and demonstrate the expected case shown in Fig. 8.1, we design three structures with different shape factors, and the corresponding results are presented in Figs. 8.4a–c. The temperature profiles outside the shells are undistorted as if there were no core-shell structures in the center. Meanwhile, the isotherms in the cores are concentrated as expected. According to the Fourier law $\boldsymbol{J} = -\kappa \nabla T$, heat fluxes are also enhanced

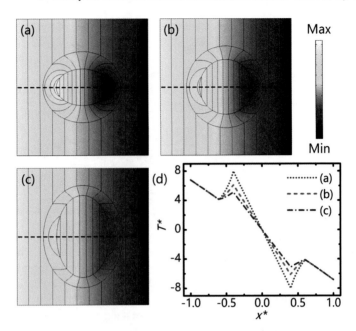

Fig. 8.4 **a–c** Simulations of the monolayer scheme with an isotropic thermal conductivity. **d** T^* as a function of x^*. Parameters: **a** $L_{c1} = 0.4$ and $\kappa_s/\kappa_c = -0.58$; **b** $L_{c1} = 0.5$ and $\kappa_s/\kappa_c = -1$; **c** $L_{c1} = 0.6$ and $\kappa_s/\kappa_c = -1.87$; and **a–c** $r_{s1}/r_{c1} = 1.5$ and $\kappa_c = \kappa_b$. Adapted from Ref. [37]

in the cores due to the larger temperature gradients. The dimensionless temperatures are plotted as a function of dimensionless position in Fig. 8.4d.

By considering the monolayer scheme with an anisotropic thermal conductivity, we confirm the theoretical prediction of Eqs. (8.14) and (8.17). Then, we design three structures with different thermal conductivities of the shells (Fig. 8.5a–c). Similar to Fig. 8.4, the temperature profiles in Fig. 8.5a–c prove the effect of thermal concentrating. Also, we draw the temperature distribution of the thermal concentrator (Fig. 8.5d) designed by transformation theory for comparison. Figure 8.5e displays the temperature distribution along the central horizontal axis. As presented in Fig. 8.2b, the temperature gradient in the core increases with the increment of κ_{srr}/κ_c, leading to the improvement of concentrating efficiency. Thus, we can control $\kappa_{srr}/\kappa_c \to 0^-$ for an extreme concentrating efficiency.

For the bilayer scheme with isotropic thermal conductivities, two coupling conditions (Eqs. (8.27) and (8.28)) are available. Similar to the structures in Figs. 8.4 and 8.5, those in Fig. 8.6 also ensure that isotherms outside the shells are straight and those in the cores are denser, thereby realizing the effect of thermal concentrating. With the coupling condition of Eq. (8.27), the efficiency changes continuously with κ_t/κ_c (Fig. 8.3b). We further design a structure to display the concentrating efficiency when $\kappa_t/\kappa_c \to 0^-$ (Fig. 8.6a). The coupling condition of Eq. (8.28) can lead to an

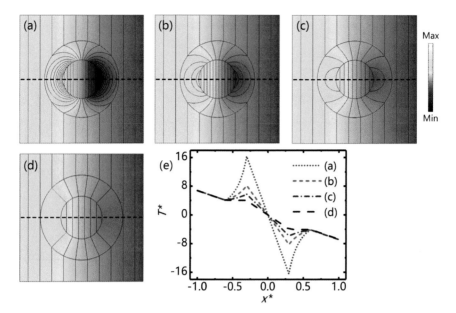

Fig. 8.5 **a–c** Simulations of the monolayer scheme with an anisotropic thermal conductivity. **d** Temperature distribution of the existing scheme based on transformation theory close to the upper limit of concentrating efficiency. **e** T^* as a function of x^*. Parameters: **a** $\kappa_{srr}/\kappa_c = -0.5$; **b** $\kappa_{srr}/\kappa_c = -1$; **c** $\kappa_{srr}/\kappa_c = -2$; **d** $\kappa_{srr}/\kappa_c = (r + 100)/r$; and **a–d** $r_s/r_c = 2$ and $\kappa_c = \kappa_b$. Adapted from Ref. [37]

infinite efficiency. That is, $\eta \to +\infty$ when $\kappa_t/\kappa_c \to 0^-$ (Fig. 8.6b) or $\kappa_t/\kappa_c \to -\infty$ (Fig. 8.6c), and $\eta \to -\infty$ for $\kappa_t/\kappa_c \to 0^+$ (Fig. 8.6d) or $\kappa_t/\kappa_c \to +\infty$ (Fig. 8.6e). As shown in Fig. 8.6f, the effect of thermal concentrating can be quantitatively observed.

8.6 Experimental Suggestion

The coupling conditions require apparent negative thermal conductivities [33–36], which cannot happen spontaneously in experiments. To achieve the equivalent effect, we can resort to external heat sources (Fig. 8.7a). According to the thermal uniqueness theorem [39, 40], as long as we realize the same boundary temperature distributions by adding external heat sources at the inner and outer boundaries of the shell, we can obtain the same temperature profiles. Since the central temperature gradient and heat flux in Fig. 8.7c are almost the same as those in Fig. 8.7b, we prove that an apparent negative thermal conductivity can be effectively achieved using external heat sources. Then we design a structure as a feasible experimental suggestion (Fig. 8.7d). We add a series of point heat sources at the inner and outer boundaries of the shell (Fig. 8.7d). The precise temperatures are presented in Tables 8.1 and 8.2, which can be

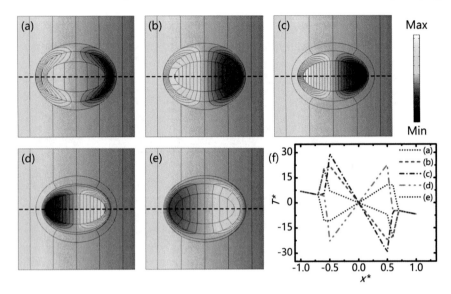

Fig. 8.6 **a–e** Simulations of the bilayer scheme with isotropic thermal conductivities. **f** T^* as a function of x^*. Parameters: **a** $\kappa_s/\kappa_c = 0.0826$ and $\kappa_t/\kappa_c = -0.05$; **b** $\kappa_s/\kappa_c = -1.14$ and $\kappa_t/\kappa_c = -0.05$; **c** $\kappa_s/\kappa_c = -0.175$ and $\kappa_t/\kappa_c = -10$; **d** $\kappa_s/\kappa_c = -0.122$ and $\kappa_t/\kappa_c = 15$; **e** $\kappa_s/\kappa_c = -2.83$ and $\kappa_t/\kappa_c = 0.05$; and **a–e** $r_{s1}/r_{c1} = 1.2$, $r_{t1}/r_{c1} = 1.4$, $L_{c1} = 1/3$, and $\kappa_c = \kappa_b$. Adapted from Ref. [37]

Table 8.1 Temperatures of point heat sources at the outer boundary of the shell in Fig. 8.7d. Adapted from Ref. [37]

Source	Temp. (K)	Source	Temp. (K)
1	293.00	13	293.00
2	290.24	14	295.76
3	287.68	15	298.32
4	285.45	16	300.55
5	283.78	17	302.22
6	282.69	18	303.30
7	282.30	19	303.70
8	282.69	20	303.30
9	283.78	21	302.22
10	285.45	22	300.55
11	287.68	23	298.32
12	290.24	24	295.76

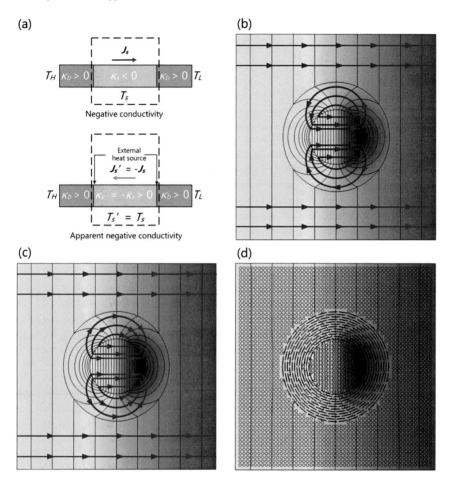

Fig. 8.7 Experimental suggestions. **a** Schematic diagram for realizing apparent negative conductivity. **b** Without temperature control. **c** Continuous temperature control. **d** Discrete point heat sources whose temperatures are shown in Tables 8.1 and 8.2. The core and background in **d** are a brass plate ($109\,\text{W}\,\text{m}^{-1}\,\text{K}^{-1}$) drilled with 2116 air circles with a radius of $0.1\,\text{cm}$. The shell is drilled with 282 air ellipses with a major (or minor) semi-axis of $0.35\,\text{cm}$ (or $0.02\,\text{cm}$). Other parameters: **b** and **d** $\kappa_s = \text{diag}(23,\ 92)\,\text{W}\,\text{m}^{-1}\,\text{K}^{-1}$; **c** $\kappa_s = \text{diag}(-23,\ -92)\,\text{W}\,\text{m}^{-1}\,\text{K}^{-1}$; and **b–d** $r_c = 1\,\text{cm}$, $r_s = 2\,\text{cm}$, $\kappa_c = \kappa_b = 46\,\text{W}\,\text{m}^{-1}\,\text{K}^{-1}$. The black lines and blue arrows in **b–d** denote isotherms and heat fluxes, respectively. Adapted from Ref. [37]

experimentally controlled by adjusting the voltages of heaters and coolers according to Eqs. (1) and (2) in Ref. [39]. The required thermal conductivity can be realized by punching air holes on a brass plate ($109\,\text{W}\,\text{m}^{-1}\,\text{K}^{-1}$), whose left and right edges are put into hot ($313\,\text{K}$) and cold ($273\,\text{K}$) sinks, respectively. To achieve the thermal conductivities of the core and background in Fig. 8.7b, 2116 air circles are drilled on the brass, leading to an effective thermal conductivity of $46\,\text{W}\,\text{m}^{-1}\,\text{K}^{-1}$ (calculated by Eq. (11) in Ref. [41]). The shell region is composed of 282 air ellipses, leading

Table 8.2 Temperatures of point heat sources at the inner boundary of the shell in Fig. 8.7d. Adapted from Ref. [37]

Source	Temp. (K)	Source	Temp. (K)
1	293.00	13	293.00
2	281.96	14	304.04
3	271.67	15	314.33
4	259.83	16	323.17
5	256.05	17	329.95
6	251.79	18	334.21
7	250.34	19	335.67
8	251.79	20	334.21
9	256.05	21	329.95
10	259.83	22	323.17
11	271.67	23	314.33
12	281.96	24	304.04

to an effective thermal conductivity of diag(23, 92) W m^{-1} K^{-1} (calculated by Eq. (11) in Ref. [41]). By comparing the temperature distributions in Fig. 8.7b–d, we can confirm that the scheme in Fig. 8.7d can realize the effect of Fig. 8.7b in experiments.

8.7   Conclusion

We break the upper limit for the concentrating efficiency of existing thermal concentrators by coupling thermal conductivities. We first explore a monolayer scheme with an isotropic thermal conductivity, which can break the upper limit but is still restricted by its geometric configuration. Then, we consider a shell with an anisotropic thermal conductivity or add the second shell with an isotropic thermal conductivity as another degree of freedom, which renders the concentrating efficiency free from geometric configurations. Apparent negative thermal conductivities are required in these three schemes, which can be effectively realized by external energy or thermoelectric materials. Since negative permeability [42–44] and negative electric conductivity [45] have been, respectively, revealed in static magnetic fields and DC fields, it is promising to extend our results to these diffusive fields due to the similar equation forms (i.e., the Laplace equation). Moreover, the present theory is applicable not only for thermal concentrators with $\eta > 1$ but also for thermal invisible sensors with $\eta = 1$ [46, 47] and thermal cloaks with $\eta = 0$ (perfect cloaking) or $\eta < 1$ (imperfect cloaking) [48, 49]. A typical feature for concentrating, sensing, or cloaking is the undistorted background temperature distribution, so these schemes may provide insights into thermal camouflage [50] for misleading infrared detection. It is also promising to extend the related mechanisms towards multi-function and micro/nano-scale.

8.8 Exercise and Solution

Exercise

1. Prove that the three-dimensional case in Sect. 8.3 is similar to two dimensions.

Solution

1. The tensorial thermal conductivity of the shell can be expressed in the spherical coordinate system (r, θ, ϕ) as $\kappa_s = \text{diag}\left(\kappa_{srr}, \kappa_{s\theta\theta}, \kappa_{s\phi\phi}\right)$. For simplicity, we assume a axial symmetry with $\kappa_{s\theta\theta} = \kappa_{s\phi\phi}$. Therefore, thermal conduction is independent of ϕ, which is dominated by

$$\frac{1}{r^2}\frac{\partial}{\partial r}\left(r^2 \kappa_{srr}\frac{\partial T}{\partial r}\right) + \frac{1}{r\sin\theta}\frac{\partial}{\partial\theta}\left(\sin\theta\kappa_{s\theta\theta}\frac{\partial T}{r\partial\theta}\right) = 0. \tag{8.37}$$

The temperature distributions in the core T_c, shell T_s, and background T_b can be written as

$$\begin{cases} T_c = A_c r \cos\theta, \\ T_s = \left(A_s r^{h_{s1}} + B_s r^{h_{s2}}\right)\cos\theta, \\ T_b = \left(A_b r + B_b r^{-2}\right)\cos\theta, \end{cases} \tag{8.38}$$

with definitions of $h_{s1} = \left(-1 + \sqrt{1 + 8\kappa_{s\theta\theta}/\kappa_{srr}}\right)/2$ and $h_{s2} = \left(-1 - \sqrt{1 + 8\kappa_{s\theta\theta}/\kappa_{srr}}\right)/2$. By substituting Eq. (8.38) into the boundary conditions, we can obtain

$$\begin{cases} A_c r_c = A_s r_c^{h_{s1}} + B_s r_c^{h_{s2}}, \\ A_s r_s^{h_{s1}} + B_s r_s^{h_{s2}} = A_b r_s + B_b r_s^{-2}, \\ \kappa_c A_c = \kappa_{srr}\left(h_{s1}A_s r_c^{h_{s1}-1} + h_{s2}B_s r_c^{h_{s2}-1}\right), \\ \kappa_{srr}\left(h_{s1}A_s r_s^{h_{s1}-1} + h_{s2}B_s r_s^{h_{s2}-1}\right) = \kappa_b\left(A_b - 2B_b r_s^{-3}\right). \end{cases} \tag{8.39}$$

We can calculate A_c, A_s, B_s, and B_b with Eq. (8.39). By taking $B_b = 0$, we can further derive

$$\kappa_b = \frac{h_{s1}\left(\kappa_c - h_{s2}\kappa_{srr}\right) - h_{s2}\left(\kappa_c - h_{s1}\kappa_{srr}\right) f^{(h_{s1}-h_{s2})/3}}{\kappa_c - h_{s2}\kappa_{srr} - \left(\kappa_c - h_{s1}\kappa_{srr}\right) f^{(h_{s1}-h_{s2})/3}}\kappa_{srr}, \tag{8.40}$$

with a definition of $f = r_c^3/r_s^3$. The concentrating efficiency is

$$\eta = \frac{A_c}{A_b} = \frac{(h_{s1} - h_{s2})\,\kappa_{srr}\,f^{(h_{s1}-1)/3}}{\kappa_c - h_{s2}\kappa_{srr} - (\kappa_c - h_{s1}\kappa_{srr})\,f^{(h_{s1}-h_{s2})/3}}. \tag{8.41}$$

For an isotropic case with $h_{s1} = 1$ and $h_{s2} = -2$, Eq. (8.14) can be simplified as

$$\eta = \frac{3\kappa_{srr}}{\kappa_c + 2\kappa_{srr} - (\kappa_c - \kappa_{srr})\,f}, \tag{8.42}$$

which has the same form as Eq. (8.7) in Sect. 8.2.

We can also derive two coupling conditions for $\kappa_c = \kappa_b$,

$$h_{s1}\kappa_{srr} = \kappa_c, \tag{8.43}$$

$$h_{s2}\kappa_{srr} = \kappa_c. \tag{8.44}$$

Equations (8.43) and (8.44) can also be unified as

$$2\kappa_{srr}\kappa_{s\theta\theta} - \kappa_c\kappa_{srr} = \kappa_c^2. \tag{8.45}$$

When Eq. (8.45) is satisfied, Eq. (8.41) can be reduced to

$$\eta = f^{-(1-\kappa_c/\kappa_{srr})/3} = (r_s/r_c)^{1-\kappa_c/\kappa_{srr}}, \tag{8.46}$$

which has the same form as two dimensions (Eq. (8.19)). Therefore, the minimum value $\eta \to 0$ occurs with $\kappa_{srr}/\kappa_c \to 0^+$, and the maximum value $\eta \to \infty$ occurs with $\kappa_{srr}/\kappa_c \to 0^-$. Moreover, we can find $\eta \to r_s/r_c$ when $\kappa_{srr}/\kappa_c \to \pm\infty$. If we consider an isotropic and nontrivial shell with $\kappa_{srr}/\kappa_c = 1/h_{s2} = -1/2$, the concentrating efficiency becomes $\eta = r_s^3/r_c^3$, which is also similar to the two-dimensional conclusion in Sect. 8.3.

References

1. Fan, C.Z., Gao, Y., Huang, J.P.: Shaped graded materials with an apparent negative thermal conductivity. Appl. Phys. Lett. **92**, 251907 (2008)
2. Chen, T.Y., Weng, C.N., Chen, J.S.: Cloak for curvilinearly anisotropic media in conduction. Appl. Phys. Lett. **93**, 114103 (2008)
3. Yang, S., Wang, J., Dai, G.L., Yang, F.B., Huang, J.P.: Controlling macroscopic heat transfer with thermal metamaterials: theory, experiment and application. Phys. Rep. **908**, 1 (2021)
4. Li, Y., Li, W., Han, T.C., Zheng, X., Li, J.X., Li, B.W., Fan, S.H., Qiu, C.-W.: Transforming heat transfer with thermal metamaterials and devices. Nat. Rev. Mater. **6**, 488 (2021)
5. Yu, G.X., Lin, Y.F., Zhang, G.Q., Yu, Z., Yu, L.L., Su, J.: Design of square-shaped heat flux cloaks and concentrators using method of coordinate transformation. Front. Phys. **6**, 70 (2011)
6. Guenneau, S., Amra, C., Veynante, D.: Transformation thermodynamics: cloaking and concentrating heat flux. Opt. Express **20**, 8207 (2012)

7. Han, T.C., Zhao, J.J., Yuan, T., Lei, D.Y., Li, B.W., Qiu, C.W.: Theoretical realization of an ultra-efficient thermal-energy harvesting cell made of natural materials. Energy Environ. Sci. **6**, 3537 (2013)

8. Moccia, M., Castaldi, G., Savo, S., Sato, Y., Galdi, V.: Independent manipulation of heat and electrical current via bifunctional metamaterials. Phys. Rev. X **4**, 021025 (2014)

9. Chen, F., Lei, D.Y.: Experimental realization of extreme heat flux concentration with easy-to-make thermal metamaterials. Sci. Rep. **5**, 11552 (2015)

10. Li, Y., Shen, X.Y., Huang, J.P., Ni, Y.S.: Temperature-dependent transformation thermotics for unsteady states: switchable concentrator for transient heat flow. Phys. Lett. A **380**, 1641 (2016)

11. Xu, G.Q., Zhang, H.C., Jin, Y.: Achieving arbitrarily polygonal thermal harvesting devices with homogeneous parameters through linear mapping function. Energy Convers. Manage. **165**, 253262 (2018)

12. Ji, Q.X., Fang, G.D., Liang, J.: Achieving thermal concentration based on fiber reinforced composite microstructures design. J. Phys. D: Appl. Phys. **51**, 315304 (2018)

13. Xu, G.Q., Zhou, X., Liu, Z.J.: Converging heat transfer in completely arbitrary profiles with unconventional thermal concentrator. Int. Commun. Heat Mass Transf. **108**, 104337 (2019)

14. Li, T.H., Yang, C.F., Li, S.B., Zhu, D.L., Han, Y., Li, Z.Q.: Design of a bifunctional thermal device exhibiting heat flux concentration and scattering amplification effects. J. Phys. D: Appl. Phys. **53**, 065503 (2020)

15. Narayana, S., Sato, Y.: Heat flux manipulation with engineered thermal materials. Phys. Rev. Lett. **108**, 214303 (2012)

16. Kapadia, R.S., Bandaru, P.R.: Heat flux concentration through polymeric thermal lenses. Appl. Phys. Lett. **105**, 233903 (2014)

17. Lan, C.W., Li, B., Zhou, J.: Simultaneously concentrated electric and thermal fields using fan-shaped structure. Opt. Express **23**, 24475 (2015)

18. Chen, T.Y., Weng, C.N., Tsai, Y.L.: Materials with constant anisotropic conductivity as a thermal cloak or concentrator. J. Appl. Phys. **117**, 054904 (2015)

19. Wang, R.Z., Xu, L.J., Ji, Q., Huang, J.P.: A thermal theory for unifying and designing transparency, concentrating and cloaking. J. Appl. Phys. **123**, 115117 (2018)

20. Shen, X.Y., Li, Y., Jiang, C.R., Ni, Y.S., Huang, J.P.: Thermal cloak-concentrator. Appl. Phys. Lett. **109**, 031907 (2016)

21. Xu, G.Q., Zhou, X., Zhang, J.Y.: Bilayer thermal harvesters for concentrating temperature distribution. Int. J. Heat Mass Transf. **142**, 118434 (2019)

22. Xu, L.J., Yang, S., Huang, J.P.: Thermal theory for heterogeneously architected structure: fundamentals and application. Phys. Rev. E **98**, 052128 (2018)

23. Fujii, G., Akimoto, Y.: Cloaking a concentrator in thermal conduction via topology optimization. Int. J. Heat Mass Transf. **159**, 120082 (2020)

24. Ji, Q.X., Chen, X.Y., Liang, J., Laude, V., Guenneau, S., Fang, G.D., Kadic, M.: Designing thermal energy harvesting devices with natural materials through optimized microstructures. Int. J. Heat Mass Transf. **169**, 120948 (2021)

25. Fujii, G., Akimoto, Y., Takahashi, M.: Exploring optimal topology of thermal cloaks by CMA-ES. Appl. Phys. Lett. **112**, 061108 (2018)

26. Fujii, G., Akimoto, Y.: Optimizing the structural topology of bifunctional invisible cloak manipulating heat flux and direct current. Appl. Phys. Lett. **115**, 174101 (2019)

27. Fujii, G., Akimoto, Y.: Topology-optimized thermal carpet cloak expressed by an immersed-boundary level-set method via a covariance matrix adaptation evolution strategy. Int. J. Heat Mass Transf. **137**, 1312 (2019)

28. Fujii, G., Akimoto, Y.: dc electric cloak concentrator via topology optimization. Phys. Rev. E **102**, 033308 (2020)

29. Xu, L.J., Yang, S., Huang, J.P.: Passive metashells with adaptive thermal conductivities: chameleonlike behavior and its origin. Phys. Rev. Appl. **11**, 054071 (2019)

30. Sun, F., Liu, Y.C., Yang, Y.B., Chen, Z.H., He, S.L.: Thermal surface transformation and its applications to heat flux manipulations. Opt. Express **27**, 33757 (2019)

31. Yang, F.B., Tian, B.Y., Xu, L.J., Huang, J.P.: Experimental demonstration of thermal chameleonlike rotators with transformation-invariant metamaterials. Phys. Rev. Appl. **14**, 054024 (2020)
32. Sedeh, H.B., Fakheri, M.H., Abdolali, A., Sun, F., Ma, Y.G.: Feasible thermodynamics devices enabled by thermal-null medium. Phys. Rev. Appl. **14**, 064034 (2020)
33. Gao, Y., Huang, J.P.: Unconventional thermal cloak hiding an object outside the cloak. EPL **104**, 44001 (2013)
34. Wegener, M.: Metamaterials beyond optics. Science **342**, 939 (2013)
35. Shen, X.Y., Huang, J.P.: Thermally hiding an object inside a cloak with feeling. Int. J. Heat Mass Transf. **78**, 1 (2014)
36. Yang, S., Xu, L.J., Huang, J.P.: Thermal magnifier and external cloak in ternary component structure. J. Appl. Phys. **125**, 055103 (2019)
37. Zhuang, P.F., Xu, L.J., Tan, P., Ouyang, X.P., Huang, J.P.: Breaking efficiency limit of thermal concentrators by conductivity couplings. Sci. China-Phys. Mech. Astron. in press (2022)
38. Milton, G.W.: The Theory of Composites. Cambridge University Press (2002)
39. Nguyen, D.M., Xu, H.Y., Zhang, Y.M., Zhang, B.L.: Active thermal cloak. Appl. Phys. Lett. **107**, 121901 (2015)
40. Xu, L.J., Huang, J.P.: Active thermal wave cloak. Chin. Phys. Lett. **37**, 120501 (2020)
41. Yang, S., Xu, L.J., Wang, R.Z., Huang, J.P.: Full control of heat transfer in single-particle structural materials. Appl. Phys. Lett. **111**, 121908 (2017)
42. Mach-Batlle, R., Parra, A., Prat-Camps, J., Laut, S., Navau, C., Sanchez, A.: Negative permeability in magnetostatics and its experimental demonstration. Phys. Rev. B **96**, 094422 (2017)
43. Mach-Batlle, R., Parra, A., Laut, S., Del-Valle, N., Navau, C., Sanchez, A.: Magnetic illusion: transforming a magnetic object into another object by negative permeability. Phys. Rev. Appl. **9**, 034007 (2018)
44. Mach-Batlle, R., Bason, M.G., Del-Valle, N., Prat-Camps, J.: Tailoring magnetic fields in inaccessible regions. Phys. Rev. Lett. **125**, 177204 (2020)
45. Chen, T.H., Zheng, B., Yang, Y.H., Shen, L., Wang, Z.J., Gao, F., Li, E.P., Luo, Y., Cui, T.J., Chen, H.S.: Direct current remote cloak for arbitrary objects. Light Sci. Appl. **8**, 30 (2019)
46. Yang, T.Z., Bai, X., Gao, D.L., Wu, L.Z., Li, B.W., Thong, J.T.L., Qiu, C.W.: Invisible sensors: simultaneous sensing and camouflaging in multiphysical fields. Adv. Mater. **27**, 7752 (2015)
47. Jin, P., Xu, L.J., Jiang, T., Zhang, L., Huang, J.P.: Making thermal sensors accurate and invisible with an anisotropic monolayer scheme. Int. J. Heat Mass Transf. **163**, 120437 (2020)
48. Li, Y., Zhu, K.J., Peng, Y.G., Li, W., Yang, T.Z., Xu, H.X., Chen, H., Zhu, X.F., Fan, S.H., Qiu, C.W.: Thermal meta-device in analogue of zero-index photonics. Nat. Mater. **18**, 48 (2019)
49. Xu, L.J., Yang, S., Huang, J.P.: Effectively infinite thermal conductivity and zero-index thermal cloak. EPL **131**, 24002 (2020)
50. Hu, R., Xi, W., Liu, Y.D., Tang, K.C., Song, J.L., Luo, X.B., Wu, J.Q., Qiu, C.-W.: Thermal camouflaging metamaterials. Mater. Today **45**, 120 (2021)

Chapter 9
Theory for Chameleonlike Thermal Rotators: Extremely Anisotropic Conductivity

Abstract In this chapter, we propose a mechanism for intelligent thermal regulation based on transformation-invariant metamaterials, which possess highly anisotropic thermal conductivities. As an application, we design intelligent thermal rotators that can guide heat flux direction with different environmental parameters. Since the adaptive behavior is similar to chameleons, the present rotators are called chameleonlike rotators. We further perform finite-element simulations and laboratory experiments to validate the scheme and demonstrate the chameleonlike behavior. These results have potential applications for implementing adaptive and adjustable thermal metamaterials. Similar behaviors can also be expected in other fields, such as hydrodynamics.

Keywords Thermal rotators · Transformation invariance · Extreme anisotropy

9.1 Opening Remarks

Transformation thermotics [1, 2] provides a fundamental and powerful method to control heat flux at will. Initial explorations mainly focused on thermal conduction, and many functions were proposed, such as cloaking, concentrating, and rotating [3]. For the sake of practical applications, convection [4–6] and radiation [7, 8] have also been considered to develop corresponding transformation theories.

Although transformation-thermotics-based metamaterials have achieved great success, the lack of intelligence remains a problem. Specifically, the key equation of transformation thermotics is $\kappa' = \mathbf{J}\kappa\mathbf{J}^\tau / \det \mathbf{J}$, where κ' is transformed thermal conductivity, κ is environmental thermal conductivity, $\mathbf{J}$ is the Jacobian matrix, and τ denotes transpose [1, 2]. The transformed parameter (κ') is crucially dependent on the environmental parameter (κ). In other words, once the environmental parameter changes, the transformed parameter should change accordingly, making the original design fail in the new environment. This limitation is fatal because one device applies to only one environment. Similar problems also exist in the fields of electromagnetism and thermal radiation, and some inspiring studies [9, 10] gave insights.

To improve intelligence, we propose a mechanism based on thermal transformation-invariant metamaterials [11, 12], whose thermal conductivities are

© The Author(s) 2023
L.-J. Xu and J.-P. Huang, *Transformation Thermotics and Extended Theories*,
https://doi.org/10.1007/978-981-19-5908-0_9

highly anisotropic [13–15], i.e., $0\,\mathrm{W\,m^{-1}\,K^{-1}}$ in one direction and $\infty\,\mathrm{W\,m^{-1}\,K^{-1}}$ in the other. Transformation-invariant (i.e., highly anisotropic) metamaterials have aroused broad interest in various fields, such as electromagnetism [16, 17] and acoustics [18, 19]. For a two-dimensional case, highly anisotropic thermal conductivities have adaptive responses to environmental changes [20, 21], just like chameleons. We perform coordinate transformations based on transformation-invariant metamaterials, which can keep the chameleonlike behavior. Therefore, the designed devices have adaptive responses to environmental changes. We take thermal rotators [22–26] as an example, which can guide heat flux direction. Existing designs only apply to a particular environment, which cannot cope with environmental changes. Here, we propose the concept of thermal chameleonlike rotators. Going beyond a normal isotropic shell with near-zero thermal conductivity (Fig. 9.1a), we start the rotation transformation from a transformation-invariant shell (Fig. 9.1b). In this way, the designed rotator can work in different environments (Fig. 9.1c and d), thus called a chameleonlike rotator. The environment denotes the regions except for the rotator, and the environmental parameter refers to its thermal conductivity.

9.2   Chameleonlike Behavior Origin

We consider a passive and stable conduction process in two dimensions, which is governed by the Fourier law,

$$\nabla \cdot (-\kappa \cdot \nabla T) = 0. \tag{9.1}$$

The whole system is divided into three regions, i.e., core, shell, and background, with tensorial thermal conductivities of $\kappa_1 = \kappa_1 I$, $\kappa_2 = \mathrm{diag}\,(\kappa_{rr}, \kappa_{\theta\theta})$, and $\kappa_3 = \kappa_3 I$, respectively. We treat the core and background as the environment, and suppose their thermal conductivities to be the same, i.e., $\kappa_1 = \kappa_3$. κ_2 is expressed in cylindrical coordinates (r, θ). By solving the Laplace equation, the effective thermal conductivity of the core and shell κ_e can be expressed as

$$\kappa_e = \kappa_{rr} \frac{n_1 (\kappa_1 - n_2\kappa_{rr}) - n_2 (\kappa_1 - n_1\kappa_{rr})\, p^{(n_1-n_2)/2}}{\kappa_1 - n_2\kappa_{rr} - (\kappa_1 - n_1\kappa_{rr})\, p^{(n_1-n_2)/2}}, \tag{9.2}$$

where $n_{1,2} = \pm\sqrt{\kappa_{\theta\theta}/\kappa_{rr}}$ and $p = (R_1/R_2)^2$. R_1 and R_2 are the inner and outer radii of the shell, respectively. The thermal conductivity of a transformation-invariant metamaterial is

$$\kappa_2 = \begin{pmatrix} \infty & 0 \\ 0 & 0 \end{pmatrix}. \tag{9.3}$$

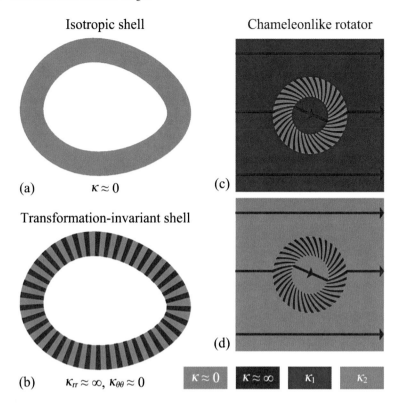

Fig. 9.1 Schematic diagram of the thermal chameleonlike rotator. **a** Isotropic shell with near-zero thermal conductivity. **b** Transformation-invariant shell with near-zero thermal conductivity in the tangential direction and near-infinite thermal conductivity in the radial direction. **c** and **d** Thermal chameleonlike rotator working in different environments. Lines with arrows indicate heat flow. The environmental thermal conductivities of **c** and **d** are κ_1 and κ_2, respectively. Adapted from Ref. [27]

The substitution of Eq. (9.3) into Eq. (9.2) yields

$$\kappa_e \approx \kappa_1, \tag{9.4}$$

which means that the effective thermal conductivity of the core-shell structure can adaptively change with the environment. In other words, two-dimensional transformation-invariant metamaterials (Eq. (9.3)) have a chameleonlike behavior (Eq. (9.4)). We then consider an arbitrary two-dimensional coordinate transformation,

$$r' = R(r, \theta), \tag{9.5a}$$
$$\theta' = \Theta(r, \theta), \tag{9.5b}$$

where (r', θ') are physical coordinates and (r, θ) are virtual coordinates. We can express the Jacobian matrix $\mathbf{J}$ as

$$\mathbf{J} = \begin{pmatrix} \dfrac{\partial r'}{\partial r} & \dfrac{\partial r'}{r\partial \theta} \\[3mm] \dfrac{r'\partial \theta'}{\partial r} & \dfrac{r'\partial \theta'}{r\partial \theta} \end{pmatrix}. \tag{9.6}$$

The transformed thermal conductivity is

$$\kappa_2' = \frac{\mathbf{J}\kappa_2\mathbf{J}^{\mathrm{T}}}{\det \mathbf{J}}, \tag{9.7}$$

which can be expressed in detail as

$$\kappa_2' = \frac{1}{\det \mathbf{J}} \begin{pmatrix} \kappa_{rr}\left(\dfrac{\partial r'}{\partial r}\right)^2 + \kappa_{\theta\theta}\left(\dfrac{\partial r'}{r\partial\theta}\right)^2 & \kappa_{rr}\left(\dfrac{\partial r'}{\partial r}\right)\left(\dfrac{r'\partial\theta'}{\partial r}\right) + \kappa_{\theta\theta}\left(\dfrac{\partial r'}{r\partial\theta}\right)\left(\dfrac{r'\partial\theta'}{r\partial\theta}\right) \\[3mm] \kappa_{rr}\left(\dfrac{\partial r'}{\partial r}\right)\left(\dfrac{r'\partial\theta'}{\partial r}\right) + \kappa_{\theta\theta}\left(\dfrac{\partial r'}{r\partial\theta}\right)\left(\dfrac{r'\partial\theta'}{r\partial\theta}\right) & \kappa_{rr}\left(\dfrac{r'\partial\theta'}{\partial r}\right)^2 + \kappa_{\theta\theta}\left(\dfrac{r'\partial\theta'}{r\partial\theta}\right)^2 \end{pmatrix}. \tag{9.8}$$

With Eq. (9.3), the eigenvalues of Eq. (9.8) are

$$\lambda_1 = \frac{\kappa_{rr}}{\det \mathbf{J}}\left[\left(\frac{\partial r'}{\partial r}\right)^2 + \left(\frac{r'\partial\theta'}{\partial r}\right)^2\right], \tag{9.9a}$$

$$\lambda_2 \approx \frac{\kappa_{\theta\theta}}{\det \mathbf{J}}. \tag{9.9b}$$

Due to $\kappa_{rr} = \infty$ and $\kappa_{\theta\theta} = 0$, Eq. (9.9) can be further reduced to

$$\lambda_1 = \infty, \tag{9.10a}$$

$$\lambda_2 = 0. \tag{9.10b}$$

An arbitrary coordinate transformation does not change the eigenvalues.

We then design a thermal chameleonlike rotator with transformation-invariant metamaterials. The coordinate transformation of rotating can be expressed as

$$r' = r, \tag{9.11a}$$

$$\theta' = \theta + \theta_0 \quad (r < R_1), \tag{9.11b}$$

$$\theta' = \theta + \theta_0 (R_2 - r)/(R_2 - R_1) \quad (R_1 < r < R_2), \tag{9.11c}$$

where θ_0 is rotation angle. With Eqs. (9.6) and (9.7), we can derive the thermal conductivity of the rotator as

$$\kappa_2' = \begin{pmatrix} \kappa_{rr} & \kappa_{rr}\dfrac{r'\theta_0}{R_2 - R_1} \\ \kappa_{rr}\dfrac{r'\theta_0}{R_2 - R_1} & \kappa_{rr}\left(\dfrac{r'\theta_0}{R_2 - R_1}\right)^2 + \kappa_{\theta\theta} \end{pmatrix}, \tag{9.12}$$

which is the key parameter for a thermal chameleonlike rotator as long as κ_2 satisfies Eq. (9.3).

9.3 Finite-Element Simulation

To verify the scheme, we first perform simulations with COMSOL Multiphysics. The system is the same as Fig. 9.1c. We compare the difference between a chameleonlike rotator and a normal rotator (Fig. 9.2). Before performing the rotation transformation, the thermal conductivities for the chameleonlike rotator and normal rotator are diag $\left(10^6, \ 10^{-3}\right)$ and $100\,\mathrm{W\,m^{-1}\,K^{-1}}$, respectively. The radial thermal conductivity of the transformation-invariant metamaterial should be much larger than the environmental thermal conductivity (at least two orders of magnitude), or the chameleonlike rotator may fail. We then change the environmental thermal conductivity from 10 to 1000 W m^{-1} K^{-1}, and the chameleonlike rotator can always work, i.e., rotating heat flux without distorting the environmental temperature profile (Fig. 9.2a–c).

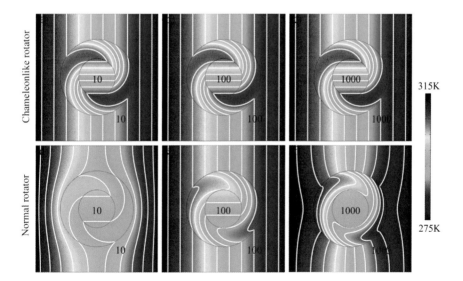

Fig. 9.2 Simulation results of **a–c** chameleonlike rotator and **d–f** normal rotator. White lines represent isotherms, and the values in each simulation are the corresponding thermal conductivities. The system size is 1×1 m^2. The outer and inner diameters of the shell are 0.3 and 0.6 m, respectively. Adapted from Ref. [27]

Therefore, the simulation results confirm the chameleonlike property. However, the normal rotator fails. When the environmental thermal conductivity is $100\,\mathrm{W\,m^{-1}\,K^{-1}}$, it behaves like a traditional rotator (Fig. 9.2e). When the environment changes, the temperature profile is distorted (Fig. 9.2d and f). Therefore, the normal rotator has no response to environmental changes.

9.4   Laboratory Experiment

For experimental verification, it is not easy to find a material in nature that satisfies Eq. (9.12). Therefore, we use the effective medium theory to realize the corresponding parameter. Drawing on the multilayered structure [22], we design the chameleonlike rotator as shown in Fig. 9.3a. As required by Eqs. (9.3) and (9.12), we choose two materials with extremely large ($\kappa_l \approx 10^6\ \mathrm{W\ m^{-1}\ K^{-1}}$) and extremely small ($\kappa_s \approx 10^{-3}\ \mathrm{W\ m^{-1}\ K^{-1}}$) thermal conductivities to approximately satisfy Eq. (9.3), and then use the helical structure to approximately satisfy Eq. (9.12). The simulation results are shown in Fig. 9.3b–g. Among them, Fig. 9.3b–d show the results of chameleonlike rotator-1, rotating heat flux 90 °C. Figure 9.3e, f presents the results of chameleonlike rotator-2, rotating heat flux 180 °C. Therefore, it is feasible to fabricate chameleonlike rotators with multilayered composite structures.

Restricted by experimental conditions, we choose copper ($\kappa_{cu} \approx 400\,\mathrm{W\,m^{-1}\,K^{-1}}$) and air ($\kappa_{air} \approx 0.026\,\mathrm{W\ m^{-1}\ K^{-1}}$) to fabricate a multilayered composite structure to realize a small-angle rotator. According to the series/parallel connection formula [28], the effective thermal conductivity of the composite structure is about

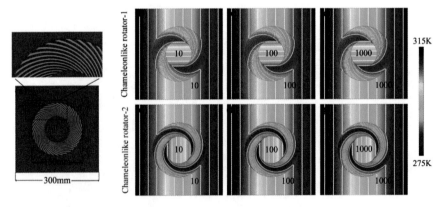

Fig. 9.3 Simulation results of chameleonlike rotators with multilayered structures. **a** Schematic diagram. The structure is composed of two kinds of material with thermal conductivities of 10^6 and 10^{-3} W m^{-1} K^{-1}, respectively. Simulations results of **b–d** chameleonlike rotator-1 and **e–g** chameleonlike rotator-2 in different environments. The composite materials in **b–g** are the same as those in **a**. Adapted from Ref. [27]

diag (200, 0.052) W m^{-1} K^{-1} before transformation, putting a limit on the variation range of the environmental thermal conductivity. We calculate κ_e with κ_1 changing from 0.1 to 50 W m^{-1} K^{-1}, and confirm that the chameleonlike rotator works well from 0.1 to 5 W m^{-1} K^{-1}, as shown in Fig. 9.4b. The difference $|\kappa_e - \kappa_1|$ is smaller than 0.05 W m^{-1} K^{-1} (denoted by the star, with a deviation smaller than 1%). There-fore, we conduct experiments with environmental thermal conductivities of 1 and 5 W m^{-1} K^{-1}. The system is designed as shown in Fig. 9.4a. The chameleonlike rota-tor is composed of air and copper, fabricated by laser cutting. The environment is colloidal materials obtained by mixing silica gel ($\kappa_{gel} = 0.15$ W m^{-1} K^{-1} and density $\rho_{gel} = 1.14 \times 10^3$ kg m^{-3}) and white copper powder ($\kappa_{wcu} = 33$ W m^{-1} K^{-1} and $\rho_{wcu} = 8.65 \times 10^3$ kg m^{-3}). The thermal conductivity of the mixture is determined by the Bruggeman formula [29],

$$p_{gel}\frac{\kappa_{gel} - \kappa_{mix}}{\kappa_{gel} + 2\kappa_{mix}} + (1 - p_{gel})\frac{\kappa_{wcu} - \kappa_{mix}}{\kappa_{wcu} + 2\kappa_{mix}} = 0, \tag{9.13}$$

where p_{gel} is the volume fraction of silica gel in the mixture. By setting $\kappa_{mix} = 1$ or 5 W m^{-1} K^{-1}, we can derive the composition ratio of silica gel, which helps us fabricate the colloidal materials. Although interface thermal conductance [30, 31] exists, the mixture in regions I and III has a little fluidity, ensuring good contact between the object and copper. We then fill two hot and ice water tanks as hot and cold sources. The FLIR E60 infrared camera measures the temperature profile of the sample. The experimental results are shown in Fig. 9.4d ($\kappa_{mix} = 1$ W m^{-1} K^{-1}) and 9.4f ($\kappa_{mix} = 5$ W m^{-1} K^{-1}). The corresponding simulation results are presented in Fig. 9.4c and e. Heat dissipation exists because the sample is connected to the hot and cold tanks with two copper plates. Moreover, the natural convection between the sample and air also results in heat dissipation. Therefore, there is a small difference between the computational and experimental values, but this does not affect the expected results. The isotherms still keep straight even though the environmental thermal conductivity changes. Meanwhile, heat flux is rotated as expected. Therefore, the experimental results are consistent with the simulation results, verifying the feasibility of chameleonlike rotators.

9.5 Discussion

The major difference of our scheme is to start the coordinate transformation from a highly anisotropic parameter, which is proved to have a chameleonlike behavior. Therefore, the designed rotator based on this parameter can also have a chameleonlike behavior. Meanwhile, no matter how the shape of the rotator changes, the chameleon-like behavior still exists. Therefore, the present scheme can also design chameleon-like rotators with arbitrary shapes. Nevertheless, a perfect transformation-invariant (i.e., highly anisotropic) shell is described by Eq. (9.3), indicating that the higher anisotropy yields a better chameleonlike behavior and a wider working range. Due to

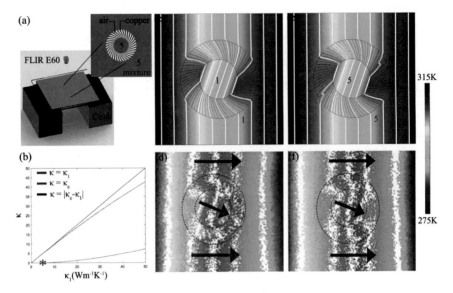

Fig. 9.4 Laboratory experiments of chameleonlike rotator. **a** Experimental setup. The structure is composed of copper ($\kappa_{cu} \approx 400$ W m^{-1} K^{-1}) and air ($\kappa_{air} \approx 0.026$ W m^{-1} K^{-1}). **b** κ as a function of κ_1. The blue (top) and red (middle) lines correspond to κ_1 and κ_e, respectively. The black (bottom) line refers to $|\kappa_e - \kappa_1|$. The coordinate of $*$ is (5, 0.047). **c** and **e** Simulation results and **d** and **f** experimental results of the samples. The arrows indicate the direction of heat flux. The inner and outer diameters of the shell are 0.075 and 0.15 m, respectively. Adapted from Ref. [27]

the lack of highly conductive materials, the working range of the fabricated rotator is from 0.1 to 5 W m^{-1} K^{-1}. More methods [32–36] can be applied to enhance thermal conductivities.

Moreover, the scheme can be extended to transient regimes by taking density and heat capacity into account [23, 37–42]. The scheme is also not limited to conductive systems. Recent studies explored convective-diffusive systems [43–45], hydrodynamic systems [46, 47], and acoustic systems [18, 19] to design functional devices. Therefore, it is also promising to design chameleonlike rotators in these fields. Although these results are obtained at the macroscopic scale described by the Fourier law, intelligence may also be helpful for heat manipulations with nanostructures [48, 49].

9.6   Conclusion

We have designed thermal chameleonlike rotators based on transformation-invariant metamaterials. With a highly anisotropic thermal conductivity, the designed rotator can work in different environments, saving time and labor. Both simulations and experiments verify the feasibility of the scheme. These results improve the intelli-

gence of traditional thermal metamaterials and have potential applications in designing intelligent metamaterials. The proposed scheme can also be extended to other
fields, such as hydrodynamics, where the critical parameter (permeability or viscosity) plays a similar role as thermal conductivity in thermotics.

9.7 Exercise and Solution

Exercise

1. Calculate the Jacobian transformation matrix of Eq. (9.11).

Solution

1. For the region $r < R_1$, $\mathbf{J} = \mathbf{I}$. For the region $R_1 < r < R_2$,

$$\mathbf{J} = \begin{pmatrix} 1 & 0 \\ -\theta_0/(R_2 - R_1) & 1 \end{pmatrix}. \tag{9.14}$$

References

1. Fan, C.Z., Gao, Y., Huang, J.P.: Shaped graded materials with an apparent negative thermal conductivity. Appl. Phys. Lett. **92**, 251907 (2008)
2. Chen, T.Y., Weng, C.-N., Chen, J.-S.: Cloak for curvilinearly anisotropic media in conduction. Appl. Phys. Lett. **93**, 114103 (2008)
3. Huang, J.P.: Theoretical Thermotics: Transformation Thermotics and Extended Theories for Thermal Metamaterials. Springer, Singapore (2020)
4. Guenneau, S., Petiteau, D., Zerrad, M., Amra, C., Puvirajesinghe, T.: Transformed Fourier and Fick equations for the control of heat and mass diffusion. AIP Adv. **5**, 053404 (2015)
5. Dai, G.L., Shang, J., Huang, J.P.: Theory of transformation thermal convection for creeping flow in porous media: cloaking, concentrating, and camouflage. Phys. Rev. E **97**, 022129 (2018)
6. Xu, L.J., Huang, J.P.: Controlling thermal waves with transformation complex thermotics. Int. J. Heat Mass Transf. **159**, 120133 (2020)
7. Xu, L.J., Dai, G.L., Huang, J.P.: Transformation multithermotics: controlling radiation and conduction simultaneously. Phys. Rev. Appl. **13**, 024063 (2020)
8. Xu, L.J., Yang, S., Dai, G.L., Huang, J.P.: Transformation omnithermotics: simultaneous manip-ulation of three basic modes of heat transfer. ES Energy Environ. **7**, 65 (2020)
9. Peng, R.G., Xiao, Z.Q., Zhao, Q., Zhang, F.L., Meng, Y.G., Li, B., Zhou, J., Fan, Y.C., Zhang, P., Shen, N.-H., Koschny, T., Soukoulis, C.M.: Temperature controlled chameleonlike cloak. Phys. Rev. X **7**, 011033 (2017)
10. Li, Y., Bai, X., Yang, T.Z., Luo, H.L., Qiu, C.-W.: Structured thermal surface for radiative camouflage. Nat. Commun. **9**, 273 (2018)
11. Liu, Y.C., Sun, F., He, S.L.: Fast adaptive thermal buffering by a passive open shell based on transformation thermodynamics. Adv. Theory Simul. **1**, 1800026 (2018)
12. Sun, F., Liu, Y.H., Yang, Y.B., Chen, Z.H., He, S.L.: Thermal surface transformation and its applications to heat flux manipulations. Opt. Express **27**, 33757 (2019)

13. Athanasopoulos, N., Siakavellas, N.J.: Heat manipulation using highly anisotropic pitch-based carbon fiber composites. Adv. Eng. Mater. **17**, 1494 (2015)
14. Wan, J.Y., Song, J.W., Yang, Z., Kirsch, D., Jia, C., Xu, R., Dai, J.Q., Zhu, M.W., Xu, L.S., Chen, C.J., Wang, Y.B., Wang, Y.L., Hitz, E., Lacey, S.D., Li, Y.F., Yang, B., Hu, L.B.: Highly anisotropic conductors. Adv. Mater. **29**, 1703331 (2017)
15. Hamed, A., Ndao, S.: High anisotropy metamaterial heat spreader. Int. J. Heat Mass Transf. **121**, 10 (2018)
16. Zhang, Y.M., Zhang, B.L.: Bending, splitting, compressing and expanding of electromagnetic waves in infinitely anisotropic media. J. Opt. **20**, 014001 (2018)
17. Zhang, Y.M., Luo, Y., Pendry, J.B., Zhang, B.L.: Transformation-invariant metamaterials. Phys. Rev. Lett. **123**, 067701 (2019)
18. Wu, L.T., Oudich, M., Cao, W.K., Jiang, H.L., Zhang, C., Ke, J.C., Yang, J., Deng, Y.C., Cheng, Q., Cui, T.J., Jing, Y.: Routing acoustic waves via a metamaterial with extreme anisotropy. Phys. Rev. Appl. **12**, 044011 (2019)
19. Fakheri, M.H., Abdolali, A., Sedeh, H.B.: Arbitrary shaped acoustic concentrators enabled by null media. Phys. Rev. Appl. **13**, 034004 (2020)
20. Xu, L.J., Yang, S., Huang, J.P.: Passive metashells with adaptive thermal conductivities: chameleonlike behavior and its origin. Phys. Rev. Appl. **11**, 054071 (2019)
21. Xu, L.J., Huang, J.P.: Chameleonlike metashells in microfluidics: a passive approach to adaptive responses. Sci. China-Phys. Mech. Astron. **63**, 228711 (2020)
22. Narayana, S., Sato, Y.: Heat flux manipulation with engineered thermal materials. Phys. Rev. Lett. **108**, 214303 (2012)
23. Guenneau, S., Amra, C.: Anisotropic conductivity rotates heat fluxes in transient regimes. Opt. Express **21**, 6578 (2013)
24. Xu, L.J., Yang, S., Huang, J.P.: Thermal theory for heterogeneously architected structure: fundamentals and application. Phys. Rev. E **98**, 052128 (2018)
25. Zhou, L.L., Huang, S.Y., Wang, M., Hu, R., Luo, X.B.: While rotating while cloaking. Phys. Lett. A **383**, 759 (2019)
26. Tsai, Y.-L., Li, J.Y., Chen, T.Y.: Simultaneous focusing and rotation of a bifunctional thermal metamaterial with constant anisotropic conductivity. J. Appl. Phys. **126**, 095103 (2019)
27. Yang, F.B., Tian, B.Y., Xu, L.J., Huang, J.P.: Experimental demonstration of thermal chameleonlike rotators with transformation-invariant metamaterials. Phys. Rev. Appl. **14**, 054024 (2020)
28. Wang, R.Z., Xu, L.J., Huang, J.P.: Thermal imitators with single directional invisibility. J. Appl. Phys. **122**, 215107 (2017)
29. Shang, J., Tian, B.Y., Jiang, C.R., Huang, J.P.: Digital thermal metasurface with arbitrary infrared thermogram. Appl. Phys. Lett. **113**, 261902 (2018)
30. Li, J.Y., Gao, Y., Huang, J.P.: A bifunctional cloak using transformation media. J. Appl. Phys. **108**, 074504 (2010)
31. Ma, D.K., Zhang, G., Zhang, L.F.: Interface thermal conductance between β-Ga_2O^3 and different substrates. J. Phys. D-Appl. Phys. **53**, 434001 (2020)
32. Li, Y., Zhu, K.-J., Peng, Y.-G., Li, W., Yang, T.Z., Xu, H.-X., Chen, H., Zhu, X.-F., Fan, S.H., Qiu, C.-W.: Thermal meta-device in analogue of zero-index photonics. Nat. Mater. **18**, 48 (2019)
33. Yang, S., Xu, L.J., Huang, J.P.: Metathermotics: nonlinear thermal responses of core-shell metamaterials. Phys. Rev. E **99**, 042144 (2019)
34. Xu, L.J., Yang, S., Huang, J.P.: Effectively infinite thermal conductivity and zero-index thermal cloak. EPL **131**, 24002 (2020)
35. Li, J.X., Li, Y., Wang, W.Y., Li, L.Q., Qiu, C.-W.: Effective medium theory for thermal scattering off rotating structures. Opt. Express **28**, 25894 (2020)
36. Li, J.X., Li, Y., Cao, P.-C., Yang, T.Z., Zhu, X.-F., Wang, W.Y., Qiu, C.-W.: A continuously tunable solid-like convective thermal metadevice on the reciprocal line. Adv. Mater. **32**, 2003823 (2020)

37. Ma, Y.G., Lan, L., Jiang, W., Sun, F., He, S.L.: A transient thermal cloak experimentally realized through a rescaled diffusion equation with anisotropic thermal diffusivity. NPJ Asia Mater. **5**, e73 (2013)

38. Yang, T.Z., Su, Y., Xu, W., Yang, X.D.: Transient thermal camouflage and heat signature control. Appl. Phys. Lett. **109**, 121905 (2016)

39. Liu, Y.X., Guo, W.L., Han, T.C.: Arbitrarily polygonal transient thermal cloaks with natural bulk materials in bilayer configurations. Int. J. Heat Mass Transf. **115**, 1 (2017)

40. He, X., Yang, T.Z., Zhang, X.W., Wu, L.Z., He, X.Q.: Transient experimental demonstration of an elliptical thermal camouflage device. Sci. Rep. **7**, 16671 (2017)

41. Han, T.C., Yang, P., Li, Y., Lei, D.Y., Li, B.W., Hippalgaonkar, K., Qiu, C.-W.: Full-parameter omnidirectional thermal metadevices of anisotropic geometry. Adv. Mater. **30**, 1804019 (2018)

42. Huang, S.Y., Zhang, J.W., Wang, M., Lan, W., Hu, R., Luo, X.B.: Macroscale thermal diode-like black box with high transient rectification ratio. ES Energy Environ. **6**, 51 (2019)

43. Yang, F.B., Xu, L.J., Huang, J.P.: Thermal illusion of porous media with convection-diffusion process: Transparency, concentrating, and cloaking. ES Energy Environ. **6**, 45 (2019)

44. Yeung, W.-S., Mai, V.-P., Yang, R.-J.: Cloaking: controlling thermal and hydrodynamic fields simultaneously. Phys. Rev. Appl. **13**, 064030 (2020)

45. Xu, L.J., Huang, J.P.: Negative thermal transport in conduction and advection. Chin. Phys. Lett. **37**, 080502 (2020)

46. Park, J., Youn, J.R., Song, Y.S.: Hydrodynamic metamaterial cloak for drag-free flow. Phys. Rev. Lett. **123**, 074502 (2019)

47. Park, J., Youn, J.R., Song, Y.S.: Fluid-flow rotator based on hydrodynamic metamaterial. Phys. Rev. Appl. **12**, 061002 (2019)

48. Ma, D.K., Wan, X., Yang, N.: Unexpected thermal conductivity enhancement in pillared graphene nanoribbon with isotopic resonance. Phys. Rev. B **98**, 245420 (2018)

49. Bao, H., Chen, J., Gu, X.K., Cao, B.Y.: A review of simulation methods in micro/nanoscale heat conduction. ES Energy Environ. **1**, 16 (2018)

Chapter 10
Theory for Invisible Thermal Sensors: Bilayer Scheme

Abstract In this chapter, we propose a bilayer scheme with isotropic materials to design invisible thermal sensors with detecting accuracy. Therefore, the original temperature fields in the sensor and matrix can keep unchanged. By solving the linear Laplace equation with a temperature-independent thermal conductivity, we derive two groups of thermal conductivities to realize invisible thermal sensors, even considering geometrically anisotropic cases. These results can be directly extended to thermally nonlinear cases with temperature-dependent thermal conductivity, as long as the ratio between the nonlinear thermal conductivities of the sensor and matrix is a temperature-independent constant. These explorations are beneficial to temperature detection and provide insights into thermal camouflage.

Keywords Invisible thermal sensors · Detecting accuracy · Bilayer scheme

10.1 Opening Remarks

Precision measurement is indispensable in many fields, so high-performance sensors are crucial. Generally, when a sensor is put in a physical field, it will distort the physical field. Therefore, the measured value is not the original one, thus making the sensor inaccurate. In addition to inaccuracy, the perturbation induced by the sensor also makes the sensor "visible", which is adverse in many practical applications. The methods of scattering cancellation [1] and transformation optics [2] were proposed to design invisible electromagnetic sensors. Invisible acoustic sensors [3–5] and invisible magnetic sensors [6] were also presented successively.

Invisible thermal sensors also attracted research interest. The methods of scattering cancellation [7–9], neutral inclusion [10], and transformation thermotics [11] were put forward to design invisible thermal sensors. Furthermore, an invisible multiphysical sensor was also fabricated for both thermal and electric detection [12]. These studies focused on thermal invisibility because it is particularly important to fight against infrared detection. For example, invisibility can protect the sensor from being discovered when a thermal sensor detects temperature. However, accuracy is almost neglected in these schemes, so the detected temperature has deviations from

© The Author(s) 2023   133
L.-J. Xu and J.-P. Huang, *Transformation Thermotics and Extended Theories*,
https://doi.org/10.1007/978-981-19-5908-0_10

the original one, thus making thermal sensors inaccurate. Meanwhile, invisible thermal sensors for nonlinear cases are still lacking, limiting practical applications. Here, "nonlinear" means that thermal conductivities are temperature-dependent.

To solve the problem, we propose a bilayer scheme to design invisible thermal sensors, even considering geometrically anisotropic and thermally nonlinear cases. These two points benefit practical applications because thermal sensors do not have to be geometrically isotropic, and nonlinear thermal conductivity is common. In fact, bilayer scheme has achieved great success in designing thermal cloaks [13–18], thermal concentrators [19], and chameleonlike metashells [20, 21]. Cloaks make the temperature gradient in the center zero; concentrators make the temperature gradient in the center steeper than that in the matrix; invisible sensors keep the same temperature gradient in the sensor and matrix. We derive two groups of thermal conductivities by solving the linear Laplace equation, making thermal sensors both accurate and invisible. Moreover, we prove that the bilayer scheme can be directly extended to thermally nonlinear cases as long as the ratio between the nonlinear thermal conductivities of the sensor and matrix is a temperature-independent constant.

10.2 Linear and Geometrically Isotropic Case

We discuss the case shown in Fig. 10.1a. The Cartesian coordinates are denoted as x_i ($i = 1$, 2 for two dimensions and $i = 1$, 2, 3 for three dimensions). The radii of the core, inner shell, and outer shell are denoted as λ_a, λ_b, and λ_c, respectively. The thermal conductivities of the core, inner shell, outer shell, and matrix are denoted as κ_a, κ_b, κ_c, and κ_d, respectively. Since the geometry is isotropic, we discuss the case in cylindrical coordinates (r, θ) or spherical coordinates (r, θ, φ). Here, two

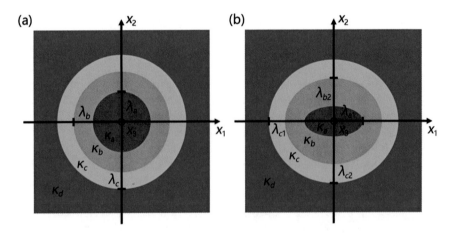

Fig. 10.1 Schematic diagrams of **a** geometrically isotropic case and **b** geometrically anisotropic case. Adapted from Ref. [22]

dimensions and three dimensions are similar because φ essentially does not matter. In the presence of an external linear thermal field G_0, the temperature profiles in different regions can be expressed as

$$T_a = u_a r \cos\theta, \tag{10.1a}$$

$$T_b = \left(u_b r + v_b r^{-\tau}\right)\cos\theta, \tag{10.1b}$$

$$T_c = \left(u_c r + v_c r^{-\tau}\right)\cos\theta, \tag{10.1c}$$

$$T_d = \left(u_d r + v_d r^{-\tau}\right)\cos\theta, \tag{10.1d}$$

where T_a, T_b, T_c, and T_d are the temperatures in the core, inner shell, outer shell, and matrix, respectively. $\tau = 1$ for two dimensions and $\tau = 2$ for three dimensions. u_a, u_b, v_b, u_c, v_c, u_d, and v_d are seven coefficients to be determined by the following boundary conditions,

$$u_a\lambda_a = u_b\lambda_a + v_b\lambda_a^{-\tau}, \tag{10.2a}$$

$$u_b\lambda_b + v_b\lambda_b^{-\tau} = u_c\lambda_b + v_c\lambda_b^{-\tau}, \tag{10.2b}$$

$$u_c\lambda_c + v_c\lambda_c^{-\tau} = u_d\lambda_c + v_d\lambda_c^{-\tau}, \tag{10.2c}$$

$$\kappa_a u_a = \kappa_b\left(u_b - \tau v_b\lambda_a^{-\tau-1}\right), \tag{10.2d}$$

$$\kappa_b\left(u_b - \tau v_b\lambda_b^{-\tau-1}\right) = \kappa_c\left(u_c - \tau v_c\lambda_b^{-\tau-1}\right), \tag{10.2e}$$

$$\kappa_c\left(u_c - \tau v_c\lambda_c^{-\tau-1}\right) = \kappa_d\left(u_d - \tau v_d\lambda_c^{-\tau-1}\right), \tag{10.2f}$$

$$u_d = G_0, \tag{10.2g}$$

$$v_d = 0, \tag{10.2h}$$

$$u_a = u_d. \tag{10.2i}$$

Equations (10.2a)–(10.2c) and (10.2d)–(10.2f) indicate the continuities of temperature and heat flux, respectively. Equations (10.2g) and (10.2h) ensure a linear thermal field in the matrix, thus making the sensor thermally invisible. Equation (10.2i) makes the temperature in the sensor the same as that in the matrix, thus ensuring accurate detection. We take κ_b and κ_c as other two coefficients which (together with the seven coefficients in Eqs. (10.1a)–(10.1d)) can be determined by the nine equations in Eq. (10.2). Therefore, κ_b and κ_c can be solved as

$$\kappa_b^{(1)} = \frac{\kappa_a\alpha_3 - \kappa_d\alpha_1 + \sqrt{(\kappa_a - \kappa_d)\left(\kappa_a\alpha_2^2 - \kappa_d\alpha_1^2\right)}}{\alpha_5}, \tag{10.3a}$$

$$\kappa_c^{(1)} = \frac{\kappa_a \alpha_2 - \kappa_d \alpha_4 - \sqrt{(\kappa_a - \kappa_d)\left(\kappa_a \alpha_2^2 - \kappa_d \alpha_1^2\right)}}{\alpha_6}, \tag{10.3b}$$

or

$$\kappa_b^{(2)} = \frac{\kappa_a \alpha_3 - \kappa_d \alpha_1 - \sqrt{(\kappa_a - \kappa_d)\left(\kappa_a \alpha_2^2 - \kappa_d \alpha_1^2\right)}}{\alpha_5}, \tag{10.4a}$$

$$\kappa_c^{(2)} = \frac{\kappa_a \alpha_2 - \kappa_d \alpha_4 + \sqrt{(\kappa_a - \kappa_d)\left(\kappa_a \alpha_2^2 - \kappa_d \alpha_1^2\right)}}{\alpha_6}, \tag{10.4b}$$

where

$$\alpha_1 = \lambda_a^{1+\tau}\left(\lambda_b^{1+\tau} + \tau\lambda_c^{1+\tau}\right) + \lambda_b^{1+\tau}\left[\tau\lambda_b^{1+\tau} - (2\tau + 1)\lambda_c^{1+\tau}\right], \tag{10.5a}$$

$$\alpha_2 = \lambda_a^{1+\tau}\left[(2\tau + 1)\lambda_b^{1+\tau} - \tau\lambda_c^{1+\tau}\right] - \lambda_b^{1+\tau}\left(\tau\lambda_b^{1+\tau} + \lambda_c^{1+\tau}\right), \tag{10.5b}$$

$$\alpha_3 = \lambda_a^{1+\tau}\left[2\tau\lambda_a^{1+\tau} - (2\tau - 1)\lambda_b^{1+\tau} - \tau\lambda_c^{1+\tau}\right] + \lambda_b^{1+\tau}\left(\tau\lambda_b^{1+\tau} - \lambda_c^{1+\tau}\right), \tag{10.5c}$$

$$\alpha_4 = \lambda_b^{1+\tau}\left(\lambda_a^{1+\tau} - \tau\lambda_b^{1+\tau}\right) + \lambda_c^{1+\tau}\left[\tau\lambda_a^{1+\tau} + (2\tau - 1)\lambda_b^{1+\tau} - 2\tau\lambda_c^{1+\tau}\right], \tag{10.5d}$$

$$\alpha_5 = 2\tau\left(\lambda_a^{1+\tau} - \lambda_b^{1+\tau}\right)\left(\lambda_a^{1+\tau} - \lambda_c^{1+\tau}\right), \tag{10.5e}$$

$$\alpha_6 = 2\tau\left(\lambda_a^{1+\tau} - \lambda_c^{1+\tau}\right)\left(\lambda_b^{1+\tau} - \lambda_c^{1+\tau}\right). \tag{10.5f}$$

When $\kappa_a < \kappa_d$, Eqs. (10.3) and (10.4) are always positive. When $\kappa_a = \kappa_d$, the sensor has the same thermal conductivity as the matrix, resulting in $\kappa_b = \kappa_c = \kappa_a = \kappa_d$, so the bilayer scheme is not necessary. When $\kappa_a > \kappa_d$, Eqs. (10.3b) and (10.4a) are negative. Negative thermal conductivity means that the direction of heat flux is from low temperature to high temperature, which can be effectively realized by introducing extra energy [23]. Also, we do not need to worry about complex values as long as the value of λ_b is appropriately chosen. Physically, when $\kappa_a < \kappa_d$, the temperature gradient in the sensor is larger than that in the matrix, and the bilayer scheme can reduce the temperature gradient to make the temperature gradients in the sensor and matrix the same. When $\kappa_a > \kappa_d$, the temperature gradient in the sensor is smaller than that in the matrix, but the bilayer scheme cannot enhance the temperature gradient with only positive thermal conductivities.

10.3 Linear and Geometrically Anisotropic Case

We discuss the case shown in Fig. 10.1b. The semi axes of the core, inner shell, and outer shell are denoted as λ_{ai}, λ_{bi}, and λ_{ci}, respectively ($i = 1$, 2 for two dimensions and $i = 1$, 2, 3 for three dimensions). Since the geometry is anisotropic, we discuss the case in elliptical coordinates (ρ, ξ) or ellipsoidal coordinates (ρ, ξ, η). Here, although two dimensions and three dimensions are different, we can remove the terms associated with η and x_3 to reduce three dimensions to two dimensions. The ellipsoidal coordinates (ρ, ξ, η) can be expressed as

$$\frac{x_1^2}{\rho + \lambda_{a1}^2} + \frac{x_2^2}{\rho + \lambda_{a2}^2} + \frac{x_3^2}{\rho + \lambda_{a3}^2} = 1, \tag{10.6a}$$

$$\frac{x_1^2}{\xi + \lambda_{a1}^2} + \frac{x_2^2}{\xi + \lambda_{a2}^2} + \frac{x_3^2}{\xi + \lambda_{a3}^2} = 1, \tag{10.6b}$$

$$\frac{x_1^2}{\eta + \lambda_{a1}^2} + \frac{x_2^2}{\eta + \lambda_{a2}^2} + \frac{x_3^2}{\eta + \lambda_{a3}^2} = 1, \tag{10.6c}$$

where $\rho = $ constant denotes an ellipsoidal surface, and λ_i is the semi axis of the ellipsoid $(\rho = $ constant$)$ along x_i axis. Accordingly, the Cartesian coordinates can be expressed as

$$x_1^2 = \frac{\left(\rho + \lambda_{a1}^2\right)\left(\xi + \lambda_{a1}^2\right)\left(\eta + \lambda_{a1}^2\right)}{\left(\lambda_{a1}^2 - \lambda_{a2}^2\right)\left(\lambda_{a1}^2 - \lambda_{a3}^2\right)}, \tag{10.7a}$$

$$x_2^2 = \frac{\left(\rho + \lambda_{a2}^2\right)\left(\xi + \lambda_{a2}^2\right)\left(\eta + \lambda_{a2}^2\right)}{\left(\lambda_{a2}^2 - \lambda_{a1}^2\right)\left(\lambda_{a2}^2 - \lambda_{a3}^2\right)}, \tag{10.7b}$$

$$x_3^2 = \frac{\left(\rho + \lambda_{a3}^2\right)\left(\xi + \lambda_{a3}^2\right)\left(\eta + \lambda_{a3}^2\right)}{\left(\lambda_{a3}^2 - \lambda_{a1}^2\right)\left(\lambda_{a3}^2 - \lambda_{a2}^2\right)}. \tag{10.7c}$$

In the presence of an external linear thermal field G_0 along x_i axis, the temperature profiles in different regions can be expressed as [24]

$$T_a = u_a x_i, \tag{10.8a}$$

$$T_b = \left[u_b + v_b \int_{\rho_a}^{\rho} \frac{d\rho}{\left(\rho + \lambda_{ai}^2\right) g(\rho)}\right] x_i, \tag{10.8b}$$

$$T_c = \left[u_c + v_c \int_{\rho_a}^{\rho} \frac{d\rho}{\left(\rho + \lambda_{ai}^2\right) g(\rho)}\right] x_i, \tag{10.8c}$$

$$T_d = \left[u_d + v_d \int_{\rho_a}^{\rho} \frac{d\rho}{\left(\rho + \lambda_{ai}^2\right) g(\rho)}\right] x_i, \tag{10.8d}$$

where $g(\rho) = \sqrt{\left(\rho + \lambda_{a1}^2\right)\left(\rho + \lambda_{a2}^2\right)\left(\rho + \lambda_{a3}^2\right)} = \lambda_1 \lambda_2 \lambda_3$, and ρ_a $(= 0)$ denotes the ellipsoidal core surface with semiaxes λ_{ai}. As explained above, $g(\rho) = \sqrt{\left(\rho + \lambda_{a1}^2\right)\left(\rho + \lambda_{a2}^2\right)} = \lambda_1 \lambda_2$ for two dimensions.

We use two mathematical skills to proceed. The first one is associated with the temperature derivations in Eq. (10.8),

$$\frac{\partial x_i}{\partial \rho} = \frac{x_i}{2\left(\rho + \lambda_{ai}^2\right)}, \tag{10.9a}$$

$$\frac{\partial}{\partial \rho}\left[x_i \int_{\rho_a}^{\rho} \frac{d\rho}{\left(\rho + \lambda_{ai}^2\right) g\left(\rho\right)}\right] = \frac{x_i}{\left(\rho + \lambda_{ai}^2\right) g\left(\rho\right)} \tag{10.9b}$$

$$+ \frac{x_i}{2\left(\rho + \lambda_{ai}^2\right)} \int_{\rho_a}^{\rho} \frac{d\rho}{\left(\rho + \lambda_{ai}^2\right) g\left(\rho\right)}.$$

The second one is related to the integrations in Eqs. (10.8b)–(10.8d) which can be rewritten as

$$\int_{\rho_a}^{\rho} \frac{d\rho}{\left(\rho + \lambda_{ai}^2\right) g\left(\rho\right)} = \int_{\rho_a}^{\infty} \frac{d\rho}{\left(\rho + \lambda_{ai}^2\right) g\left(\rho\right)} - \int_{\rho}^{\infty} \frac{d\rho}{\left(\rho + \lambda_{ai}^2\right) g\left(\rho\right)} \tag{10.10}$$

$$= \frac{2L_{ai}}{g\left(\rho_a\right)} - \frac{2L_i}{g\left(\rho\right)},$$

where L_{ai} and L_i are shape factors along x_i axis,

$$L_{ai} = \frac{g\left(\rho_a\right)}{2} \int_{\rho_a}^{\infty} \frac{d\rho}{\left(\rho + \lambda_{ai}^2\right) g\left(\rho\right)}, \tag{10.11a}$$

$$L_i = \frac{g\left(\rho\right)}{2} \int_{\rho}^{\infty} \frac{d\rho}{\left(\rho + \lambda_{ai}^2\right) g\left(\rho\right)}. \tag{10.11b}$$

Then, the boundary conditions can be expressed as

$$u_a = u_b, \tag{10.12a}$$

$$u_b + v_b \int_{\rho_a}^{\rho_b} \frac{d\rho}{\left(\rho + \lambda_{ai}^2\right) g\left(\rho\right)} = u_c + v_c \int_{\rho_a}^{\rho_b} \frac{d\rho}{\left(\rho + \lambda_{ai}^2\right) g\left(\rho\right)}, \tag{10.12b}$$

$$u_c + v_c \int_{\rho_a}^{\rho_c} \frac{d\rho}{\left(\rho + \lambda_{ai}^2\right) g\left(\rho\right)} = u_d + v_d \int_{\rho_a}^{\rho_c} \frac{d\rho}{\left(\rho + \lambda_{ai}^2\right) g\left(\rho\right)}, \tag{10.12c}$$

$$\kappa_a u_a = \kappa_b \left[u_b + \frac{2v_b}{g\left(\rho_a\right)}\right], \tag{10.12d}$$

$$\kappa_b \left[u_b + \frac{2v_b}{g\left(\rho_b\right)} + v_b \int_{\rho_a}^{\rho_b} \frac{d\rho}{\left(\rho + \lambda_{ai}^2\right) g\left(\rho\right)}\right] = \kappa_c \left[u_c + \frac{2v_c}{g\left(\rho_b\right)} + v_c \int_{\rho_a}^{\rho_b} \frac{d\rho}{\left(\rho + \lambda_{ai}^2\right) g\left(\rho\right)}\right], \tag{10.12e}$$

$$\kappa_c \left[u_c + \frac{2v_c}{g\left(\rho_c\right)} + v_c \int_{\rho_a}^{\rho_c} \frac{d\rho}{\left(\rho + \lambda_{ai}^2\right) g\left(\rho\right)} \right] = \kappa_d \left[u_d + \frac{2v_d}{g\left(\rho_c\right)} + v_d \int_{\rho_a}^{\rho_c} \frac{d\rho}{\left(\rho + \lambda_{ai}^2\right) g\left(\rho\right)} \right],$$

$$\tag{10.12f}$$

$$u_d = G_0, \tag{10.12g}$$

$$v_d = 0, \tag{10.12h}$$

$$u_a = u_d. \tag{10.12i}$$

The physical understanding of Eq. (10.12) is similar to Eq. (10.2). Similarly, we can derive two groups of thermal conductivities as

$$\kappa_b^{(1)} = \beta^{(1)} \left(\kappa_a, \kappa_d, \lambda_{ai}, \lambda_{bi}, \lambda_{ci}\right), \tag{10.13a}$$

$$\kappa_c^{(1)} = \gamma^{(1)} \left(\kappa_a, \kappa_d, \lambda_{ai}, \lambda_{bi}, \lambda_{ci}\right), \tag{10.13b}$$

or

$$\kappa_b^{(2)} = \beta^{(2)} \left(\kappa_a, \kappa_d, \lambda_{ai}, \lambda_{bi}, \lambda_{ci}\right), \tag{10.14a}$$

$$\kappa_c^{(2)} = \gamma^{(2)} \left(\kappa_a, \kappa_d, \lambda_{ai}, \lambda_{bi}, \lambda_{ci}\right), \tag{10.14b}$$

where $\beta^{(1)}$, $\gamma^{(1)} \left[< \beta^{(1)}\right]$, $\beta^{(2)}$, and $\gamma^{(2)} \left[> \beta^{(2)}\right]$ are four functions determined by Eq. (10.12). The physical understanding of Eqs. (10.13) and (10.14) are consistent with that of Eqs. (10.3) and (10.4). The isotropic case with Eqs. (10.3) and (10.4) is very complicated, let alone the anisotropic case with Eqs. (10.13) and (10.14). Therefore, we use Mathematica to calculate thermal conductivities with determined $\left(\kappa_a, \kappa_d, \lambda_{ai}, \lambda_{bi}, \lambda_{ci}\right)$ when performing simulations. Certainly, the anisotropic case with Eqs. (10.13) and (10.14) can be reduced to the isotropic case with Eqs. (10.3) and (10.4). We do not start from Eqs. (10.13) and (10.14) to derive Eqs. (10.3) and (10.4) because Eqs. (10.13) and (10.14) are too complicated to simplify.

10.4 Nonlinear Case

We discuss the thermally nonlinear case where thermal conductivities are dependent on temperature. This consideration is necessary because many common materials, such as silicon and germanium, are nonlinear. We suppose the thermal conductivity of the matrix to be $\kappa_d\left(T\right) = \kappa_d f\left(T\right)$, where $f\left(T\right)$ can be any temperature-dependent functions. Then, we prove that the bilayer scheme can also be applied for thermally nonlinear cases as long as the ratio between the nonlinear thermal conductivities of core and matrix is a temperature-independent constant, namely $\kappa_d\left(T\right)/\kappa_a\left(T\right) = \kappa_d/\kappa_a$. Therefore, the thermal conductivity of the core should be $\kappa_a\left(T\right) = \kappa_a f\left(T\right)$.

We directly substitute $\kappa_d(T)$ and $\kappa_a(T)$ into Eqs. (10.3) and (10.4). Then, we can also derive two groups of $\kappa_b(T)$ and $\kappa_c(T)$ which satisfy

$$\kappa_b(T) = \kappa_b f(T), \qquad (10.15\mathrm{a})$$

$$\kappa_c(T) = \kappa_c f(T). \qquad (10.15\mathrm{b})$$

Here, superscripts are omitted because both two groups of thermal conductivities satisfy this property. More generally, we substitute $\kappa_d(T)$ and $\kappa_a(T)$ into Eqs. (10.13) and (10.14). $\kappa_b(T)$ and $\kappa_c(T)$ also satisfy

$$\kappa_b(T) = \beta\left[\kappa_a f(T), \kappa_d f(T), \lambda_{ai}, \lambda_{bi}, \lambda_{ci}\right]$$
$$= \beta\left[\kappa_a, \kappa_d, \lambda_{ai}, \lambda_{bi}, \lambda_{ci}\right] f(T) = \kappa_b f(T), \qquad (10.16\mathrm{a})$$
$$\kappa_c(T) = \gamma\left[\kappa_a f(T), \kappa_d f(T), \lambda_{ai}, \lambda_{bi}, \lambda_{ci}\right]$$
$$= \gamma\left[\kappa_a, \kappa_d, \lambda_{ai}, \lambda_{bi}, \lambda_{ci}\right] f(T) = \kappa_c f(T). \qquad (10.16\mathrm{b})$$

Such a property allows us to transform the nonlinear Laplace equation into the linear Laplace equation. Meanwhile, general solutions are consistent in different regions. The nonlinear Laplace equation in different regions can be expressed as

$$\nabla \cdot \left[-\kappa_{a,b,c,d}(T)\nabla T\right] = \nabla \cdot \left[-\kappa_{a,b,c,d} f(T)\nabla T\right]$$
$$= \nabla \cdot \left[-\kappa_{a,b,c,d}\nabla h(T)\right] = 0, \qquad (10.17\mathrm{a})$$

where $\partial h(T)/\partial T = f(T)$. In other words, as long as we replace T with $h(T)$, the nonlinear Laplace equation can be transformed into the linear Laplace equation. Therefore, the above theories can be applied without any correction. The only assumption is that the ratio between the nonlinear thermal conductivities of sensor and matrix is a temperature-independent constant.

10.5   Finite-Element Simulation

We use the template of solid heat transfer in COMSOL Multiphysics to confirm these theoretical analyses. Without loss of generality, we perform simulations in two dimensions. Although interfacial thermal resistance may exist in practice [25, 26], its macroscopic effect is usually small. Therefore, we neglect the interfacial thermal resistance in simulations.

Firstly, we discuss the geometrically isotropic case in Fig. 10.2. A thermal sensor is embedded in the matrix for temperature detection. Since the thermal conductivity of the sensor is different from that of the matrix, the whole temperature profile is distorted (Fig. 10.2a). Therefore, the sensor is not only thermally visible but also inaccurate. When a pioneering monolayer scheme [12] is applied, it can ensure thermal invisibility. However, it does not perform well in detecting accuracy because the

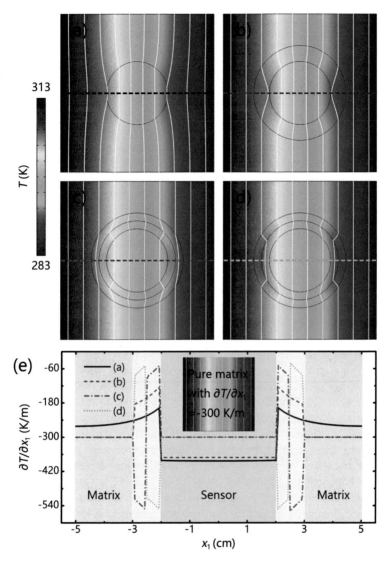

Fig. 10.2 Simulations of geometrically isotropic case. **a** Sensor embedded in the matrix. **b** Sensor coated by the monolayer scheme proposed in Ref. [12] with inner and outer radii of λ_a and λ_c, respectively. The thermal conductivity of the single layer is $161.1\,\mathrm{W\,m^{-1}\,K^{-1}}$. **c** Sensor coated by two layers designed with Eq. (10.3). **d** Sensor coated by two layers designed with Eq. (10.4). **e** Temperature gradients on the dashed lines in **a–d** as a function of x_1. The simulation size is $10 \times 10\,\mathrm{cm^2}$. The temperatures of the left and right boundaries are set at 313 and 283 K. Other boundaries are insulated. $\lambda_a = 2$, $\lambda_b = 2.5$, $\lambda_c = 3$ cm, and $\kappa_a = 50$, $\kappa_d = 100\,\mathrm{W\,m^{-1}\,K^{-1}}$. $\kappa_b^{(1)} = 378.5$, $\kappa_c^{(1)} = 58.5$, and $\kappa_b^{(2)} = 26.7$, $\kappa_c^{(2)} = 346.3\,\mathrm{W\,m^{-1}\,K^{-1}}$. Adapted from Ref. [22]

temperature in the sensor is still different from the original one (Fig. 10.2b). Then, we resort to the bilayer scheme. We coat the sensor with the bilayer scheme whose thermal conductivities are designed according to Eq. (10.3), and the simulation result is shown in Fig. 10.2c. The temperature in the matrix becomes linear again, thus making the sensor thermally invisible. Meanwhile, the temperature in the sensor is the same as the original one, thus ensuring accurate detection. We also design the thermal conductivities of the two layers according to Eq. (10.4), and the same effect can be obtained (Fig. 10.2d). For quantitative comparison, we export the data on the dashed lines in Fig. 10.2a–d. Since the temperature difference is not significant enough to observe, we export temperature gradient $\partial T / \partial x_1$ for comparison. The result is presented in Fig. 10.2e, indicating that the bilayer scheme can indeed simultaneously ensure thermal invisibility and accurate detection. The inset of Fig. 10.2e shows the temperature profile of a pure matrix with a linear thermal field of -300 K/m.

Then, we discuss the geometrically anisotropic case in Fig. 10.3, which is more practical. The results are similar to the geometrically isotropic case. Figure 10.3a and b demonstrate the temperature profiles without and with a sensor embedded in the matrix, respectively. The sensor distorts the whole temperature profile, resulting in thermal visibility and inaccurate sensor detection. When the monolayer scheme [16] is applied, it can ensure thermal invisibility, but the temperature in the sensor is still changed (Fig. 10.3c). Figure 10.3d and e shows the results coated by two layers designed with Eqs. (10.13) and (10.14), respectively. Again, the temperatures in the matrix and sensor become the same. Therefore, the sensor is thermally invisible and accurate. For clarity, we plot the temperature difference Δ with the temperature in Fig. 10.3c (Fig. 10.3d or e) minus that in Fig. 10.3a, which is shown in Fig. 10.3f (Fig. 10.3g or h). Our scheme ensures that the temperature difference Δ in the matrix and sensor is always zero, confirming an accurate and thermally invisible sensor.

Finally, we discuss the thermally nonlinear case in Fig. 10.4. Nonlinear (temperature-dependent) thermal conductivities, whether weak or strong, are common in nature. Here, "strong" (or "weak") means that the nonlinear (or linear) term of thermal conductivity is dominant. Therefore, it is necessary to extend our scheme to thermally nonlinear cases. To make nonlinear properties clear, we discuss strong nonlinearity directly. A typical case of strong nonlinearity is the thermal radiation described by the Rosseland diffusion approximation, which is proportional to T^3 [27–29]. Therefore, we take on $f(T) = \mu + \nu T^3$ with μ and ν being two constants. We set a high temperature at 2283 K K, and aerogel (or ceramic), with excellent tolerance to high temperatures, can be applied to observe thermal nonlinearity. As proved in Eq. (10.16), we can directly multiply the original thermal conductivities with $f(T)$ to proceed.

Since the thermal conductivity of the matrix is nonlinear, the temperature gradient is no longer a constant (Fig. 10.4a). When an elliptical sensor is embedded in the matrix, the straight isotherms are distorted (Fig. 10.4b). Then, we coat the sensor with two layers designed with Eq. (10.16). The simulation results are presented in Fig. 10.4c and d, respectively. The distorted isotherms in the matrix and sensor restore. Similarly, we also plot the temperature difference Δ with the temperature in Fig. 10.4c (or Fig. 10.4d) minus that in Fig. 10.4a, and the results are shown in

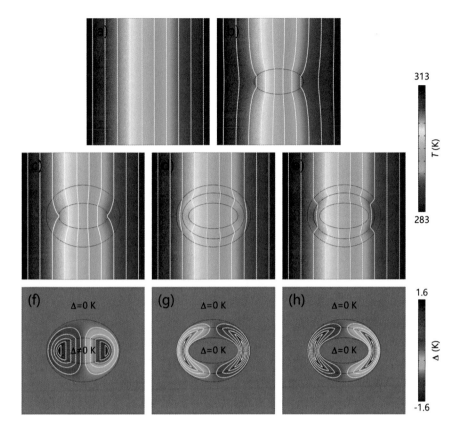

Fig. 10.3 Simulations of geometrically anisotropic case. **a** Pure matrix. **b** Sensor embedded in the matrix. **c** Sensor coated by the monolayer scheme proposed in Ref. [16] whose thermal conductivity is 149.5 W m^{-1} K^{-1}. **d** Sensor coated by two layers designed with Eq. (10.13). **e** Sensor coated by two layers designed with Eq. (10.14). **f** Temperature difference with the temperature in **c** minus that in **a**. **g** Temperature difference with the temperature in **d** minus that in **a**. **h** Temperature difference with the temperature in **e** minus that in **a**. $\lambda_{a1} = 2, \lambda_{a2} = 1, \lambda_{b1} = 2.5, \lambda_{b2} = 1.8, \lambda_{c1} = 3, \lambda_{c2} = 2.45$ cm, and $\kappa_a = 5, \kappa_d = 100$ W m^{-1} K^{-1}. $\kappa_b^{(1)} = 274.5, \kappa_c^{(1)} = 61.8$, and $\kappa_b^{(2)} = 2.4$, $\kappa_c^{(2)} = 342.1$ W m^{-1} K^{-1}. Adapted from Ref. [22]

Fig. 10.4e (or Fig. 10.4f). We can observe zero temperature difference Δ in the matrix and sensor, so the bilayer scheme performs satisfactorily.

The bilayer scheme can be extended to transient regimes by considering heat capacity and density [30–32]. Since invisibility is a special case of camouflage, these results may guide thermal camouflage [33–40]. The present scheme is dependent on elliptical/ellipsoid shapes because the Laplace equation can be analytically handled. Therefore, other methods remain to be explored for complex shapes [41, 42], such as combining neutral inclusion [10] and transformation thermotics [43, 44].

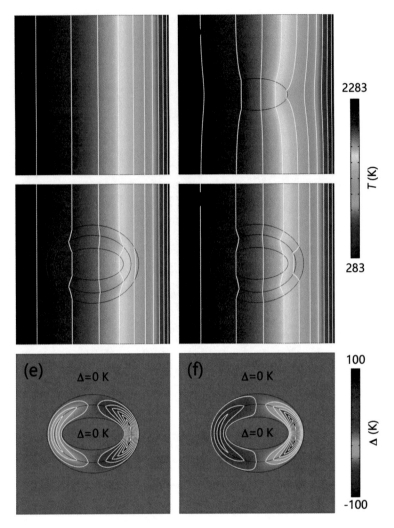

Fig. 10.4 Simulations of thermally nonlinear case. The temperatures of the left and right boundaries are set at 2283 and 283 K, respectively. $f(T) = 1 + 10^{-9}T^3$. The other parameters are the same as those for Fig. 10.3. Adapted from Ref. [22]

10.6   Conclusion

We have proposed a bilayer scheme to design invisible thermal sensors. Compared with existing schemes, the present one is accurate and applicable for geometrically anisotropic and thermally nonlinear cases. Thermal invisibility can protect sensors from being detected, and accurate detection benefits practical applications. The extensions to geometric anisotropy and thermal nonlinearity make thermal sensors more widely applicable. Moreover, we unify two/three-dimensional cases, isotropic/anisotropic cases, and linear/nonlinear cases with a single theoretical framework, laying a solid foundation for designing thermal metamaterials under different conditions.

10.7 Exercise and Solution

Exercise

1. Take the interfacial thermal resistance discussed in Ref. [26] into account and derive the corresponding parameters for invisible thermal sensors.

Solution

1. When interfacial thermal resistance is taken into consideration, temperature jumps will occur at the interfaces of the system. Therefore, the boundary conditions associated with the continuity of temperatures (Eqs. (10.2a)–(10.2c)) should be rewritten as [26]

$$u_a \lambda_a + R_{ab} \kappa_a u_a = u_b \lambda_a + v_b \lambda_a^{-\tau}, \tag{10.18a}$$

$$u_b \lambda_b + v_b \lambda_b^{-\tau} + R_{bc} \kappa_b \left(u_b - \tau v_b \lambda_b^{-\tau-1} \right) = u_c \lambda_b + v_c \lambda_b^{-\tau}, \tag{10.18b}$$

$$u_c \lambda_c + v_c \lambda_c^{-\tau} + R_{cd} \kappa_c \left(u_c - \tau v_c \lambda_c^{-\tau-1} \right) = u_d \lambda_c + v_d \lambda_c^{-\tau}, \tag{10.18c}$$

where R_{ab}, R_{bc}, and R_{cd} are the interfacial thermal resistances of the sensor and first layer, the first layer and second layer, and the second layer and matrix, respectively. The other boundary conditions are unchanged.

References

1. Alù, A., Engheta, N.: Cloaking a sensor. Phys. Rev. Lett. **102**, 233901 (2009)
2. Greenleaf, A., Kurylev, Y., Lassas, M., Uhlmann, G.: Cloaking a sensor via transformation optics. Phys. Rev. E **83**, 016603 (2011)
3. Zhu, X.F., Liang, B., Kan, W.W., Zou, X.Y., Cheng, J.C.: Acoustic cloaking by a superlens with single-negative materials. Phys. Rev. Lett. **106**, 014301 (2011)
4. Fleury, R., Soric, J., Alù, A.: Physical bounds on absorption and scattering for cloaked sensors. Phys. Rev. B **89**, 045122 (2014)
5. Fleury, R., Sounas, D., Alù, A.: An invisible acoustic sensor based on parity-time symmetry. Nat. Commun. **6**, 5905 (2015)
6. Mach-Batlle, R., Navau, C., Sanchez, A.: Invisible magnetic sensors. Appl. Phys. Lett. **112**, 162406 (2018)

7. Chen, P.-Y., Soric, J., Alù, A.: Invisibility cloaking: invisibility and cloaking based on scattering cancellation. Adv. Mater. **24**, OP281 (2012)
8. Farhat, M., Chen, P.-Y., Bagci, H., Amra, C., Guenneau, S., Alù, A.: Thermal invisibility based on scattering cancellation and mantle cloaking. Sci. Rep. **5**, 9876 (2015)
9. Farhat, M., Guenneau, S., Chen, P.-Y., Alù, A., Salama, K.N.: Scattering cancellation-based cloaking for the Maxwell-Cattaneo heat waves. Phys. Rev. Appl. **11**, 044089 (2019)
10. He, X., Wu, L.Z.: Thermal transparency with the concept of neutral inclusion. Phys. Rev. E **88**, 033201 (2013)
11. Shen, X.Y., Huang, J.P.: Thermally hiding an object inside a cloak with feeling. Int. J. Heat Mass Transf. **78**, 1 (2014)
12. Yang, T.Z., Bai, X., Gao, D.L., Wu, L.Z., Li, B.W., Thong, J.T.L., Qiu, C.-W.: Invisible sensors: simultaneous sensing and camouflaging in multiphysical fields. Adv. Mater. **27**, 7752 (2015)
13. Xu, H.Y., Shi, X.H., Gao, F., Sun, H.D., Zhang, B.L.: Ultrathin three-dimensional thermal cloak. Phys. Rev. Lett. **112**, 054301 (2014)
14. Han, T.C., Bai, X., Gao, D.L., Thong, J.T.L., Li, B.W., Qiu, C.-W.: Experimental demonstration of a bilayer thermal cloak. Phys. Rev. Lett. **112**, 054302 (2014)
15. Ma, Y.G., Liu, Y.C., Raza, M., Wang, Y.D., He, S.L.: Experimental demonstration of a multiphysics cloak: manipulating heat flux and electric current simultaneously. Phys. Rev. Lett. **113**, 205501 (2014)
16. Han, T.C., Yang, P., Li, Y., Lei, D.Y., Li, B.W., Hippalgaonkar, K., Qiu, C.-W.: Full-parameter omnidirectional thermal metadevices of anisotropic geometry. Adv. Mater. **30**, 1804019 (2018)
17. Li, Y., Zhu, K.-J., Peng, Y.-G., Li, W., Yang, T.Z., Xu, H.X., Chen, H., Zhu, X.-F., Fan, S.H., Qiu, C.-W.: Thermal meta-device in analogue of zero-index photonics. Nat. Mater. **18**, 48 (2019)
18. Qin, J., Luo, W., Yang, P., Wang, B., Deng, T., Han, T.C.: Experimental demonstration of irregular thermal carpet cloaks with natural bulk material. Int. J. Heat Mass Transf. **141**, 487 (2019)
19. Xu, G.Q., Zhou, X., Zhang, J.Y.: Bilayer thermal harvesters for concentrating temperature distribution. Int. J. Heat Mass Transf. **142**, 118434 (2019)
20. Xu, L.J., Huang, J.P.: Magnetostatic chameleonlike metashells with negative permeabilities. EPL **125**, 64001 (2019)
21. Yang, S., Xu, L.J., Huang, J.P.: Two exact schemes to realize thermal chameleonlike metashells. EPL **128**, 34002 (2019)
22. Xu, L.J., Huang, J.P., Jiang, T., Zhang, L., Huang, J.P.: Thermally invisible sensors. EPL **132**, 14002 (2020)
23. Yang, S., Xu, L.J., Huang, J.P.: Intelligence thermotics: correlated self-fixing behavior of thermal metamaterials. EPL **126**, 54001 (2019)
24. Milton, G.W.: The Theory of Composites. Cambridge University Press, Cambridge (2004)
25. Li, J.Y., Gao, Y., Huang, J.P.: A bifunctional cloak using transformation media. J. Appl. Phys. **108**, 074504 (2010)
26. Zheng, X., Li, B.W.: Effect of interfacial thermal resistance in a thermal cloak. Phys. Rev. Appl. **13**, 024071 (2020)
27. Xu, L.J., Huang, J.P.: Metamaterials for manipulating thermal radiation: Transparency, cloak, and expander. Phys. Rev. Appl. **12**, 044048 (2019)
28. Xu, L.J., Dai, G.L., Huang, J.P.: Transformation multithermotics: controlling radiation and conduction simultaneously. Phys. Rev. Appl. **13**, 024063 (2020)
29. Su, C., Xu, L.J., Huang, J.P.: Nonlinear thermal conductivities of core-shell metamaterials: rigorous theory and intelligent application. EPL **130**, 34001 (2020)
30. Yang, T.Z., Su, Y.S., Xu, W.K., Yang, X.D.: Transient thermal camouflage and heat signature control. Appl. Phys. Lett. **109**, 121905 (2016)
31. Liu, Y.X., Guo, W.L., Han, T.C.: Arbitrarily polygonal transient thermal cloaks with natural bulk materials in bilayer configurations. Int. J. Heat Mass Transf. **115**, 1 (2017)
32. He, X., Yang, T.Z., Zhang, X.W., Wu, L.Z., He, X.Q.: Transient experimental demonstration of an elliptical thermal camouflage device. Sci. Rep. **7**, 16671 (2017)

33. Hu, R., Zhou, S.L., Li, Y., Lei, D.Y., Luo, X.B., Qiu, C.-W.: Illusion thermotics. Adv. Mater. **30**, 1707237 (2018)
34. Li, Y., Bai, X., Yang, T.Z., Luo, H., Qiu, C.-W.: Structured thermal surface for radiative camouflage. Nat. Commun. **9**, 273 (2018)
35. Xu, L.J., Yang, S., Huang, J.P.: Thermal transparency induced by periodic interparticle interaction. Phys. Rev. Appl. **11**, 034056 (2019)
36. Xu, L.J., Yang, S., Huang, J.P.: Passive metashells with adaptive thermal conductivities: chameleonlike behavior and its origin. Phys. Rev. Appl. **11**, 054071 (2019)
37. Xu, L.J., Yang, S., Huang, J.P.: Dipole-assisted thermotics: experimental demonstration of dipole-driven thermal invisibility. Phys. Rev. E **100**, 062108 (2019)
38. Xu, L.J., Yang, S., Huang, J.P.: Designing effective thermal conductivity of materials of core-shell structure: theory and simulation. Phys. Rev. E **99**, 022107 (2019)
39. Yang, F.B., Xu, L.J., Huang, J.P.: Thermal illusion of porous media with convection-diffusion process: transparency, concentrating, and cloaking. ES Energy Environ. **6**, 45 (2019)
40. Xu, L.J., Yang, S., Dai, G.L., Huang, J.P.: Transformation omnithermotics: simultaneous manipulation of three basic modes of heat transfer. ES Energy Environ. **7**, 65 (2020)
41. Ji, Q., Chen, X., Fang, G., Liang, J., Yan, X., Laude, V., Kadic, M.: Thermal cloaking of complex objects with the neutral inclusion and the coordinate transformation methods. AIP Adv. **9**, 045029 (2019)
42. Jin, P., Yang, S., Xu, L.J., Dai, G.L., Huang, J.P., Ouyang, X.P.: Particle swarm optimization for realizing bilayer thermal sensors with bulk isotropic materials. Int. J. Heat Mass Transf. **172**, 121177 (2021)
43. Fan, C.Z., Gao, Y., Huang, J.P.: Shaped graded materials with an apparent negative thermal conductivity. Appl. Phys. Lett. **92**, 251907 (2008)
44. Chen, T.Y., Weng, C.-N., Chen, J.-S.: Cloak for curvilinearly anisotropic media in conduction. Appl. Phys. Lett. **93**, 114103 (2008)

Chapter 11
Theory for Invisible Thermal Sensors: Monolayer Scheme

Abstract In this chapter, we propose an anisotropic monolayer scheme to prevent thermal sensors from distorting local and background temperature profiles, making them accurate and thermally invisible. We design metashells with anisotropic thermal conductivity and perform finite-element simulations in two or three dimensions for arbitrarily given thermal conductivity of sensors and backgrounds. We further experimentally fabricate a metashell with an anisotropic thermal conductivity based on the effective medium theory, which confirms the feasibility of our scheme. Our results are beneficial to improving the performance of thermal detection and may also guide other diffusive physical fields.

Keywords Invisible thermal sensors · Monolayer scheme · Anisotropic thermal conductivity

11.1 Opening Remarks

Temperature measurement has broad applications, requiring high sensitivity for thermal sensors. However, the distortion of temperature profiles resulting from thermal sensors cannot be avoided by only improving sensitivity. A severe problem lies in the thermal-conductivity mismatch between sensors and backgrounds. Similar problem (parametric mismatch) also occurs in some other fields, and promotes relevant researches in electromagnetism [1, 2], magnetics [3], and acoustics [4–6].

Many schemes were proposed based on neutral inclusion [7] or transformation thermotics [8]. Furthermore, a multiphysical scheme was proposed by coating a sensor with an isotropic shell [9]. Though these schemes improve the performance of thermal detection, they are also faced with problems of complex parameters and technological difficulty. For the scheme based on transformation thermotics [8], anisotropic, inhomogeneous, and even negative thermal conductivity is required, which makes its experimental realization extremely difficult. For the scheme based on an isotropic shell [7, 9–11], local temperature profiles are still different from the original ones, making thermal detection inaccurate. Here, "local" indicates the region occupied by a sensor. To improve accuracy, one should minimize the thickness

© The Author(s) 2023
L.-J. Xu and J.-P. Huang, *Transformation Thermotics and Extended Theories*,
https://doi.org/10.1007/978-981-19-5908-0_11

of the isotropic shell, which, however, still cannot completely remove inaccuracy. In this sense, to date, thermal sensors with both accuracy and invisibility are still experimentally lacking. Here, accuracy and invisibility respectively indicate that local and background temperature profiles are not distorted.

Different from existing methods [7–11], we propose an anisotropic monolayer scheme that can accurately measure local temperature profiles without disturbing background thermal fields. It is worth noting that a similar scheme has been successfully applied to design other thermal functions, such as cloaking [12–14], concentrating [13–15], and chameleon [16, 17]. The present scheme is applicable for arbitrarily given thermal conductivities of backgrounds and sensors, which is confirmed by finite-element simulations in two or three dimensions. Furthermore, we experimentally fabricate a metashell with anisotropic thermal conductivity based on the effective medium theory [18–22], and the experimental results agree well with the theory and finite-element simulations.

11.2   Theoretical Foundation

We start by discussing a two-dimensional system shown in Fig. 11.1. The system is divided by a metashell (Area II) into three areas, whose thermal conductivities are κ_1 for sensor (Area I), $\kappa_2 = \mathrm{diag}\,(\kappa_{rr}, \kappa_{\theta\theta})$ for metashell (Area II), and κ_3 for background (Area III). κ_2 is expressed in cylindrical coordinates (r, θ), where κ_{rr} and $\kappa_{\theta\theta}$ are radial and tangential thermal conductivities, respectively. We consider the known equation describing passive heat conduction at steady states,

$$\nabla \cdot (-\kappa \cdot \nabla T) = 0, \tag{11.1}$$

where κ and T are thermal conductivity and temperature, respectively.

Equation (11.1) can be expanded in cylindrical coordinates as

$$\frac{1}{r}\frac{\partial}{\partial r}\left(r\kappa_{rr}\frac{\partial T}{\partial r}\right) + \frac{1}{r}\frac{\partial}{\partial \theta}\left(\kappa_{\theta\theta}\frac{\partial T}{r\partial \theta}\right) = 0. \tag{11.2}$$

The general solution to Eq. (11.2) is

$$T = A_0 + B_0 \ln r + \sum_{i=1}^{\infty}[A_i \sin(i\theta) + B_i \cos(i\theta)]\,r^{im_1} \tag{11.3}$$

$$+ \sum_{j=1}^{\infty}[C_j \sin(j\theta) + D_j \cos(j\theta)]\,r^{jm_2},$$

where $m_1 = \sqrt{\kappa_{\theta\theta}/\kappa_{rr}}$ and $m_2 = -\sqrt{\kappa_{\theta\theta}/\kappa_{rr}}$, representing anisotropy degree.

Fig. 11.1 Schematic diagram of the anisotropic monolayer scheme. Adapted from Ref. [23]

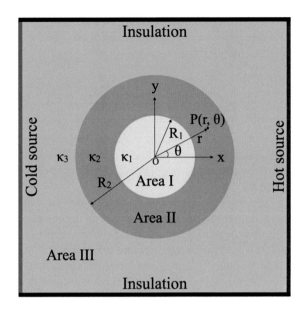

The temperature profiles of sensor (Area I), metashell (Area II), and background (Area III) are respectively denoted as T_1, T_2, and T_3, which satisfy the general solution in Eq. (11.3). Especially, T_1 and T_3 can be given by the right side of Eq. (11.3) with $m_1 = 1$ and $m_2 = -1$. Boundary conditions are determined by the continuities of temperature and normal heat flux,

$$\begin{cases} T_1\left(R_1\right) = T_2\left(R_1\right), \\ T_2\left(R_2\right) = T_3\left(R_2\right), \\ \left(-\kappa_1 \dfrac{\partial T_1}{\partial r}\right)_{R_1} = \left(-\kappa_{rr} \dfrac{\partial T_2}{\partial r}\right)_{R_1}, \\ \left(-\kappa_{rr} \dfrac{\partial T_2}{\partial r}\right)_{R_2} = \left(-\kappa_3 \dfrac{\partial T_3}{\partial r}\right)_{R_2}. \end{cases} \tag{11.4}$$

Considering the symmetry of boundary conditions, we only keep certain terms in Eq. (11.3) as the temperature profiles of three areas,

$$T_1 = A_0 + Ar\cos\theta, \tag{11.5a}$$

$$T_2 = A_0 + Br^{m_1}\cos\theta + Cr^{m_2}\cos\theta, \tag{11.5b}$$

$$T_3 = A_0 + Dr\cos\theta + Er^{-1}\cos\theta, \tag{11.5c}$$

where the temperature at $\theta = \pm\pi/2$ is defined as A_0, and $D = |\nabla T_0|$ is the modulus of an external linear thermal field ∇T_0.

We have six undetermined coefficients (i.e., A, B, C, E, κ_{rr}, and $\kappa_{\theta\theta}$) and only four equations (Eq. (11.4)). The other two equations are to make thermal sensors accurate and thermally invisible,

$$\begin{cases} A = D, \\ E = 0, \end{cases} \tag{11.6}$$

where $A = D$ indicates that the local temperature profile is the same as the background temperature profile, making a thermal sensor accurate; and $E = 0$ indicates that the background temperature profile is undistorted, making a thermal sensor invisible.

Then, the six unknown coefficients can be uniquely determined by six equations (Eqs. (11.4) and (11.6)), including the anisotropic thermal conductivity of the metashell $\kappa_2 = \mathrm{diag}\,(\kappa_{rr},\ \kappa_{\theta\theta})$. Therefore, thermal sensors with accuracy and thermal invisibility in two dimensions are designed.

On the same footing, we extend the two-dimensional theory to three dimensions. Accordingly, the thermal conductivity of the metashell is denoted as $\kappa_2 = \mathrm{diag}\left(\kappa_{rr},\ \kappa_{\theta\theta},\ \kappa_{\varphi\varphi}\right)$ with $\kappa_{\theta\theta} = \kappa_{\varphi\varphi}$. Then, Eq. (11.1) can be expanded in spherical coordinates $(r,\ \theta,\ \varphi)$ as

$$\frac{1}{r^2}\frac{\partial}{\partial r}\left(r^2 \kappa_{rr}\frac{\partial T}{\partial r}\right) + \frac{1}{r}\frac{1}{\sin\theta}\frac{\partial}{\partial\theta}\left(\sin\theta\,\kappa_{\theta\theta}\frac{\partial T}{r\partial\theta}\right) = 0, \tag{11.7}$$

where φ is neglected because we consider a rotational symmetric case.

The general solution to Eq. (11.7) is

$$T = \sum_{i=0}^{\infty}\left(A_i r^{s_1} + B_i r^{s_2}\right)P_i\left(\cos\theta\right), \tag{11.8}$$

where $s_1 = \left[-1 + \sqrt{1 + 4i\,(i+1)\,\kappa_{\theta\theta}/\kappa_{rr}}\right]/2$, $s_2 = \left[-1 - \sqrt{1 + 4i\,(i+1)\,\kappa_{\theta\theta}/\kappa_{rr}}\right]/2$, i is summation index, and P_i is the Legendre polynomials. Since three-dimensional boundary conditions are the same as two-dimensional ones, the temperature profiles for three areas can be expressed as

$$T_1 = A_0 + Ar\cos\theta, \tag{11.9a}$$

$$T_2 = A_0 + Br^{s_1}\cos\theta + Cr^{s_2}\cos\theta, \tag{11.9b}$$

$$T_3 = A_0 + Dr\cos\theta + Er^{-2}\cos\theta, \tag{11.9c}$$

where $s_1 = \left[-1 + \sqrt{1 + 8\kappa_{\theta\theta}/\kappa_{rr}}\right]/2$, $s_2 = \left[-1 - \sqrt{1 + 8\kappa_{\theta\theta}/\kappa_{rr}}\right]/2$, and the six unknown coefficients (i.e., A, B, C, E, κ_{rr}, and $\kappa_{\theta\theta}$) can also be solved by Eqs. (11.4) and (11.6). Therefore, we also make three-dimensional thermal sensors accurate and invisible. We use Mathematica to solve the six equations (Eqs. (11.4) and (11.6)) and obtain the required thermal conductivity of the metashell.

11.3   Finite-Element Simulation

To confirm our theory, we further perform finite-element simulations with COM-SOL Multiphysics. We can derive the anisotropic thermal conductivity of metashells and perform finite-element simulations for arbitrarily given thermal conductivity of backgrounds and sensors.

For comparison, we show a reference with uniform thermal conductivity in Fig. 11.2a. The finite-element simulations with a reference shell (a bare sensor embedded in the background, i.e., bareness), an isotropic shell (calculated with the theory in Ref. [9]), and an anisotropic metashell (calculated by our theory) are pre-

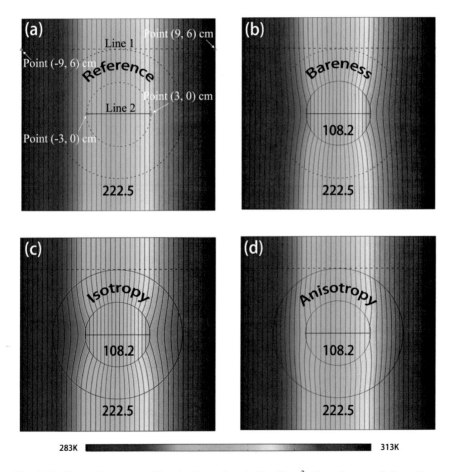

Fig. 11.2 Case with $\kappa_1 < \kappa_3$. The simulation box is $18 \times 18\,\text{cm}^2$, $R_1 = 3$ cm, and $R_2 = 6$ cm. Black lines represent isotherms. The temperatures of cold source (left boundary) and hot source (right boundary) are set at 283 and 313 K, respectively. The thermal conductivities of **a** reference (all areas) and **b** reference shell (Area II) are set to be the same, i.e., 222.5 W m^{-1} K^{-1}. The thermal conductivities of sensor (Area I) and background (Area III) in **b**–**d** are set to be 108.2 and 222.5 W m^{-1} K^{-1}, respectively. The thermal conductivities of **c** isotropic shell and **d** anisotropic metashell are 277.3 and diag (178.0, 349.0) W m^{-1} K^{-1}, respectively. Adapted from Ref. [23]

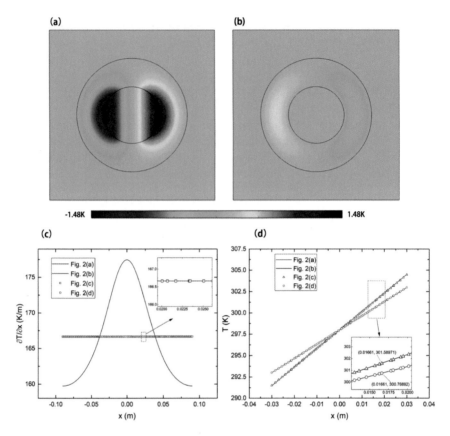

Fig. 11.3 Quantitative comparison of Fig. 11.2. Temperature-difference distributions with **a** the temperature in Fig. 11.2c minus that in Fig. 11.2a and b the temperature in Fig. 11.2d minus that in Fig. 11.2a. **c** Temperature-gradient distributions on Line 1 in Fig. 11.2. **d** Temperature distributions on Line 2 in Fig. 11.2. Adapted from Ref. [23]

sented in Fig. 11.2b–d, respectively. Figure 11.2b indicates that a bare thermal sensor indeed distorts local (Area I) and background (Area III) temperature profiles. Comparing Fig. 11.2a and c, although the isotropic shell keeps the background temperature profile undistorted, the local temperature profile is still changed. Fortunately, our scheme (Fig. 11.2d) makes local and background temperature profiles the same as those in Fig. 11.2a, thus making the sensor accurate and thermally invisible.

To compare different schemes quantitatively, we also plot temperature-difference profiles with the temperature in Fig. 11.2c minus that in Fig. 11.2a (Fig. 11.3a) and the temperature in Fig. 11.2d minus that in Fig. 11.2a (Fig. 11.3b). The temperature difference in the local region of Fig. 11.3a is nonzero, indicating that the detected temperature is not the original one. The temperature difference in the local region of Fig. 11.3b is zero, making a thermal sensor accurate. Certainly, the temperature profile of the metashell has a small difference. We export the temperature-gradient distributions along x axis on Line 1 in Fig. 11.2, as shown in Fig. 11.3c. The results indicate that the isotropic and anisotropic shells (Fig. 11.2c and d) show the same

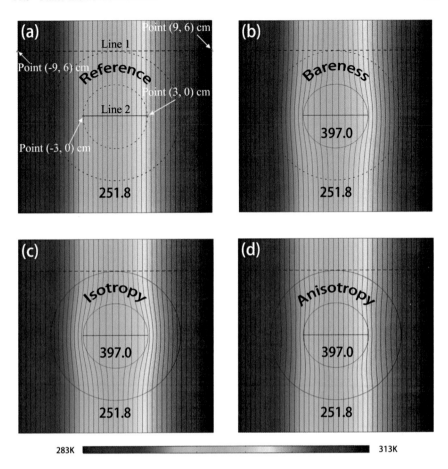

Fig. 11.4 Case with $\kappa_1 > \kappa_3$. The thermal conductivities of **a** reference (all areas) and **b** reference shell (Area II) are set to be the same, i.e., 251.8 W m^{-1} K^{-1}. The thermal conductivities of sensor (Area I) and background (Area III) in **b–d** are set to be 397.0 and 251.8 W m^{-1} K^{-1}, respectively. The thermal conductivities of **c** isotropic shell and **d** anisotropic metashell are 217.5 and diag (308.0, 104.0) W m^{-1} K^{-1}, respectively. Other parameters are the same as those for Fig. 11.2. Adapted from Ref. [23]

advantage of thermal invisibility (i.e., the constant temperature-gradient distribution as that in reference). However, due to the thermal-conductivity mismatch between the sensor and background, a bare sensor distorts the heat flow of the original background thermal field, so the temperature gradient at the position of Line 1 in Fig. 11.2b is not a constant. We also export the temperature distributions on Line 2 in Fig. 11.2, as shown in Fig. 11.3d. The results show that the anisotropic scheme is better than the isotropic one because what the thermal sensor in Fig. 11.2d detects is completely consistent with the reference. However, the detection in Fig. 11.2c deviates from the reference.

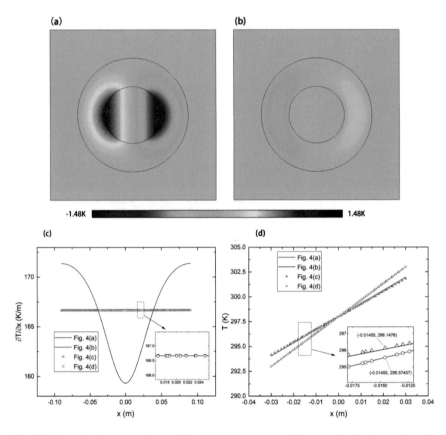

Fig. 11.5 Quantitative comparison of Fig. 11.4. Temperature-difference distributions with **a** the temperature in Fig. 11.4c minus that in Fig. 11.4a and **b** the temperature in Fig. 11.4d minus that in Fig. 11.4a. **c** Temperature-gradient distributions on Line 1 in Fig. 11.4. **d** Temperature distribution on Line 2 in Fig. 11.4. Adapted from Ref. [23]

Moreover, we also discuss the case that the thermal conductivity of the sensor (Area I) is larger than that of the background (Area III) to ensure completeness; see Figs. 11.4 and 11.5. The results are similar to Figs. 11.2 and 11.3. That is, a bare thermal sensor will distort temperature profiles of all areas (Fig. 11.4b). The existing isotropic scheme can keep the background temperature profile undistorted, but the local one is changed (Fig. 11.4c). The present scheme can ensure both local and background temperature profiles undistorted (Fig. 11.4d). Factual data can be found in Fig. 11.5.

To go further, we also perform three-dimensional finite-element simulations. The temperature profiles in the sensor (Area I) in Fig. 11.6b and c are different from that in Fig. 11.6a, which means that the sensor cannot accurately measure local temperature distributions. Fortunately, the temperature profiles in the sensor (Area I) and background (Area III) in Fig. 11.6d are identical to those in Fig. 11.6a. We also perform quantitative analyses on "accurate" and "invisible" properties of the sensor (Fig. 11.6e–h), and the results are what we expect, just like the two-dimensional case.

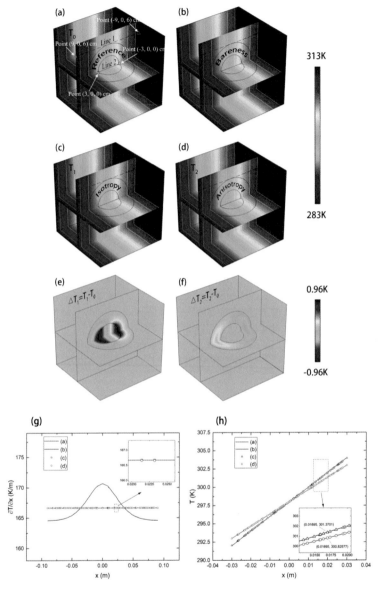

Fig. 11.6 Three-dimensional simulations. The simulation box is $18 \times 18 \times 18 \, \text{cm}^3$, $R_1 = 3$ cm, and $R_2 = 6$ cm. The thermal conductivities of **a** reference (all areas) and **b** reference shell (Area II) are set to be the same, i.e., $222.5 \, \text{W m}^{-1} \, \text{K}^{-1}$. The thermal conductivities of sensor (Area I) and background (Area III) in **b**-**d** are set to be 108.2 and $222.5 \, \text{W m}^{-1} \, \text{K}^{-1}$, respectively. The thermal conductivities of **c** isotropic shell and **d** anisotropic metashell are 242.5 and diag (183.7, 269.0, 269.0) $\text{W m}^{-1} \, \text{K}^{-1}$, respectively. Temperature-difference distributions with **e** the temperature in **c** minus that in **a** (i.e., $\Delta T_1 = T_1 - T_0$) and **f** the temperature in **d** minus that in **a** (i.e., $\Delta T_2 = T_2 - T_0$). **g** Temperature-gradient distributions on Line 1 in **a**–**d**. **h** Temperature distributions on Line 2 in **a**–**d**. Adapted from Ref. [23]

To sum up, the accuracy and invisibility of three-dimensional thermal sensors are also confirmed by simulations.

11.4 Laboratory Experiment

To experimentally validate the finite-element simulations in Fig. 11.2b and d, we set up a device shown in Fig. 11.7a. By utilizing laser cutting, we fabricate two samples (Fig. 11.7b and c) based on ellipse-embedded structures [18]. The holes in Areas I and III are uniformly distributed with circular shapes, ensuring that the effective thermal conductivity is isotropic. The holes in Area II have anisotropic (elliptical) geometry, so the effective thermal conductivity is also anisotropic. Therefore, the perforated structure indeed follows the theory. To eliminate infrared reflection and thermal convection as much as possible, we also apply transparent and foamed plastic films (insulating materials) on the upper and lower surfaces of the two samples, respectively. Then, we measure the temperature profiles of these two samples with the infrared camera Flir E60. The measured results with a reference shell and with an anisotropic metashell are shown in Fig. 11.7d and e, respectively. We also perform finite-element simulations based on these two samples, as shown in Fig. 11.7f and g. Both finite-element simulations (Fig. 11.2b and d, and Fig. 11.7f and g) and experiments (Fig. 11.7d and e) prove that with the present scheme, a thermal sensor does accurately detect the local temperature profile without disturbing the background thermal field.

We have discussed the scheme in steady heat conduction, and extending it to transient states is promising, which should consider density and heat capacity. Furthermore, topology optimization is also a powerful method to design metamaterials [24–27] beyond transformation method [28–31], which could be applied to design accurate and invisible sensors.

11.5 Conclusion

We have proposed an anisotropic monolayer scheme to make thermal sensors accurate and thermally invisible. By coating a thermal sensor with a metashell with anisotropic thermal conductivity, the thermal sensor can accurately measure local temperature profiles without disturbing surrounding thermal fields. The present scheme is validated by two-dimensional simulations and experiments, which also apply to three dimensions. These results may advance the performance of thermal detection and provide guidance to thermal camouflage [32–39]. On the same basis, this work also offers hints to obtaining counterparts in other diffusive fields.

11.6 Exercise and Solution

Exercise

1. Solve the unknown numbers in Eqs. (11.5) and (11.9).

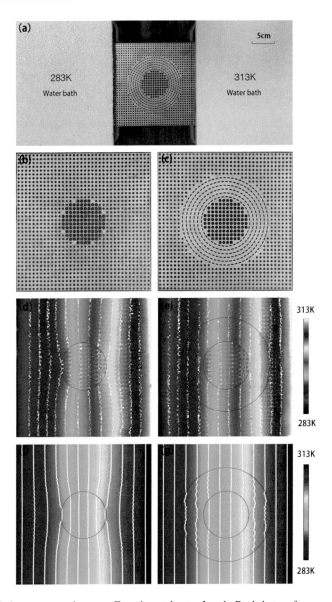

Fig. 11.7 Laboratory experiments. **a** Experimental setup. **b** and **c** Real photos of two samples. **d** and **e** (or **f** and **g**) are measured results (or finite-element simulations) corresponding to the two samples shown in **b** and **c**, respectively. White lines represent isotherms. The sensor (or background) in **b** and **c** is carved with air circles with radius 0.21 cm (or 0.15 cm). The anisotropic metashell is carved with air ellipses with major (or minor) semiaxis of 0.21 cm (or 0.044 cm). The thermal conductivities of copper and air are 397 and 0.026 W m^{-1} K^{-1}, respectively. These parameters cause the tensorial thermal conductivity of the anisotropic metashell in **c** to be diag (178.0, 349.0) W m^{-1} K^{-1}, and the thermal conductivities of sensor (or background) in **b** and **c** to be 108.2 W m^{-1} K^{-1} (or 222.5 W m^{-1} K^{-1}). The sample size in **b** and **c** is the same as that for Fig. 11.2b and d. Adapted from Ref. [23]

Solution

1. Since κ_{rr} and $\kappa_{\theta\theta}$ appear in the exponent, it is difficult to analytically express them. To solve the problem, we treat κ_1 and κ_3 as two undetermined coefficients, which together with other coefficients can be respectively expressed in two and three dimensions as

$$
\begin{cases}
A = |\nabla T_0|, \\[2ex]
B = \dfrac{R_1^{m_2} R_2 - R_1 R_2^{m_2}}{R_1^{m_2} R_2^{m_1} - R_1^{m_1} R_2^{m_2}} |\nabla T_0|, \\[3ex]
C = -\dfrac{R_1^{m_1} R_2 - R_1 R_2^{m_1}}{R_1^{m_2} R_2^{m_1} - R_1^{m_1} R_2^{m_2}} |\nabla T_0|, \\[3ex]
E = 0, \\[2ex]
\kappa_1 = \kappa_{rr} \dfrac{m_1 \left(-R_1^{m_1} R_2^{m_2} + R_1^{m_1+m_2-1} R_2 \right) + m_2 \left(R_1^{m_2} R_2^{m_1} - R_1^{m_1+m_2-1} R_2 \right)}{R_1^{m_2} R_2^{m_1} - R_1^{m_1} R_2^{m_2}}, \\[4ex]
\kappa_3 = \kappa_{rr} \dfrac{m_1 \left(R_1^{m_2} R_2^{m_1} - R_1 R_2^{m_1+m_2-1} \right) + m_2 \left(-R_1^{m_1} R_2^{m_2} + R_1 R_2^{m_1+m_2-1} \right)}{R_1^{m_2} R_2^{m_1} - R_1^{m_1} R_2^{m_2}},
\end{cases}
$$

$$(11.10)$$

$$
\begin{cases}
A = |\nabla T_0|, \\[2ex]
B = \dfrac{R_1^{s_2} R_2 - R_1 R_2^{s_2}}{R_1^{s_2} R_2^{s_1} - R_1^{s_1} R_2^{s_2}} |\nabla T_0|, \\[3ex]
C = -\dfrac{R_1^{s_1} R_2 - R_1 R_2^{s_1}}{R_1^{s_2} R_2^{s_1} - R_1^{s_1} R_2^{s_2}} |\nabla T_0|, \\[3ex]
E = 0, \\[2ex]
\kappa_1 = \kappa_{rr} \dfrac{s_1 \left(-R_1^{s_1} R_2^{s_2} + R_1^{s_1+s_2-1} R_2 \right) + s_2 \left(R_1^{s_2} R_2^{s_1} - R_1^{s_1+s_2-1} R_2 \right)}{R_1^{s_2} R_2^{s_1} - R_1^{s_1} R_2^{s_2}}, \\[4ex]
\kappa_3 = \kappa_{rr} \dfrac{s_1 \left(R_1^{s_2} R_2^{s_1} - R_1 R_2^{s_1+s_2-1} \right) + s_2 \left(-R_1^{s_1} R_2^{s_2} + R_1 R_2^{s_1+s_2-1} \right)}{R_1^{s_2} R_2^{s_1} - R_1^{s_1} R_2^{s_2}},
\end{cases}
$$

$$(11.11)$$

Equations (11.10) and (11.11) have similar forms with only different $m_{1,2}$ and $s_{1,2}$.

References

1. Alù, A., Engheta, N.: Cloaking a sensor. Phys. Rev. Lett. **102**, 233901 (2009)
2. Greenleaf, A., Kurylev, Y., Lassas, M., Uhlmann, G.: Cloaking a sensor via transformation optics. Phys. Rev. E **83**, 016603 (2011)
3. Mach-Batlle, R., Navau, C., Sanchez, A.: Invisible magnetic sensors. Appl. Phys. Lett. **112**, 162406 (2018)
4. Zhu, X.F., Liang, B., Kan, W.W., Zou, X.Y., Cheng, J.C.: Acoustic cloaking by a superlens with single-negative materials. Phys. Rev. Lett. **106**, 014301 (2011)
5. Fleury, R., Soric, J., Alù, A.: Physical bounds on absorption and scattering for cloaked sensors. Phys. Rev. B **89**, 045122 (2014)
6. Fleury, R., Sounas, D., Alù, A.: An invisible acoustic sensor based on parity-time symmetry. Nat. Commun. **6**, 5905 (2015)
7. He, X., Wu, L.Z.: Thermal transparency with the concept of neutral inclusion. Phys. Rev. E **88**, 033201 (2013)
8. Shen, X.Y., Huang, J.P.: Thermally hiding an object inside a cloak with feeling. Int. J. Heat Mass Transf. **78**, 1 (2014)
9. Yang, T.Z., Bai, X., Gao, D.L., Wu, L.Z., Li, B.W., Thong, J.T.L., Qiu, C.-W.: Invisible sensors: simultaneous sensing and camouflaging in multiphysical fields. Adv. Mater. **27**, 7752 (2015)
10. Yang, T.Z., Su, Y.S., Xu, W.K., Yang, X.D.: Transient thermal camouflage and heat signature control. Appl. Phys. Lett. **109**, 121905 (2016)
11. Han, T.C., Yang, P., Li, Y., Lei, D.Y., Li, B.W., Hippalgaonkar, K., Qiu, C.-W.: Full-parameter omnidirectional thermal metadevices of anisotropic geometry. Adv. Mater. **30**, 1804019 (2018)
12. Han, T.C., Yuan, T., Li, B.W., Qiu, C.-W.: Homogeneous thermal cloak with constant conductivity and tunable heat localization. Sci. Rep. **3**, 1593 (2013)
13. Chen, T.Y., Weng, C.N., Tsai, Y.L.: Materials with constant anisotropic conductivity as a thermal cloak or concentrator. J. Appl. Phys. **117**, 054904 (2015)
14. Hu, R., Huang, S.Y., Wang, M., Zhou, L.L., Peng, X.Y., Luo, X.B.: Binary thermal encoding by energy shielding and harvesting units. Phys. Rev. Appl. **10**, 054032 (2018)
15. Han, T.C., Zhao, J.J., Yuan, T., Lei, D.Y., Li, B.W., Qiu, C.-W.: Theoretical realization of an ultra-efficient thermal-energy harvesting cell made of natural materials. Energ. Environ. Sci. **6**, 3537 (2013)
16. Xu, L.J., Yang, S., Huang, J.P.: Passive metashells with adaptive thermal conductivities: chameleonlike behavior and its origin. Phys. Rev. Appl. **11**, 054071 (2019)
17. Xu, L.J., Huang, J.P.: Chameleonlike metashells in microfluidics: a passive approach to adaptive responses. Sci. China-Phys. Mech. Astron. **63**, 228711 (2020)
18. Yang, S., Xu, L.J., Wang, R.Z., Huang, J.P.: Full control of heat transfer in single-particle structural materials. Appl. Phys. Lett. **111**, 121908 (2017)
19. Shang, J., Wang, R.Z., Xin, C., Dai, G.L., Huang, J.P.: Macroscopic networks of thermal conduction: failure tolerance and switching processes. Int. J. Heat Mass Transf. **121**, 321 (2018)
20. Xu, L.J., Yang, S., Huang, J.P.: Thermal transparency induced by periodic interparticle interaction. Phys. Rev. Appl. **11**, 034056 (2019)
21. Li, J.X., Li, Y., Li, T.L., Wang, W.Y., Li, L.Q., Qiu, C.-W.: Doublet thermal metadevice. Phys. Rev. Appl. **11**, 044021 (2019)
22. Dai, G.L., Huang, J.P.: Nonlinear thermal conductivity of periodic composites. Int. J. Heat Mass Transf. **147**, 118917 (2020)
23. Jin, P., Xu, L.J., Jiang, T., Zhang, L., Huang, J.P.: Making thermal sensors accurate and invisible with an anisotropic monolayer scheme. Int. J. Heat Mass Transf. **163**, 120437 (2020)

24. Fujii, G., Akimoto, Y., Takahashi, M.: Exploring optimal topology of thermal cloaks by CMA-ES. Appl. Phys. Lett. **112**, 061108 (2018)
25. Fujii, G., Akimoto, Y.: Optimizing the structural topology of bifunctional invisible cloak manipulating heat flux and direct current. Appl. Phys. Lett. **115**, 174101 (2019)
26. Fujii, G., Akimoto, Y.: Topology-optimized thermal carpet cloak expressed by an immersed-boundary level-set method via a covariance matrix adaptation evolution strategy. Int. J. Heat Mass Transf. **137**, 1312 (2019)
27. Fujii, G., Akimoto, Y.: Cloaking a concentrator in thermal conduction via topology optimization. Int. J. Heat Mass Transf. **159**, 120082 (2020)
28. Fan, C.Z., Gao, Y., Huang, J.P.: Shaped graded materials with an apparent negative thermal conductivity. Appl. Phys. Lett. **92**, 251907 (2008)
29. Chen, T.Y., Weng, C.N., Chen, J.S.: Cloak for curvilinearly anisotropic media in conduction. Appl. Phys. Lett. **93**, 114103 (2008)
30. Xu, L.J., Dai, G.L., Huang, J.P.: Transformation multithermotics: Controlling radiation and conduction simultaneously. Phys. Rev. Appl. **13**, 024063 (2020)
31. Xu, L.J., Yang, S., Dai, G.L., Huang, J.P.: Transformation omnithermotics: simultaneous manipulation of three basic modes of heat transfer. ES Energy Environ. **7**, 65 (2020)
32. Li, Y., Bai, X., Yang, T.Z., Luo, H., Qiu, C.-W.: Structured thermal surface for radiative camouflage. Nat. Commun. **9**, 273 (2018)
33. Hu, R., Zhou, S.L., Li, Y., Lei, D.Y., Luo, X.B., Qiu, C.-W.: Illusion thermotics. Adv. Mater. **30**, 1707237 (2018)
34. Zhou, S.L., Hu, R., Luo, X.B.: Thermal illusion with twinborn-like heat signatures. Int. J. Heat Mass Transf. **127**, 607 (2018)
35. Guo, J., Qu, Z.G.: Thermal cloak with adaptive heat source to proactively manipulate temperature field in heat conduction process. Int. J. Heat Mass Transf. **127**, 1212 (2018)
36. Hu, R., Huang, S.Y., Wang, M., Luo, X.L., Shiomi, J., Qiu, C.-W.: Encrypted thermal printing with regionalization transformation. Adv. Mater. **31**, 1807849 (2019)
37. Qin, J., Luo, W., Yang, P., Wang, B., Deng, T., Han, T.C.: Experimental demonstration of irregular thermal carpet cloaks with natural bulk material. Int. J. Heat Mass Transf. **141**, 487 (2019)
38. Xu, L.J., Yang, S., Huang, J.P.: Dipole-assisted thermotics: experimental demonstration of dipole-driven thermal invisibility. Phys. Rev. E **100**, 062108 (2019)
39. Peng, X.Y., Hu, R.: Three-dimensional illusion thermotics with separated thermal illusions. ES Energy Environ. **6**, 39 (2019)

Chapter 12
Theory for Invisible Thermal Sensors: Optimization Scheme

Abstract Metamaterial-based devices have been extensively explored for their intriguing functions, such as cloaking, concentrating, rotating, and sensing. However, they are usually achieved by employing metamaterials with extreme parameters, critically restricting engineering preparation. In this chapter, we propose an optimization model with particle swarm algorithms to simplify parametric designs to realize bilayer thermal sensors composed of bulk isotropic materials (circular structure). For this purpose, the fitness function is defined to evaluate the difference between the actual and expected temperatures. By choosing suitable materials for different regions and treating the sensor, inner shell, and outer shell radii as design variables, we finally minimize the fitness function via particle swarm optimization. The designed scheme is easy to implement in applications and shows excellent performances in detective accuracy and thermal invisibility, which are confirmed by finite-element simulations and laboratory experiments. The optimization model can also be flexibly extended to a square case. This method can calculate numerical solutions for difficult analytical theories (circular structure) and optimal solutions for problems without analytical theories (such as square structure), providing new inspiration for simplifying the design of metamaterials in various communities.

Keywords Invisible thermal sensors · Particle swarm optimization · Irregular shapes

12.1 Opening Remarks

Novel meta-devices [1–24] have been researched continuously over the decades in various fields since the pioneering theoretical proposals of transformation theory [1–4]. Recently, many fruitful strategies have been proposed for offering new avenue of devising thermal meta-devices such as neutral inclusion [6], bilayer schemes [10], illusion thermotics [13], regionalization transformation [14], and many-particle thermal invisibility [21]. However, most experimental devices are prepared by employing metamaterials with unconventional thermal conductivities (i.e., anisotropic, graded or singular), which remain to be overcomed for engineering applications. New

© The Author(s) 2023

L.-J. Xu and J.-P. Huang, *Transformation Thermotics and Extended Theories*,
https://doi.org/10.1007/978-981-19-5908-0_12

schemes deserve exploring for purpose of simplifying engineering preparation and developing novel functional meta-devices.

Optimization method has been conjectured as an effective tool for the design of metamaterials in the macro [25–33] or micro [34–36] scale, which is applied comprehensively in recent years. A gradient-based numerical optimization algorithm was used to control the heat flow in the printed circuit board [27]. Also, non-gradient-based black-box algorithms, including evolutionary algorithms, have been used in the reverse structural design of thermal metamaterials [31]. Based on topology optimization [29, 30, 32, 33], thermal meta-devices are reversely designed for specific objective functions, providing excellent performance. Topology optimization usually involves a change in structural topology, resulting in complex structural parameters. This feature inspires us to explore simpler structural meta-devices with equally high performance by employing optimization algorithms.

Here, we propose an optimization model with particle swarm algorithms [37] (PSA) for designing bilayer thermal sensors composed of bulk isotropic materials. For example, we design a circular bilayer thermal sensor with detective accuracy and thermal invisibility. For this purpose, two objective functions are constructed simultaneously, one for detecting the temperature distribution of the region occupied by the sensor with accuracy and the other for undisturbing the temperature distribution of the original background. When choosing suitable material analogies for different regions, we treat the radius of the sensor, inner shell, and outer shell as design variables. By adopting PSA, the characteristics of the prescribed optimized structure are precisely and efficiently found. The designed scheme not only simplifies practical fabrication but shows almost perfect performance, as both simulation and experimental results exhibit. The optimization model can also be flexibly extended to a square case.

12.2 Theoretical Foundation

The scheme for a bilayer thermal sensor is shown in Fig. 12.1a. A sensor (with radius of R_1, Material (1) coated with a bilayer shell (inner shell with radius of R_2, Material (2); outer shell with radius of R_3, Material (3) is put in the center of background (Material 4) for detection of the temperature distribution of region occupied by it. Hot source and cold source are, respectively, set at the right-most and left-most boundaries. The up-most and down-most boundaries are thermally insulated. Temperature distributions follow the Laplace equation with passive heat conduction at steady state,

$$\nabla \cdot (-\kappa (\boldsymbol{x}) \cdot \nabla T) = 0, \tag{12.1}$$

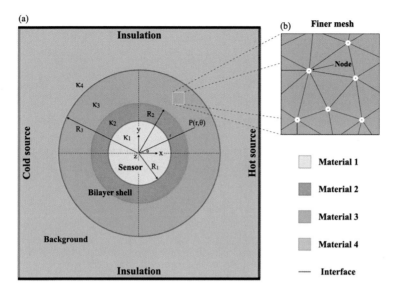

Fig. 12.1 **a** Schematic diagram for bilayer thermal sensors. **b** Discretized finer mesh for optimization. Adapted from Ref. [38]

where $\kappa\left(\boldsymbol{x}\right)$ is thermal conductivity, denoted by

$$\kappa\left(\boldsymbol{x}\right) = \begin{cases} \kappa_1 & \text{for } \boldsymbol{x} \text{ in region of Material 1,} \\ \kappa_2 & \text{for } \boldsymbol{x} \text{ in region of Material 2,} \\ \kappa_3 & \text{for } \boldsymbol{x} \text{ in region of Material 3,} \\ \kappa_4 & \text{for } \boldsymbol{x} \text{ in region of Material 4.} \end{cases} \quad (12.2)$$

Considering expanding Eq. (12.1) in cylindrical coordinates and symmetry of boundary conditions, the general solution of the temperature distribution in four regions can be expressed as

$$T_1 = A_0 + Ar\cos\theta, \quad (12.3)$$

$$T_2 = A_0 + Br\cos\theta + Cr^{-1}\cos\theta, \quad (12.4)$$

$$T_3 = A_0 + Dr\cos\theta + Er^{-1}\cos\theta, \quad (12.5)$$

$$T_4 = A_0 + Fr\cos\theta + Gr^{-1}\cos\theta, \quad (12.6)$$

where A_0 is the temperature at $\theta = \pm\pi/2$, and $F = |\nabla T_0|$ represents the modulus of an external linear thermal field ∇T_0.

Boundary continuity conditions of temperature and normal heat flow should be satisfied,

$$\begin{cases} T_1\,(R_1) = T_2\,(R_1)\,, \\ T_2\,(R_2) = T_3\,(R_2)\,, \\ T_3\,(R_3) = T_4\,(R_3)\,, \\ \left(-\kappa_1 \frac{\partial T_1}{\partial r}\right)_{R_1} = \left(-\kappa_2 \frac{\partial T_2}{\partial r}\right)_{R_1}\,, \\ \left(-\kappa_2 \frac{\partial T_2}{\partial r}\right)_{R_2} = \left(-\kappa_3 \frac{\partial T_3}{\partial r}\right)_{R_2}\,, \\ \left(-\kappa_3 \frac{\partial T_3}{\partial r}\right)_{R_3} = \left(-\kappa_4 \frac{\partial T_4}{\partial r}\right)_{R_3}\,. \end{cases} \tag{12.7}$$

When thermal sensor works, there is no thermal disturbances in sensor and background regions, which means we have following two equations,

$$\begin{cases} A = F, \\ G = 0. \end{cases} \tag{12.8}$$

We simplify the form of equations that are composed of Eqs. (12.7) and (12.8)

$$\begin{cases} F R_1 = B R_1 + C R_1^{-1}, \\ B R_2 + C R_2^{-1} = D R_2 + E R_2^{-1}, \\ D R_3 + E R_3^{-1} = F R_3, \\ -\kappa_1 F = -\kappa_2 \left(B - C R_1^{-2}\right), \\ -\kappa_2 \left(B - C R_2^{-2}\right) = -\kappa_3 \left(D - E R_2^{-2}\right), \\ -\kappa_3 \left(D - E R_3^{-2}\right) = -\kappa_4 F. \end{cases} \tag{12.9}$$

For given R_1, κ_1, κ_2, κ_3 and κ_4, we do have six unknown coefficients (B, C, D, E, R_2 and R_3) determined by six equations (Eq. (12.9)) uniquely. However, due to the nonlinear coupling of multiple unknowns, it is difficult to find an explicit analytical expression for any one radius. When the thermal conductivity of the shell (circular or elliptic structure) is taken as unknown coefficients, analytic expressions can be obtained, which have great limitations. In this way, the physical image of the influence of geometric size on the performance of thermal sensors cannot be given intuitively from the analytical theory. There is a mapping relationship between the radius of the circles (say, R_1, R_2 and R_3) and the performance of the thermal sensor. We turn it into an optimization problem and reversely design the geometry size according to the performance.

12.3 Optimization Problem Description

In principle, a thermal sensor should have the ability of reproducing temperature distributions in sensor and background regions, which are the same as those in corresponding regions of original background [39]. The heat field to be studied is discretized for numerical optimization by using finer mesh in COMSOL Multiphysics, as shown in Fig. 12.1b. Considering optimization problem, two objective functions for bilayer thermal sensors with accuracy and invisibility are, respectively, defined as

$$\Psi_s = \frac{1}{N_s} \sum_{i=1}^{N_s} \left| T(i) - T_{ref}(i) \right|, \qquad (12.10)$$

$$\Psi_b = \frac{1}{N_b} \sum_{i=1}^{N_b} \left| T(i) - T_{ref}(i) \right|, \qquad (12.11)$$

where i, T, T_{ref}, N_s, and N_b represent sequence number of nodes, temperature distribution controlled by bilayer thermal sensor, temperature distribution in pure background, number of nodes in sensor and background regions after discretization, respectively. Then we add Eqs. (12.10) and (12.11) to represent the fitness function,

$$\Psi = \Psi_s + \Psi_b. \qquad (12.12)$$

As a swarm intelligence optimization algorithm, PSA (Fig. 12.2) has a very high convergence rate, adopted extensively for inverse problems. PSA gets the optimal solution through the coordination of particles in the solution space, and particles constantly follow the current optimal particle. To solve the optimal problem mentioned above, we first initialize N particles R_j^0 $j = 1, 2, ..., N$, in the feasible solution space K, given as

$$K = \left\{ R = (R_1, R_2, R_3) : R_{min} \leq R_i < R_j \leq R_{max}, \ i < j; \ i, \ j \in \{1, 2, 3\} \right\}. \qquad (12.13)$$

The characteristics of each particle are represented by position, velocity and fitness function in K space. Particles move constantly in the solution space, updating the position and velocity of individuals by tracking individual and group extremum points. Here, individual and group extremum points are the positions with the minimum fitness function among all the positions experienced by the individual and group particles. Each time positions of particles are updated, the fitness function of which are calculated. During each iteration, the updating formula of particle velocity and position are

$$V_j^{i+1} = wV_j^i + c_1 d_1 \left(P_j^i - R_j^i \right) + c_2 d_2 \left(P_g - R_j^i \right), \qquad (12.14)$$

$$R_j^{i+1} = R_j^i + V_j^{i+1}, \qquad (12.15)$$

where i is iteration number, j is sequence number of each particle, P_j^i is individual extremum point of j-th particle at the i-th iteration, and P_g is group extremum point. w is inertia weight usually taken as a linear decreasing function, denoted by

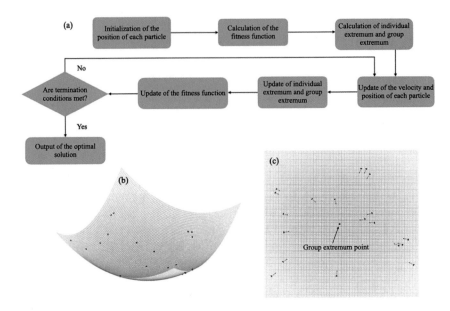

Fig. 12.2 a Illustration of the algorithm of PSA. **b** Schematic diagram of a particle swarm in search of an optimal solution. **c** Contour map of (**b**). Adapted from Ref. [38]

Table 12.1 Parameter setting of optimization model

N	w_s	w_e	I_{max}	c_1	c_2
50	0.9	0.4	150	1.49	1.49

$$w = w_s - (w_s - w_e)\,\frac{i}{I_{max}}, \tag{12.16}$$

where w_s, w_e, and I_{max} are respectively initial inertia weight (for global search), final inertia weight (for local search), and total number of iteration. c_1 and c_2 are empirical constant. d_1 and d_2 are random numbers between 0 and 1. Relevant parameters of the optimization model are shown in Table 12.1. When finishing total iterations, termination conditions are met (or say minimum fitness function converged). Therefore, we get the optimal solution for design variables. Also, we have finished the design of bilayer thermal sensors.

On the same footing, we extend bilayer thermal sensors to a square case, where there is no strict analytical theory of bilayer thermal sensors. For a given set of four different bulk isotropic materials, the sides of three squares (L_1, L_2 and L_3, from inside to outside) are selected as design variables. Using PSA, we can obtain the geometrical size of the bilayer thermal sensor with the best performance of accuracy and invisibility.

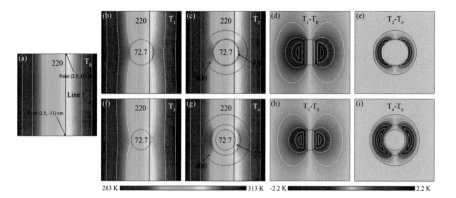

Fig. 12.3 Finite-element simulation results of circular bilayer thermal sensors. The simulation box is $22 \times 22\,\text{cm}^2$. White lines represent isotherms. **a** Pure background for reference. **b** and **c** First case for bare sensor with $R_1 = 3.42$ cm and bilayer thermal sensor with $R_1 = 3.42$ cm, $R_2 = 3.64$ cm, and $R_3 = 5.95$ cm. **f** and **g** Second case for bare sensor with $R_1 = 2.73$ cm and bilayer thermal sensor with $R_1 = 2.73$ cm, $R_2 = 4.11$ cm, and $R_3 = 6.85$ cm. **d** and **e** Temperature difference between (**b**), (**c**), and (**a**). **h** and **i** Temperature difference between (**f**), (**g**), and (**a**). Adapted from Ref. [38]

12.4 Finite-Element Simulation

With each suitable set of selected materials, we obtain the optimal solution R of the design variables, representing the sensor, inner shell, and outer shell radii. For numerical demonstrations, we choose two different materials for the inner shell (Inconel alloy 625 and Stainless steel 436 with thermal conductivity of 9.8 and $30\,\text{W}\,\text{m}^{-1}\,\text{K}^{-1}$) to perform finite-element simulations with COMSOL MULTIPHYSICS. Sensor, outer shell, and background are Magnesium alloy, Copper, and Aluminum with thermal conductivity of 72.7, 400, and $220\,\text{W}\,\text{m}^{-1}\,\text{K}^{-1}$. Thus, we get two sets of design variables, parameterizing two cases of bilayer thermal sensors.

Before discussing the results of bilayer thermal sensors, we first show two reference schemes; one for pure background (Fig. 12.3a), the other for bare sensor (Fig. 12.3b and f). The presence of a bare sensor disturbs the thermal field of the pure background, making the thermal field in the sensor region distorted. Figure 12.3c and g show the simulation results of two cases of bilayer thermal sensors designed by PSA. External isotherms in two cases are both vertical. In two cases, the interval between internal isotherms (in the sensor region) is almost identical to the pure background. We plot the temperature-difference distributions between various schemes and pure background to accentuate the temperature difference. From Fig. 12.3d and h, we get a significant temperature deviation in sensor and background regions imposed by bare sensors. On the contrary, the temperature difference in sensor and background regions in Fig. 12.3e and i is almost zero. Furthermore, the temperature distributions along the y axis on Line 1 in Fig. 12.3 are exported to contrast the performance of these schemes quantitatively; see Fig. 12.4. Each bilayer thermal sensor maintains

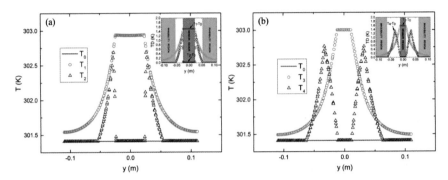

Fig. 12.4 Quantitative comparison of Fig. 12.3. **a** Temperature distributions (T_0, T_1, and T_2) on Line 1 in Fig. 12.3a, b, and c. Upper-right inset shows the temperature difference on Line 1 between T_1 (T_2) and T_0. **b** Temperature distributions (T_0, T_3, and T_4) on Line 1 in Fig. 12.3a, f, and g. Upper-right inset shows the temperature difference on Line 1 between T_3 (T_4) and T_0. Adapted from Ref. [38]

the same temperature distributions in the sensor and background regions as those in the pure background, demonstrating its excellent performance. However, a bare sensor not only measures the sensor region inaccurately but also distorts the thermal field in the background region, whose temperature difference (TD) is shown in the upper right inset of Fig. 12.4.

Moreover, we also perform finite-element simulations of square bilayer thermal sensors with different geometrical sizes (structure with single shell, structure with bilayer shell of random size, and structure with bilayer shell of optimized size). As expected, optimized size dramatically improves the performance of the thermal sensor, reproducing temperature distributions in sensor and background regions from the original thermal field; see Fig. 12.5.

12.5   Laboratory Experiment

To test the performance of the bilayer thermal sensor, we select one case (Fig. 12.3g) to experiment with a setup shown in Fig. 12.6a. For comparison, we prepare two samples, one for background with pure Aluminum (Fig. 12.6c); another for the bilayer scheme made of Magnesium alloy (AZ91D), Stainless steel (ASTM 436), Copper, and Aluminum (from inside to outside) (Fig. 12.6e). Both of them have the size of 22 $\times$ 22 $\times$ 0.5 cm^3, each with two tentacles 5 cm high on the left and right, respectively. Sensor (R_1), inner shell (R_1 and R_2), and outer shell (R_2 and R_3) have the circular boundaries with the radii of $R_1 = 2.73$ cm, $R_2 = 4.11$ cm, and $R_3 = 6.85$ cm. We process the structure of four materials by laser cutting and combine these parts using mechanics enchases craft, as shown in Fig. 12.6b. The left and right tentacles of samples are immersed in 283 and 313 K water baths, and an infrared camera FLIR

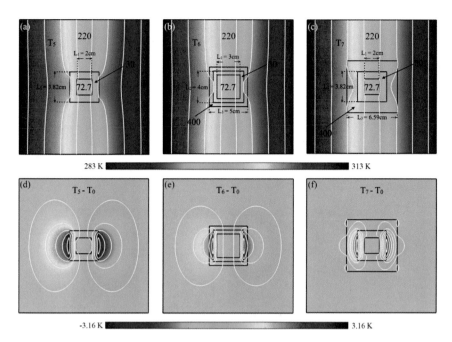

Fig. 12.5 Simulation results of square bilayer thermal sensors. The simulation box is $17 \times 17\,\mathrm{cm}^2$. White lines represent isotherms. Sensor, inner shell, outer shell, and background are respectively expanded Magnesium alloy, Stainless steel, Copper, and Aluminum with conductivity of 72.7, 30, 400, and 220 $\mathrm{W\,m^{-1}\,K^{-1}}$. **a** Thermal sensor of single layer size with $L_1 = 2$ cm and $L_2 = 3.82$ cm. **b** Bilayer thermal sensor of random size with $L_1 = 3$ cm, $L_2 = 4$ cm and $L_3 = 5$ cm. **c** Bilayer thermal sensor of optimized size with $L_1 = 2$ cm, $L_2 = 3.82$ cm and $L_3 = 6.59$ cm. **d–f** Temperature difference between **a–c** and Fig. 12.3a. Adapted from Ref. [38]

E60 is used to measure the temperature distributions of two samples at the steady state (after 10 min). Figure 12.6d and f show the measured results of Fig. 12.6c and e, respectively. Though thermal contact resistance exists in the interface of different materials, our scheme exhibits excellent properties in both accuracy and thermal invisibility, which is well consistent with the simulation result of Fig. 12.3g.

12.6 Conclusion

In summary, we have proposed an optimization model with particle swarm algorithms for designing bilayer thermal sensors composed of bulk isotropic materials. For example, we design and fabricate a circular bilayer thermal sensor with high performance, as both simulation result and experimental result exhibit. Such a scheme removes the need for extreme parameters (anisotropic, graded, or singular), making engineering applications readily and efficiently. The optimization model can also

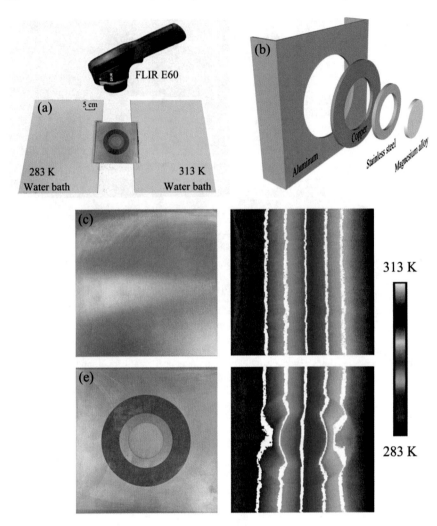

Fig. 12.6 Laboratory experiments. **a** Experimental setup. **b** Composition of bilayer thermal sensor. **c** and **e** Real photos of reference and bilayer thermal sensor. **d** and **f** Measured results for (**c**) and (**e**). White lines represent isotherms. The sample size in (**c**) and (**e**) is the same as that for Fig. 12.3a and g. Adapted from Ref. [38]

be flexibly extended to a square case. Finally, an intelligent method of simplifying structures and materials can calculate numerical solutions for difficult analytical theories (such as circular structure) and optimal solutions for problems without analytical theories (such as square structure). This property provides an insight into the development of metamaterials in a wide range of communities.

12.7 Exercise and Solution

Exercise

1. Let $y = x^2$, we initialize two particles whose x coordinates are $x_1 = 1$ and $x_2 = 2$, respectively. Using particle swarm optimization, we can get the minimum value of y. Please write down the $x_1^{(2)}$ and $x_2^{(2)}$ coordinates of the two particles after the first two iterations.

$$V_j^{(i+1)} = wV_j^{(i)} + \left(P_j^{(i)} - R_j^{(i)} \right) + \left(P_g - R_j^{(i)} \right), R_j^{(i+1)} = R_j^{(i)} + V_j^{(i+1)}, \tag{12.17}$$

where $w = 0.5$ and $V_j^{(0)} = 0$.

Solution

1. First iteration of particle j, we have

$$V_j^{(1)} = wV_j^{(0)} + \left(P_j^{(0)} - x_j^{(0)} \right) + \left(P_g - x_j^{(0)} \right), x_j^{(1)} = x_j^{(0)} + V_j^{(1)}. \tag{12.18}$$

Second iteration of particle j, we have

$$V_j^{(2)} = wV_j^{(1)} + \left(P_j^{(1)} - x_j^{(1)} \right) + \left(P_g - x_j^{(1)} \right), x_j^{(2)} = x_j^{(1)} + V_j^{(2)}. \tag{12.19}$$

After substituting the values, we get $x_1^{(2)} = 1$ and $x_2^{(2)} = 0.5$.

References

1. Leonhardt, U.: Optical conformal mapping. Science **312**, 1777 (2006)
2. Pendry, J.B., Schurig, D., Smith, D.R.: Controlling electromagnetic fields. Science **312**, 1780 (2006)
3. Fan, C.Z., Gao, Y., Huang, J.P.: Shaped graded materials with an apparent negative thermal conductivity. Appl. Phys. Lett. **92**, 251907 (2008)
4. Chen, T.Y., Weng, C.N., Chen, J.S.: Cloak for curvilinearly anisotropic media in conduction. Appl. Phys. Lett. **93**, 114103 (2008)
5. Narayana, S., Sato, Y.: Heat flux manipulation with engineered thermal materials. Phys. Rev. Lett. **108**, 214303 (2012)
6. He, X., Wu, L.Z.: Thermal transparency with the concept of neutral inclusion. Phys. Rev. E **88**, 033201 (2013)
7. Yu, C.J., Li, Y.H., Zhang, X., Huang, X., Malyarchuk, V., Wang, S.D., Shi, Y., Gao, L., Su, Y.W., Zhang, Y.H., Xu, H.X., Hanlon, R.T., Huang, Y.G., Rogers, J.A.: Adaptive optoelectronic camouflage systems with designs inspired by cephalopod skins. Proc. Natl. Acad. Sci. U. S. A. **111**, 12998 (2014)
8. Xu, H.Y., Shi, X.H., Gao, F., Sun, H.D., Zhang, B.L.: Ultrathin three-dimensional thermal cloak. Phys. Rev. Lett. **112**, 054301 (2014)

9. Ma, Y.G., Liu, Y.C., Raza, M., Wang, Y.D., He, S.L.: Experimental demonstration of a multiphysics cloak: manipulating heat flux and electric current simultaneously. Phys. Rev. Lett. **113**, 205501 (2014)

10. Han, T.C., Bai, X., Gao, D.L., Thong, J.T.L., Li, B.W., Qiu, C.W.: Experimental demonstration of a bilayer thermal cloak. Phys. Rev. Lett. **112**, 054302 (2014)

11. Yang, T.Z., Bai, X., Gao, D.L., Wu, L.Z., Li, B.W., Thong, J.T.L., Qiu, C.W.: Invisible sensors: simultaneous sensing and camouflaging in multiphysical fields. Adv. Mater. **27**, 7752 (2015)

12. Han, T.C., Yang, P., Li, Y., Lei, D.Y., Li, B.W., Hippalgaonkar, K., Qiu, C.W.: Full-parameter omnidirectional thermal metadevices of anisotropic geometry. Adv. Mater. **30**, 1804019 (2018)

13. Hu, R., Zhou, S.L., Li, Y., Lei, D.Y., Luo, X.B., Qiu, C.W.: Illusion thermotics. Adv. Mater. **30**, 1707237 (2018)

14. Hu, R., Huang, S.Y., Wang, M., Luo, X.B., Shiomi, J., Qiu, C.W.: Encrypted thermal printing with regionalization transformation. Adv. Mater. **31**, 1807849 (2019)

15. Shang, J., Wang, R.Z., Xin, C., Dai, G.L., Huang, J.P.: Macroscopic networks of thermal conduction: failure tolerance and switching processes. Int. J. Heat Mass Transf. **121**, 321 (2018)

16. Qin, J., Luo, W., Yang, P., Wang, B., Deng, T., Han, T.C.: Experimental demonstration of irregular thermal carpet cloaks with natural bulk material. Int. J. Heat Mass Transf. **141**, 487 (2019)

17. Dai, G.L., Huang, J.P.: Nonlinear thermal conductivity of periodic composites. Int. J. Heat Mass Transf. **147**, 118917 (2020)

18. Xu, L.J., Huang, J.P.: Controlling thermal waves with transformation complex thermotics. Int. J. Heat Mass Transf. **159**, 120133 (2020)

19. Zhou, L.L., Huang, S.Y., Wang, M., Hu, R., Luo, X.B.: While rotating while cloaking. Phys. Lett. A **383**, 759 (2019)

20. Hu, R., Huang, S.Y., Wang, M., Zhou, L.L., Peng, X.Y., Luo, X.B.: Binary thermal encoding by energy shielding and harvesting units. Phys. Rev. Appl. **10**, 054032 (2018)

21. Xu, L.J., Yang, S., Huang, J.P.: Thermal transparency induced by periodic interparticle interaction. Phys. Rev. Appl. **11**, 034056 (2019)

22. Li, J.X., Li, Y., Li, T.L., Wang, W.Y., Li, L.Q., Qiu, C.W.: Doublet thermal metadevice. Phys. Rev. Appl. **11**, 044021 (2019)

23. Xu, L.J., Yang, S., Huang, J.P.: Passive metashells with adaptive thermal conductivities: chameleonlike behavior and its origin. Phys. Rev. Appl. **11**, 054071 (2019)

24. Xu, L.J., Huang, J.P.: Metamaterials for manipulating thermal radiation: Transparency, cloak, and expander. Phys. Rev. Appl. **12**, 044048 (2019)

25. Popa, B.I., Cummer, S.A.: Cloaking with optimized homogeneous anisotropic layers. Phys. Rev. A **79**, 023806 (2009)

26. Dede, E.M., Nomura, T., Lee, J.: Thermal-composite design optimization for heat flux shielding, focusing, and reversal. Struct. Multidisc. Optim. **49**, 59 (2014)

27. Dede, E.M., Schmalenberg, P., Nomura, T., Ishigaki, M.: Design of anisotropic thermal conductivity in multilayer printed circuit boards. IEEE Trans. Compon. Packag. Manuf. Technol. **5**, 1763 (2015)

28. Peralta, I., Fachinotti, V.D.: Optimization-based design of heat flux manipulation devices with emphasis on fabricability. Sci. Rep. **7**, 6261 (2017)

29. Fujii, G., Akimoto, Y., Takahashi, M.: Exploring optimal topology of thermal cloaks by CMA-ES. Appl. Phys. Lett. **112**, 061108 (2018)

30. Fujii, G., Akimoto, Y.: Optimizing the structural topology of bifunctional invisible cloak manipulating heat flux and direct current. Appl. Phys. Lett. **115**, 174101 (2019)

31. Alekseev, G.V., Tereshko, D.A.: Particle swarm optimization-based algorithms for solving inverse problems of designing thermal cloaking and shielding devices. Int. J. Heat Mass Transf. **135**, 1269 (2019)

32. Fujii, G., Akimoto, Y.: Topology-optimized thermal carpet cloak expressed by an immersed-boundary level-set method via a covariance matrix adaptation evolution strategy. Int. J. Heat Mass Transf. **137**, 1312 (2019)

33. Fujii, G., Akimoto, Y.: Cloaking a concentrator in thermal conduction via topology optimization. Int. J. Heat Mass Transf. **159**, 120082 (2020)
34. Hu, R., Luo, X.B.: Two-dimensional phonon engineering triggers microscale thermal functionalities. Natl. Sci. Rev. **6**, 1071 (2019)
35. Hu, R., Iwamoto, S., Feng, L., Ju, S.H., Hu, S.Q., Ohnishi, M., Nagai, N., Hirakawa, K., Shiomi, J.: Machine-learning-optimized aperiodic superlattice minimizes coherent phonon heat conduction. Phys. Rev. X **10**, 021050 (2020)
36. Hu, R., Song, J.L., Liu, Y.D., Xi, W., Zhao, Y.T., Yu, X.J., Cheng, Q., Tao, G.M., Luo, X.B.: Machine learning-optimized Tamm emitter for high-performance thermophotovoltaic system with detailed balance analysis. Nano Energy **72**, 104687 (2020)
37. Poli, R., Kennedy, J., Blackwel, T.: Particle swarm optimization: an overview. Swarm Intel. **1**, 33 (2007)
38. Jin, P., Yang, S., Xu, L.J., Dai, G.L., Huang, J.P., Ouyang, X.P.: Particle swarm optimization for realizing bilayer thermal sensors with bulk isotropic materials. Int. J. Heat Mass Transf. **172**, 121177 (2021)
39. Jin, P., Xu, L.J., Jiang, T., Zhang, L., Huang, J.P.: Making thermal sensors accurate and invisible with an anisotropic monolayer scheme. Int. J. Heat Mass Transf. **163**, 120437 (2020)

Chapter 13
Theory for Omnithermal Illusion Metasurfaces: Cavity Effect

Abstract In this chapter, we consider multifold heat-transfer modes and propose a class of restructurable metasurfaces to show illusions in infrared and similarity in visible-light view. We consider the three basic modes of heat transfer (omnithermotics) in theoretical designs and adopt radiation-cavity effects in experimental manufacture. We also make it feasible to tune surface temperature and emissivity synergistically. Besides, such metasurfaces can work in temperature-varying backgrounds and transient states. This scheme may provide a platform for the design of adaptable thermal illusion and show robustness under multifrequency detections.

Keywords Omnithermal illusion metasurface · Cavity effect · Multifrequency detection

13.1 Opening Remarks

The temperature signals of macroscopic objects can be observed by infrared imaging because all objects with nonzero temperatures emit electromagnetic energy, known as thermal radiation [1–3]. The Wien law [4] implies that within an extensive temperature range ($10^0 \sim 10^3$ K), the radiation-spectrum peak of an ideal black body locates in the infrared region. This intrinsic property is extensively applied in industry reconnoiter, military detection, and daily life. Naturally, the technology of thermal illusion [5–9] has attracted much attention due to its promising prospect in illusion or camouflage, namely, misleading or camouflaging thermal signals. The former ("illusion") means that an existing object exists in infrared imaging, replacing another non-existing object [5–7]. In contrast, the latter ("camouflage") represents that the thermal infrared pattern of an existing object blends into the background as if the object does not exist [8, 9]. Meanwhile, various challenges arise in designing infrared illusion, mainly resulting from complex surroundings, multifold heat-transfer modes, and fabrication difficulties.

Recent progress on infrared illusion focuses on regulating surface temperatures T_{sur} and designing surface emissivities ε_{sur}, which play two key roles in infrared imaging. On the one hand, with the successful development of thermal metamaterials [10–

L.-J. Xu and J.-P. Huang, *Transformation Thermotics and Extended Theories*,
https://doi.org/10.1007/978-981-19-5908-0_13

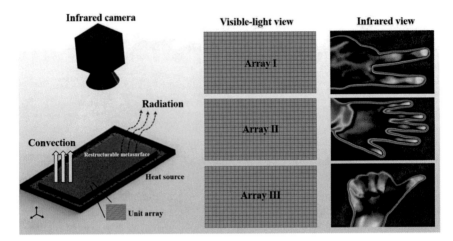

Fig. 13.1 Schematic diagram showing the proposed thermal metasurface. The units are arranged in three arrays (Array I, Array II, and Array III), which can form three different images (specific gestures) in the infrared camera (the third column). Meanwhile, they are similar in the visible-light view (the second column). Adapted from Ref. [29]

17], temperature distributions can be tailored at will with elaborate microstructure designs. Based on it, the thermal illusion has been achieved within fixed or varying backgrounds. However, there are two weaknesses of this method in the existing studies: firstly, most of them are confined to conductive systems [18–21], neglecting thermal convection and radiation; secondly, the surface structures are still identifiable from the background in the visible-light view [22–24], which make them hard to be concealed under multiband detections. On the other hand, tuning emissivities can disguise an actual object into a fake one in the infrared camera. For self-adapting control, phase-change materials are widely adopted [25–28]. But these materials are not common and usually call for additional installations to input stimulus, adapting to changing circumstances (say, changing temperatures). Besides, if ambient temperatures vary sharply or even out of the region of phase-change temperature, its effect will become invalid. So both of these two methods of infrared illusion have some limitations. Furthermore, these two tailoring methods are mutually independent and scarcely coupled due to the lack of a practical and synergistic platform.

To overcome the limitations and promote the integration of tuning T_{sur} and ε_{sur} in a single platform, we design an omnithermal restructurable thermal metasurface for infrared illusion; see Fig. 13.1. We can achieve characteristic infrared patterns by tailoring each block unit and assembling them in a specific array. We consider the three heat transfer modes, conduction, convection, and radiation (omnithermotics), which dominate surface temperatures. With the radiation-cavity effect, say, the dependence of effective emissivity on the sizes, shapes, and proportion of surface cavities, the specific emissivity can be achieved on each unit within a wide temperature range. Therefore, this single platform can tailor the surface temperature and emissivity synergistically. The unit-discretization operation in the $x - y$ plane not only provides flexibility in designing fake temperature signals (hence yielding infrared illusion) but also makes different arrays almost identical (thus causing similarity in visible light) despite different properties (T_{sur} and ε_{sur}). As a result, both illusions in infrared view and similarity in visible-light view are achieved simultaneously, as schematically shown in the middle and right columns of Fig. 13.1.

13.2 Theoretical Foundation

According to the Stefan−Boltzmann law [30], the total thermal radiative energy density I_{bb} of a black body is related to the biquadrate of surface temperature T_{sur},

$$I_{bb} = \int_0^\infty u_{bb}\,(\lambda,\ T_{sur})\,d\lambda = \int_0^\infty \frac{2\pi hc^2}{\lambda^5} \frac{1}{e^{\frac{hc}{\lambda k_B T_{sur}}} - 1} d\lambda = \left(\frac{2\pi^5 k_B^4}{15 c^2 h^3}\right) T_{sur}^4 = \sigma T_{sur}^4,$$
(13.1)

where λ is radiative wavelength and $u_{bb}\,(\lambda,\ T_{sur})$ is the black-body spectral radiance, described by the Plank law. Here, h is the Plank constant, c is the velocity of light in vacuum, k_B is the Boltzmann constant, and σ is the Stefan−Boltzmann constant. We consider a scene that an infrared camera captures the infrared signals of an object in a far field for identification, the actually received spectral radiance deviates from the result described by Eq. (13.1). Spectral directional emissivity $\varepsilon_{sur}(\lambda,\ T_{sur},\ \theta,\ \phi)$ can describe this deviation, which is defined as the spectral-radiance ratio of actual objects to black bodies at temperature T_{sur}, wavelength λ, and direction angles θ and ϕ. But in most practical situations without elaborate directed thermal emission, the diffuse-emitter approximation is reasonable enough. So we can simplify the surface emissivity to $\varepsilon_{sur}(\lambda,\ T_{sur})$. Then the actual radiative energy density I_{ac} can be written as

$$I_{ac} = \int_0^\infty \varepsilon_{sur}\,(\lambda,\ T_{sur})\,u_{bb}\,(\lambda,\ T_{sur})\,d\lambda = \int_0^\infty \varepsilon_{sur}\,(\lambda,\ T_{sur}) \frac{2\pi hc^2}{\lambda^5} \frac{1}{e^{\frac{hc}{\lambda k_B T_{sur}}} - 1} d\lambda.$$
(13.2)

As we concern the total thermal radiative energy instead of the spectral radiance, the full wavelength emissivity $\varepsilon_{sur}(T_{sur})$ makes sense. It can be defined as

$$\varepsilon_{sur}(T_{sur}) = \frac{I_{ac}}{I_{bb}} = \frac{\int_0^\infty \varepsilon_{sur}\,(\lambda,\ T_{sur})\,u_{bb}\,(\lambda,\ T_{sur})\,d\lambda}{\int_0^\infty u_{bb}\,(\lambda,\ T_{sur})\,d\lambda} = \frac{\int_0^\infty \varepsilon_{sur}\,(\lambda,\ T_{sur})\,u_{bb}\,(\lambda,\ T_{sur})\,d\lambda}{\sigma T_{sur}^4}.$$
(13.3)

Except for the intrinsic emissivity affects thermal radiation, both the signal collection range and resolution of the infrared camera should also be considered. According to the practical situations, the signal collection range $(\lambda_1,\ \lambda_2)$ covers the main emission band. Then the full wavelength emissivity $\varepsilon_{sur}(T_{sur})$ can be adopted in this scene.

Combing with Eqs. (13.2) and (13.3), the reading temperature T_{read} is given as [8]

$$
\begin{aligned}
T_{read} = C \times I_{ac} &= C \int_{\lambda_1}^{\lambda_2} \varepsilon_{sur}\,(\lambda,\,T_{sur})\,u_{bb}\,(\lambda,\,T_{sur})\,d\lambda \\[2mm]
&\approx C\varepsilon_{sur}\,(T_{sur}) \int_{\lambda_1}^{\lambda_2} u_{bb}\,(\lambda,\,T_{sur})\,d\lambda \qquad (13.4) \\[2mm]
&\approx C\varepsilon_{sur}\,(T_{sur}) \int_{\lambda_1}^{\lambda_2} \frac{2\pi hc^2}{\lambda^5} \frac{1}{e^{\frac{hc}{\lambda k_B T_{sur}}} - 1}\,d\lambda,
\end{aligned}
$$

where C is a built-in conversion parameter of the infrared camera. Equation (13.4) indicates that the two factors dominate the infrared imaging, namely, the camera capacity $[C,\,(\lambda_1,\,\lambda_2)]$ and the surface properties $(T_{sur},\,\varepsilon_{sur})$. Here, we focus on modulating the characteristic radiative spectrum, which depends on the surface properties $(T_{sur},\,\varepsilon_{sur})$. Within a limited surface temperature region, the full wavelength emissivity $\varepsilon_{sur}(T_{sur})$ is regarded as ε_{sur}, independent on T_{sur}. It is noted that if the surface temperature varies sharply, the coupling relation between ε_{sur} and λ should be underlined. And while the surface temperature difference is large enough between units, the coupling relation between ε_{sur} and T_{sur} should also be taken into consideration. Our strategy for controllable infrared illusion consists of tuning T_{sur} and ε_{sur} individually and assembling them in any specific way.

For the first step, let us consider a three-dimensional bulk as a unit, as illustrated in Fig. 13.2a. We set its sides to be thermally insulated and place a homothermal source at the bottom. The heat flows in the bulk along the z axis and dissipates into the surroundings from the top surface due to convection and radiation. This process includes the three basic modes of heat transfer. In a steady state, the temperature of the top surface T_{sur} can be determined by the conservation law of heat flow,

$$
\boldsymbol{J}_{cond} = \boldsymbol{J}_{conv} + \boldsymbol{J}_{rad}, \qquad (13.5)
$$

where $\boldsymbol{J}_{cond}$, $\boldsymbol{J}_{conv}$, and $\boldsymbol{J}_{rad}$ are conductive, convective and radiative heat flow density, respectively. We set the unit's height as H_b and thermal conductivity as κ_b. The convective coefficient and radiative emissivity of the surface are h_b and ε_b, respectively. Besides, the source and room temperatures are given as T_0 and T_{air}. We can write down the expressions of $\boldsymbol{J}_{cond}$, $\boldsymbol{J}_{conv}$, and $\boldsymbol{J}_{rad}$ as

$$
J_{cond} = \kappa_b \nabla T|_{bulk} = \kappa_b \frac{T_0 - T_{sur}}{H_b}, \qquad (13.6a)
$$

$$
J_{conv} = h_b(T_{sur} - T_{air}), \qquad (13.6b)
$$

$$
J_{rad} = \varepsilon_b \sigma (T_{sur}^4 - T_{air}^4) = \varepsilon_b \sigma (T_{sur}^2 + T_{air}^2)(T_{sur} + T_{air})(T_{sur} - T_{air})
$$

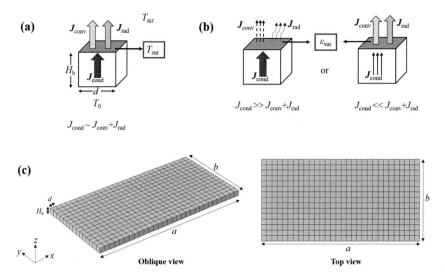

Fig. 13.2 Different tuning methods. **a** and **b** A cuboid as a block unit. Conductive flow is comparable with convective and radiative flow in (**a**), but dramatically different in (**b**). **c** Assembly of the units, which construct the whole metasurface. Adapted from Ref. [29]

$$= R_b(T_{\text{sur}})(T_{\text{sur}} - T_{\text{air}}), \tag{13.6c}$$

where $R_b(T) = \varepsilon_b \sigma (T_{\text{sur}}^2 + T_{\text{air}}^2)(T_{\text{sur}} + T_{\text{air}})$, representing the radiative ability of the surface. Combining Eqs. (13.5)–(13.6c), we can deduce the temperature of the top surface T_{sur} as

$$T_{\text{sur}} = \frac{\kappa_b T_0 / H_b + [h_b + R_b(T_{\text{sur}})] T_{\text{air}}}{\kappa_b / H_b + h_b + R_b(T_{\text{sur}})}. \tag{13.7}$$

Hereto, we obtain the general solution of the top-surface temperature of a unit. To obtain the value of T_{sur}, an iteration of $R_b(T_{\text{sur}})$ should be executed by calculator. Compared with the method reported in Ref. [19] where only κ_b is tuned, the present scheme has four parametric freedoms for handling. They are κ_b, h_b, ε_b, and H_b, involving the three basic modes of heat transfer. κ_b and H_b play the roles in controlling conductive flow. h_b and ε_b correspond to convective and radiative flows, respectively. These four parameters can be expressed as

$$\kappa_b = \frac{H_b \left[h_b(T_{\text{sur}} - T_{\text{air}}) + \varepsilon_b \sigma \left(T_{\text{sur}}^4 - T_{\text{air}}^4 \right) \right]}{T_0 - T_{\text{sur}}}, \tag{13.8a}$$

$$H_b = \frac{\kappa_b(T_0 - T_{\text{sur}})}{h_b(T_{\text{sur}} - T_{\text{air}}) + \varepsilon_b \sigma \left(T_{\text{sur}}^4 - T_{\text{air}}^4 \right)}, \tag{13.8b}$$

$$h_b = \frac{\kappa_b(T_0 - T_{\text{sur}})/H_b - \varepsilon_b \sigma \left(T_{\text{sur}}^4 - T_{\text{air}}^4 \right)}{T_{\text{sur}} - T_{\text{air}}}, \tag{13.8c}$$

$$\varepsilon_b = \frac{\kappa_b(T_0 - T_{\text{sur}})/H_b - h_b(T_{\text{sur}} - T_{\text{air}})}{\sigma\left(T_{\text{sur}}^4 - T_{\text{air}}^4\right)}. \tag{13.8d}$$

We can see if the surface temperature T_{sur} of each unit is preset to create specific infrared illusion, only three of them are independent. Also, these four parameters can be tuned arbitrarily and simultaneously to achieve the designed T_{sur} of each unit. Thus, the tuning strategy is flexible. We suppose that ε_b (equivalent to ε_{sur}) is uniform in each unit and approximate to that of a black body, the reading temperature can be estimated by Eq. (13.4) as

$$T_{\text{read1}}(x, y) \approx C \times \int_{\lambda_1}^{\lambda_2} \frac{2\pi hc^2}{\lambda^5} \frac{1}{e^{\frac{hc}{\lambda k_B T_{\text{sur}}}} - 1} d\lambda \approx T_{\text{sur}}(x, y), \tag{13.9}$$

where (x, y) refers to the central position of each unit.

It is noted that tuning ε_b plays a limited role in controlling T_{sur} due to its maximum value 1, especially in low temperature regions. However, when T_{sur} is nearly uniform in each unit under some circumstances, tuning surface emissivity is another effective method for creating illusion because ε_{sur} becomes a major impact beyond T_{sur} in Eq. (13.4). For example, according to Eq. (13.7), if κ_b is far greater than h_b and $R_b(T_{\text{sur}})$, T_{sur} will reach T_0. Inversely, it will reach T_{air}, as shown in Fig. 13.2b. Then, tailoring emissivity is the only way for creating infrared illusion in the infrared imaging. On the basis of Eq. (13.4), the reading temperature in this case is

$$T_{\text{read2}}(x, y) \approx \varepsilon_{\text{sur}}(x, y) \times C \times \int_{\lambda_1}^{\lambda_2} \frac{2\pi hc^2}{\lambda^5} \frac{1}{e^{\frac{hc}{\lambda k_B T_{\text{sur}}}} - 1} d\lambda \approx \varepsilon_{\text{sur}}(x, y) \cdot T_{\text{sur}}. \tag{13.10}$$

The final step is to assemble these units in a specific array to create the infrared illusion in infrared imaging, see Fig. 13.2c. Each unit can be regarded as a pixel. The fake surface temperature of each pixel should be distinguishable enough to make the illusion valid in infrared imaging. Therefore, the contrast ratio should be larger than the intrinsic resolution of the infrared camera under any conditions. The contrast ratio of imaging is based on the maximum and minimum values of reading temperatures. We can define the contrast ratio C as

$$C = \frac{T_{\text{read}}|_{\text{max}} - T_{\text{read}}|_{\text{min}}}{T_{\text{read}}|_{\text{max}} + T_{\text{read}}|_{\text{min}}}. \tag{13.11}$$

We have two ways for tailoring T_{read}. If the three modes of heat transfer are comparable, tuning T_{sur} solely is enough. According to Eq. (13.9), Eq. (13.11) can be written as

$$C_1 = \frac{T_{\text{sur}}|_{\text{max}} - T_{\text{sur}}|_{\text{min}}}{T_{\text{sur}}|_{\text{max}} + T_{\text{sur}}|_{\text{min}}}. \tag{13.12}$$

Otherwise, tuning ε_{sur} is necessary to present a distinguishable temperature distribution in the infrared camera. So from Eq. (13.10), Eq. (13.11) can be written as

$$C_2 = \frac{\varepsilon_{sur}|_{max} - \varepsilon_{sur}|_{min}}{\varepsilon_{sur}|_{max} + \varepsilon_{sur}|_{min}}. \tag{13.13}$$

The contrast ratio is related to the ratio of the surface temperature or the extremum difference of the effective emissivity, representing an intrinsic character of a sort of specifically-designed thermal metasurfaces. The flexible combination of units contributes to the reconfigurability, and does not affect the contrast ratio C. So, once we design the units completely, the thermal metasurface will always meet the resolution requirement of the detector.

13.3 Finite-Element Simulation

We perform finite-element simulations based on the commercial software COMSOL Multiphysics. The simulations focus on tuning the temperature T_{sur}. Here, we keep H_b fixed and tailor κ_b, h_b, and ε_b not to break the geometric construction of metasurfaces. Firstly, the metasurface is constituted with 15×30 units, as shown in Fig. 13.2c. They are cubes of 1 cm length. Then, we classify the total 450 units into 6 groups, as demonstrated in Fig. 13.3a. Each group is designed independently to obtain six patterns of T_{sur}. Here, we expect to create an illusion of "FUDAN". When tuning κ_b, we keep h_b and ε_b as constants. So as the other two parameters. Then, the six groups are assembled, as shown in Fig. 13.3a. For simplification, we heat the entire lower surface with a homothermal heat source T_0 and keep the room temperature T_{air} at 300 K. The laterals of the surface are thermally contacted with neighboring units to mimic the real situation. Figure 13.3b–d, respectively, show the results of tuning κ_b, h_b, and ε_b at $T_0 = 350$ K, while Fig. 13.3e–g are those at $T_0 = 700$ K. We can see that convection and radiation play minor roles under low-temperature surroundings. In particular, the effect of thermal radiation is nearly indistinguishable. When T_0 goes higher, they make sense gradually. We calculate the contrast ratio C with the simulation data at 350 and 700 K. It is 2.96 and 15.23% when tuning κ_b, and 0.20 and 2.28% when tuning ε_b. The amplification of tuning ε_b is about twice bigger than that of tuning κ_b, confirming that radiation plays an increasingly important role with the temperature rising.

It has been proved that the expected patterns can be observed by tuning the three heat-transfer modes individually. Convection and radiation dominate at low and high temperatures, affecting the contrast ratio of the pattern in the infrared camera. Figure 13.3h–j show the comparisons of T_{sur} between theoretical data and simulation results under three tuning modes. They echo well at low temperatures and show a little shift when they are high because the thermal interaction between different units appears. More heat exchange in the $x - y$ plane impacts T_{sur}. When the condition goes to extremes (say, T_{sur} reaches T_0 or T_{air}), we have to tune the effective emissivity.

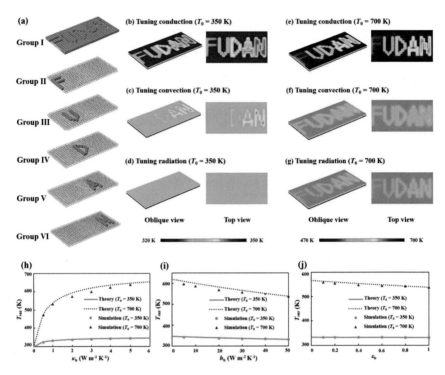

Fig. 13.3 Simulation results of tuning temperature T_{sur}. **a** Six groups and arrays with letters "FUDAN". **b–g** Temperature distributions with different tuning methods. T_0 is set at 350 K and 700 K. For tuning thermal conduction, κ_b is set as 0.5, 1, 2, 3, 4, and 5 W m^{-1} K^{-1} for six groups while h_b is 50 W m^{-2} K^{-1} and ε_b is 1. For tuning thermal convection, h_b is 5, 10, 20, 30, 40, and 50 W m^{-2} K^{-1} while κ_b is 1 W m^{-1} K^{-1} and ε_b is 1. For tuning thermal radiation, ε_b is 0.1, 0.2, 0.4, 0.6, 0.8, and 1 while κ_b is 1 W m^{-1} K^{-1} and h_b is 50 W m^{-2} K^{-1}. **h–j** Comparisons between theoretical values and simulation values of T_{sur}, corresponding to the data extracted from **b–g**. Adapted from Ref. [29]

13.4   Laboratory Experiment

As shown in Fig. 13.3d, tuning radiation with emissivity at low-temperature conditions has little effect on infrared illusion. However, the engineered emissivities can impact apparent temperature distribution. Here, we resort to the surface-cavity effect [31, 32] to modulate ε_{sur}. The cavity structures on the surface promote the block to a higher radiant exitance. Hence, the apparent temperature in infrared imaging will be deviated from the actual, forming an illusion pattern. Now, we are in the position to design a surface cavity structure. For simplification, we adopt the cylindrical structure as it is easy to manufacture, as demonstrated in Fig. 13.4a. The heat transfer process is between the surface cavity and the free space, in which the angle factor of the cavity can be omitted. According to Ref. [31], the effective emissivity of an isolated cylindrical cavity ε_e depends on its area ratio of mouth and inwall, which can be expressed as

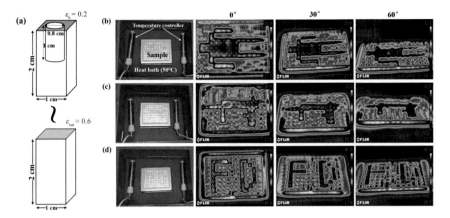

Fig. 13.4 Experimental measurements for different effective emissivities ε_{sur}. **a** Cavity structure (upper panel) and effective emissivity principle. The effective emissivity of a flat surface with cavity (upper panel) is equivalent to ε_{sur} of another flat surface (lower panel), which is quantitatively expressed in Eqs. (13.14) and (13.15). The first column of We performfinite, **c**, and **d** shows the photo of experimental apparatus for a human pattern, a machine-gun pattern, and an "FD" pattern, respectively. And the other three columns display the experimental measurements, each for one observation angle (0°, 30°, or 60°). Note that the experimental apparatus is placed in a heat bath of 50 °C. The unit of numerical values in the color bars is °C. Adapted from Ref. [29]

$$\varepsilon_e = \left[1 + \frac{S_0}{S_1}\left(\frac{1}{\varepsilon_b} - 1\right)\right]^{-1}, \tag{13.14}$$

where S_0 and S_1 are the area of mouth and inwall, respectively, and ε_{sur} is the intrinsic surface emissivity. Owing to the high thermal conductivity and regular shape of the blocks, the surface temperature can be considered a constant. The plat surface allows the energy to transfer into the environment, so the thermal interaction between cavities occurs only. Thus, a quantitative emissivity expression of the whole surface of the block can be derived as

$$\varepsilon_{sur} = \varepsilon_e' \approx f\varepsilon_e + (1-f)\varepsilon_0 = f\left[1 + \frac{S_0}{S_1}\frac{1}{\varepsilon_b}(\frac{1}{\varepsilon_b} - 1)\right]^{-1} + (1-f)\varepsilon_b. \tag{13.15}$$

The area proportion of the cavity f and inherent area ratio S_0/S_1 enable us to tailor the effective emissivity of the surface to form specific apparent temperature distribution in infrared imaging.

We examine the practical effects directly with an infrared camera FLIR E60, whose resolution is 0.1 K. We use a 10×15 array and two groups of tailored units for designing feature patterns for simplification. Copper cubes with 2 cm in length are employed as block units. The thermal conductivity of copper is about $397\,W\,m^{-1}\,K^{-1}$, to homogenize T_{sur}. Group I is not hollow with an intrinsic emissivity of 0.2, while group II is trepanned with a cylindrical hole. The hole is 0.4 cm in radius and 1 cm in depth. According to Eq. (13.15), the effective emissivity is about 0.6. Besides, we design an acrylic plat with 15×20 square holes for encoding the block units. They can be inserted in the holes for fixation. We design infrared patterns of a human, a machine gun, and the letters "FD", respectively, as shown in Fig. 13.4b–d by manually rearranging these units. This operation can also be mechanically executed with additional active installations, thus forming an active restructurable metasurface. We place the encoded surface in a water bath with a temperature of 50°C. The room temperature is about 20°C. After the system reaches a steady state, the infrared camera helps to detect the feature patterns. The metasurfaces of different arrangement ways in visible-light view are hard to distinguish (similarity). At different angles to observe, we find its robustness in both infrared and visible-light views, see Fig. 13.4b–d. It is worth mentioning that when the surface is coated with an anti-reflection film, we find that the feature pattern disappears. The reading temperatures get a little higher than the previous, confirming that the cavity engineering method helps change the imaging.

13.5 Discussion

Object emissivity and surface temperature determine the imaging pattern of infrared cameras. We have demonstrated two tuning methods by simulation or experiment of emissivity and temperature on the same platform to achieve infrared illusion and visible-light similarity. Reference [19] has given a feasible way to tune temperature by manipulating conduction processes. In addition to this, how to practically control convective and radiative flows need further study to satisfy theoretical predictions by Eqs. (13.8a)–(13.8d). Tuning T_{sur} only works with the system in the steady state, while tuning ε_{sur} works in both steady and transient states. We should note that emissivity plays two roles in the tailoring process. On the one hand, it guides the radiative flow to change the surface temperature. On the other hand, it helps conceal the actual temperature T_{sur} to cheat the infrared camera by displaying an apparent temperature. So, we perform the variable-controlling method in the above simulations and experiments. This platform is a flexible and applicable tool for infrared illusion. In different temperature regions, targeted tuning methods are available. Besides, the encoding and assembling process on unit cells is non-invasive and repeatable. Its flexibility with block assembly makes the illusion applicative to diverse situations. Moreover, infrared cameras usually have some limitations in dimensional resolution; the illusion pattern quality can be improved when the sizes of units are comparable with dimensional resolution. The proposed restructurability is essentially distinguished from the common reconfigurability or adjustability [33]. The former is property-invariant but structurally rearrangeable, while the latter is structure-invariant but property-adjustable. The proposed restructurable metasurface exhibits both illusions in infrared light and similarity in visible light. The "similarity" can be upgraded to "indistinguishability" as long as the surface is structured carefully, as implied by Fig. 13.4b–d, which should be useful for real applications.

Also, as a direct application, we suggest using our scheme to realize infrared anticounterfeiting. As we know, anticounterfeiting is extensively applied in industry, military, and daily life. The common strategies are based on optical holograms [34–36], which naked eyes or detectors can find. Nevertheless, such technologies tend to be defeated because the typical pattern can be forged. Recently, flourishing research on optical metasurfaces has been involved in this traditional field [37–40]. Light's amplitude, phase, and polarization can be tailored arbitrarily with carefully designed two-dimensional microstructures. So, its intrinsic signal is characteristic and hard to be replicated. However, we only need to capture emissive electromagnetic-wave information for identification. Intuitional insight is to tailor the characteristic radiative signals for anticounterfeiting, which does not need additional incident lights. The encryption process can be executed on our proposed metasurfaces, while decoding is achieved by using infrared imaging. The key secret is hard to be forged because of its similarity in visible-light view. Moreover, restructurability raises the difficulty level for falsifying. This kind of anticounterfeiting strategy has applicability in non-invasive and quick-recognition scenes.

13.6 Conclusion

We have proposed a practical scheme for achieving infrared-light illusion and visible-light similarity. The tuning of surface temperature and emissivity can be executed synergistically. Compared with existing thermal metamaterials, our scheme considers all the three basic modes of heat transfer (omnithermotics), thus expanding the scope of applications. Also, we have introduced the cavity effect to tailor the emissivity, simplifying the manufacture. We hope this scheme can not only overcome some challenges in designing infrared illusion but also has direct applications in industry and commerce.

13.7 Exercise and Solution

Exercise

1. Discuss the effective emissivity of a cylindrical cavity with radius r and depth h.

Solution

1. According to Eq. (13.14), we can derive

$$\varepsilon_e = \left[1 + \frac{\pi r^2}{2\pi r h + \pi r^2}\left(\frac{1}{\varepsilon_b} - 1\right)\right]^{-1} = \left[1 + \frac{1}{2h/r + 1}\left(\frac{1}{\varepsilon_b} - 1\right)\right]^{-1}$$

$$= \left[1 + \frac{1}{2\delta + 1}\left(\frac{1}{\varepsilon_b} - 1\right)\right]^{-1}, \tag{13.16}$$

where $\delta = h/r$ is the depth-radius ratio.

Then, we can define a cavity factor as $F = \varepsilon_e/\varepsilon_b$,

$$F = \frac{\varepsilon_e}{\varepsilon_b} = \left[\varepsilon_b + \frac{S_0}{S_1}(1 - \varepsilon_b)\right]^{-1} = \left[\varepsilon_b + \frac{1}{2\delta + 1}(1 - \varepsilon_b)\right]^{-1}. \tag{13.17}$$

For the same ε_b, the larger δ, the larger F. For the same δ, the smaller ε_b, the larger F.

References

1. Baranov, D.G., Xiao, Y.Z., Nechepurenko, I.A., Krasnok, A., Alù, A., Kats, M.A.: Nanophotonic engineering of far-field thermal emitters. Nat. Mater. **18**, 920 (2019)
2. Cuevas, J.C.: Thermal radiation from subwavelength objects and the violation of Planck's law. Nat. Commun. **10**, 3342 (2019)
3. Li, W., Fan, S.H.: Nanophotonic control of thermal radiation for energy applications. Opt. Express **26**, 15995 (2018)
4. Stewart, S.M.: Spectral peaks and Wien's displacement law. J. Thermophys. Heat Transf. **26**, 689 (2012)
5. He, X., Wu, L.Z.: Illusion thermodynamics: A camouflage technique changing an object into another one with arbitrary cross section. Appl. Phys. Lett. **105**, 221904 (2014)
6. Hu, R., Zhou, S.L., Li, Y., Lei, D.-Y., Luo, X.B., Qiu, C.-W.: Illusion thermotics. Adv. Mater. **30**, 1707237 (2018)
7. Yang, F.B., Xu, L.J., Huang, J.P.: Thermal illusion of porous media with convection-diffusion process: transparency, concentrating, and cloaking. ES Energy Environ. **6**, 45 (2019)
8. Qu, Y.R., Li, Q., Cai, L., Pan, M.Y., Ghosh, P., Du, K.K., Qiu, M.: Thermal camouflage based on the phase-changing material GST. Light-Sci. Appl. **7**, 26 (2018)
9. Shang, J., Jiang, C.R., Xu, L.J., Huang, J.P.: Many-particle thermal invisibility and diode from effective media. J. Heat Transf. **140**, 092004 (2018)
10. Fan, C.Z., Gao, Y., Huang, J.P.: Shaped graded materials with an apparent negative thermal conductivity. Appl. Phys. Lett. **92**, 251907 (2008)
11. Chen, T.Y., Weng, C.-N., Chen, J.-S.: Cloak for curvilinearly anisotropic media in conduction. Appl. Phys. Lett. **93**, 114103 (2008)
12. Narayana, S., Sato, Y.: Heat flux manipulation with engineered thermal materials. Phys. Rev. Lett. **108**, 214303 (2012)
13. Han, T.C., Bai, X., Gao, D.L., Thong, J.T.L., Li, B.W., Qiu, C.-W.: Experimental demonstration of a bilayer thermal cloak. Phys. Rev. Lett. **112**, 054302 (2014)
14. Li, Y., Shen, X.Y., Wu, Z.H., Huang, J.Y., Chen, Y.X., Ni, Y.S., Huang, J.P.: Temperature-dependent transformation thermotics: from switchable thermal cloaks to macroscopic thermal diodes. Phys. Rev. Lett. **115**, 195503 (2015)

15. Shen, X.Y., Li, Y., Jiang, C.R., Huang, J.P.: Temperature trapping: energy-free maintenance of constant temperatures as ambient temperature gradients change. Phys. Rev. Lett. **117**, 055501 (2016)
16. Wang, J., Shang, J., Huang, J.P.: Negative energy consumption of thermostats at ambient temperature: electricity generation with zero energy maintenance. Phys. Rev. Appl. **11**, 024053 (2019)
17. Xu, L.J., Yang, S., Huang, J.P.: Passive metashells with adaptive thermal conductivities: chameleonlike behavior and its origin. Phys. Rev. Appl. **11**, 054071 (2019)
18. Yang, T.Z., Bai, X., Gao, D.L., Wu, L.Z., Li, B.W., Thong, J.T., Qiu, C.-W.: Invisiable sensor: simultaneous camouflaging and sensing in multiphysical fields. Adv. Mater. **27**, 7752 (2015)
19. Shang, J., Tian, B.Y., Jiang, C.R., Huang, J.P.: Digital thermal metasurface with arbitrary infrared thermogram. Appl. Phys. Lett. **113**, 261902 (2018)
20. Hu, R., Huang, S.Y., Wang, M., Luo, X.B., Shiomi, J., Qiu, C.-W.: Encrypted thermal printing with regionalization transformation. Adv. Mater. **31**, 1807849 (2019)
21. Hu, R., Huang, S.Y., Wang, M., Zhou, L.L., Peng, X.Y., Luo, X.B.: Binary thermal encoding by energy shielding and harvesting units. Phys. Rev. Appl. **10**, 054032 (2018)
22. Li, Y., Bai, X., Yang, T.Z., Luo, H.L., Qiu, C.-W.: Structured thermal surface for radiative camouflage. Nat. Commun. **9**, 273 (2018)
23. Xu, L.J., Yang, S., Huang, J.P.: Dipole-assisted thermotics: experimental demonstration of dipole-driven thermal invisibility. Phys. Rev. E **100**, 062108 (2019)
24. Xu, L.J., Yang, S., Huang, J.P.: Thermal illusion with the concept of equivalent thermal dipole. Eur. Phys. J. B **92**, 264 (2019)
25. Xu, C.Y., Stiubianu, G.T., Gorodetsky, A.A.: Adaptive infrared-reflecting systems inspired by cephalopods. Science **359**, 1495 (2019)
26. Lee, N., Kim, T., Lim, J.-S., Chang, I., Cho, H.H.: Metamaterial-selective emitter for maximizing infrared camouflage performance with energy dissipation. ACS Appl. Mater. Interfaces **11**, 21250 (2019)
27. Kats, M.A., Blanchard, R., Zhang, S.Y., Genevet, P., Ko, C., Ramanathan, S., Capasso, F.: Vanadium dioxide as a natural disordered metamaterial: perfect thermal emission and large broadband negative differential thermal emittance. Phys. Rev. X **3**, 041004 (2013)
28. Xiao, L., Ma, H., Liu, J.K., Zhao, W., Jia, Y., Zhao, Q., Liu, K., Wu, Y., Wei, Y., Fan, S.S., Jiang, K.L.: Fast adaptive thermal camouflage based on flexible VO_2/graphene/CNT thin films. Nano Lett. **15**, 8365 (2015)
29. Wang, J., Yang, F.B., Xu, L.J., Huang, J.P.: Omnithermal restructurable metasurfaces for both infrared-light illusion and visible-light similarity. Phys. Rev. Appl. **14**, 014008 (2020)
30. Cuevas, J.C., García-Vidal, F.J.: Radiative heat transfer. ACS Photonics **5**, 3896 (2018)
31. Ohwada, Y.: Calculation of the effective emissivity of a cavity having non-Lambertian isothermal surfaces. J. Opt. Soc. Am. A **16**, 1059 (1999)
32. Mei, G., Zhang, J., Zhao, S., Xie, Z.: Simple method for calculating the local effective emissivity of the blackbody cavity as a temperature sensor. Infrared Phys. Technol. **85**, 372 (2017)
33. Bao, L., Cui, T.J.: Tunable, reconfigurable, and programmable metamaterials. Microw. Opt. Technol. Lett. **62**, 9 (2020)
34. Javidi, B., Horner, J.L.: Optical-pattern recognition for validation and security verification. Opt. Eng. **33**, 1752 (1994)
35. Zhang, X.S., Dalsgaard, E., Liu, S., Lai, H.K., Chen, J.Z.: Concealed holographic coding for security applications lay using a moire technique. Appl. Opt. **36**, 8096 (1997)
36. Aggarwal, A.K., Kaura, S.K., Chhachhia, D.P., Sharma, A.K.: Concealed moire pattern encoded security holograms readable by a key hologram. Opt. Laser Technol. **38**, 117 (2006)
37. Huang, L., Chen, X., Mühlenbernd, H., Zhang, H., Chen, S., Bai, B., Tan, Q., Jin, G., Cheah, K.-W., Qiu, C.-W., Li, J., Zentgraf, T., Zhang, S.: Three-dimensional optical holography using a plasmonic metasurface. Nat. Commun. **4**, 2808 (2013)

38. Wen, D., Yue, F., Li, G., Zheng, G., Chan, K., Chen, S., Chen, M., Li, K.F., Wong, P.W.H., Cheah, K.W., Pun, E.Y.B., Zhang, S., Chen, X.: Helicity multiplexed broadband metasurface holograms. Nat. Commun. **6**, 8241 (2015)
39. Zhang, C.M., Dong, F.L., Intaravanne, Y., Zang, X.F., Xu, L.H., Song, Z.W., Zheng, G.X., Wang, W., Chu, W.G., Chen, X.Z.: Multichannel metasurfaces for anticounterfeiting. Phy. Rev. Appl. **12**, 034028 (2019)
40. Sung, J., Lee, G.-V., Lee, B.: Progresses in the practical metasurface for holography and lens. Nanophotonics **8**, 1701 (2019)

Chapter 14
Theory for Effective Advection Effect: Spatiotemporal Modulation

Abstract In this chapter, we introduce spatiotemporal modulation to realize thermal wave nonreciprocity. The major mechanism is the effective advection effect of spatiotemporal modulation in an open thermal system. We further analyze the phase difference between two spatiotemporally modulated parameters, which offers a tunable parameter to control nonreciprocity. We further define a rectification ratio based on the reciprocal of spatial decay rates and discuss the nonreciprocity conditions accordingly. Finite-element simulations are performed to confirm theoretical predictions, and experimental suggestions are provided to ensure the feasibility of spatiotemporal modulation. These results have potential applications in realizing thermal detection and thermal stabilization simultaneously.

Keywords Effective advection effect · Spatiotemporal modulation · Thermal wave nonreciprocity

14.1 Opening Remarks

Ever since the concept of spatiotemporal modulation was proposed [1], intensive studies have been conducted not only in wave systems [2–16] including photonics [2–5], acoustics [6–9], and metasurfaces [10–12] but also in diffusion systems [17–19]. A direct application of spatiotemporal modulation is to realize nonreciprocity which refers to asymmetric propagation in opposite directions. Although many different kinds of waves have been studied to achieve nonreciprocity based on spatiotemporal modulation, thermal waves have received little attention despite being an important phenomenon. In terms of mechanism, thermal waves are a special kind of wave that is dominated by a diffusion equation (i.e., the Fourier equation), thus also called diffusion waves [20]. In terms of application, thermal waves can realize nondestructive detection (i.e., thermal wave imaging), widely applied in aerospace, machinery, and electricity [21–23]. Some recent studies also focused on diffusion waves to realize anti-parity-time symmetry [24–27], negative thermal transport [28], cloaks [20, 29–31], and crystals [32–34].

© The Author(s) 2023
L.-J. Xu and J.-P. Huang, *Transformation Thermotics and Extended Theories*,
https://doi.org/10.1007/978-981-19-5908-0_14

However, a mechanism to achieve thermal wave nonreciprocity is still lacking. Thermal waves can be treated as periodic temperature fluctuations, usually a double-edged sword. On the one hand, they are desirable for thermal detection. On the other hand, they are unwanted for thermal stabilization. Therefore, it is crucially important to realize thermal wave nonreciprocity. For this purpose, we explore spatiotemporal modulation to achieve thermal wave nonreciprocity, inspired by pioneering studies on nonreciprocal thermal materials [18]. It has been revealed that an advection term appears in the conduction equation at quasi-steady states if thermal conductivity and mass density are spatiotemporally modulated, thus achieving nonreciprocity. However, the applicability of thermal waves was not discussed. On the one hand, thermal waves feature completely transient states where the Willis term should be considered. On the other hand, the phase difference between two spatiotemporally modulated parameters remains explored.

Here, we thoroughly discuss thermal wave nonreciprocity based on spatiotemporal modulation. Since there is a phase difference between two spatiotemporally modulated parameters, we construct two backward cases (Fig. 14.1) with different nonreciprocity conditions. The results demonstrate that the phase difference offers a flexible and tunable parameter to control nonreciprocity. We also discuss the heat flux to reveal the feature of spatiotemporal modulation.

14.2   Theoretical Foundation

We consider a passive thermal conduction process in one dimension, dominated by

$$\rho (x - ut) \frac{\partial T}{\partial t} + \frac{\partial}{\partial x} \left[-\sigma (x - ut) \frac{\partial T}{\partial x} \right] = 0, \tag{14.1}$$

where $\sigma (x - ut)$ is thermal conductivity and $\rho (x - ut)$ is the product of mass density and heat capacity. The spatiotemporally modulated parameters in Fig. 14.1a take the form of

$$\sigma (x - ut) = \sigma_A + \sigma_B \cos [K (x - ut)], \tag{14.2a}$$

$$\rho (x - ut) = \rho_A + \rho_B \cos [K (x - ut) + \alpha], \tag{14.2b}$$

where σ_A, σ_B, ρ_A, and ρ_B are four constants. $K = 2\pi/\gamma$ is wave number, γ is wavelength, u is modulation speed, and α is phase difference. Since $\sigma (x - ut)$ and $\rho (x - ut)$ are periodic functions, the Bloch theorem is applicable and the temperature solution can be expressed as

$$T = \phi (x - ut) e^{i(kx - \omega t)}, \tag{14.3}$$

where k and ω are, respectively, the wave number and circular frequency of a thermal wave. $\phi (x - ut)$ is an amplitude modulation function that has the same periodicity as $\sigma (x - ut)$ and $\rho (x - ut)$. Equation (14.1) can then be homogenized with the approximations of $k \ll K$ and $\omega \ll uK$ [18],

$$\tilde{\rho}\frac{\partial \tilde{T}}{\partial t} + C\frac{\partial \tilde{T}}{\partial x} - \tilde{\sigma}\frac{\partial^2 \tilde{T}}{\partial x^2} - S\frac{\partial^2 \tilde{T}}{\partial x \partial t} = 0, \tag{14.4}$$

where the homogenized parameters can be expressed as

$$\tilde{\sigma} \approx \sigma_A \left(1 - \frac{\sigma_B^2}{2\sigma_A^2}\frac{1}{1+\Gamma^2}\right), \tag{14.5a}$$

$$\tilde{\rho} \approx \rho_A \left(1 - \frac{\rho_B^2}{2\rho_A^2}\frac{\Gamma^2}{1+\Gamma^2}\right), \tag{14.5b}$$

$$C \approx u\frac{\sigma_B \rho_B}{2\sigma_A}\frac{1}{1+\Gamma^2}P(\alpha), \tag{14.5c}$$

$$S \approx \frac{1}{u}\frac{\sigma_B \rho_B}{2\rho_A}\frac{\Gamma^2}{1+\Gamma^2}Q(\alpha), \tag{14.5d}$$

with $\Gamma = \rho_A u\gamma / (2\pi\sigma_A)$, $P(\alpha) = \cos\alpha + \Gamma\sin\alpha$, and $Q(\alpha) = \cos\alpha + \Gamma^{-1}\sin\alpha$. $\tilde{T}$ can be treated as the envelope line of the actual temperature T. Here, we extend the results reported in Ref. [18] by additionally considering a phase difference of α. $\tilde{\sigma}$ and $\tilde{\rho}$ are irrelevant to α, but C and S are dependent on α, offering a tunable parameter.

We then qualitatively discuss the nonreciprocity induced by spatiotemporal modulation. In what follows, the subscripts of f, $b1$, and $b2$ denote the parameters related to the forward case in Fig. 14.1a, the backward-1 case in Fig. 14.1b, and the backward-2 case in Fig. 14.1c, respectively. The two backward cases are equivalent only when $\alpha = 0$. Since $\tilde{\sigma}$ and $\tilde{\rho}$ do not contribute to nonreciprocity, we mainly discuss C and S in detail.

For the forward case, we know $C_f = C$ and $S_f = S$. For the backward-1 case, we can derive $C_{b1} = -C$ and $S_{b1} = -S$. Nonreciprocity requires $C_f \neq C_{b1}$ (or $S_f \neq S_{b1}$). Therefore, as long as $C \neq 0$ (or $S \neq 0$), nonreciprocity will occur and a larger C (or S) yields larger nonreciprocity. For clarity, we plot the functions of $C(\alpha)$ and $S(\alpha)$ in Fig. 14.2 with $\Gamma = 0.5, 1, 2$. The maximum and minimum values of C appear at $\alpha = -\text{arccot}\,\Gamma + \pi/2$ and $\alpha = -\text{arccot}\,\Gamma - \pi/2$, respectively; and the zero value occurs at $\alpha = \arctan\Gamma \pm \pi/2$. The maximum and minimum values of S appear at $\alpha = -\arctan\Gamma + \pi/2$ and $\alpha = -\arctan\Gamma - \pi/2$, respectively; and the zero value occurs at $\alpha = \text{arccot}\,\Gamma \pm \pi/2$. For the backward-2 case, we can obtain $C_{b2} = C(-u)$ and $S_{b2} = S(-u)$. Nonreciprocity requires $C_f \neq C_{b2}$ (or $S_f \neq S_{b2}$). Therefore, as long as $C(u) \neq C(-u)$ [or $S(u) \neq S(-u)$], nonreciprocity will occur. We can also observe that $\alpha = \pm\pi/2$ makes $P(\alpha)$ and $Q(\alpha)$ two odd functions of u. C and S then become two even functions of u, so nonreciprocity disappears. In one word,

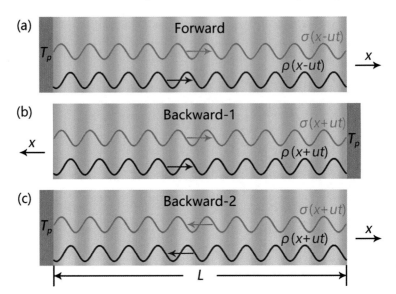

Fig. 14.1 Thermal wave nonreciprocity. **a** Forward case. **b** Backward-1 case by changing the source position. **c** Backward-2 case by changing the modulation direction. Adapted from Ref. [35]

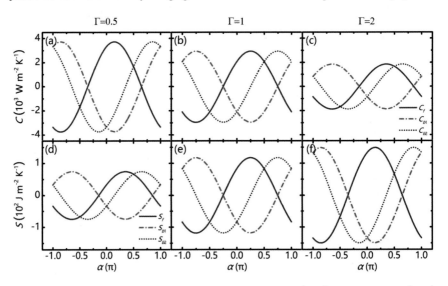

Fig. 14.2 C and S as functions of α. Parameters: $\sigma_A = 300$ W m^{-1} K^{-1}, $\sigma_B = 100$ W m^{-1} K^{-1}, $\rho_A = 3 \times 10^6$ J m^{-3} K^{-1}, $\rho_B = 5 \times 10^5$ J m^{-3} K^{-1}, and $u = 0.05$ m/s. Adapted from Ref. [35]

the nonreciprocity condition for the backward-1 case is $C \neq 0$ (or $S \neq 0$), and that for the backward-2 case is $C(u) \neq C(-u)$ [or $S(u) \neq S(-u)$]. Especially when $\alpha = 0$, $C(-u) = -C(u)$ [or $S(-u) = -S(u)$], the nonreciprocity condition for the backward-2 case can then be reduced to $C \neq 0$ (or $S \neq 0$), which is the same as that for the backward-1 case.

We then consider a transient case that can support thermal waves' propagation. Qualitative analysis is insufficient since both C and S can contribute to nonreciprocity. Therefore, we quantitatively discuss a rectification ratio. For this purpose, we apply a periodic temperature at the left side of the structure in Fig. 14.1a to generate a forward thermal wave described by Eq. (14.3). The periodic temperature has a form of $T_p = \phi_0 e^{-i\omega t} + T_0$ where ϕ_0 denote the temperature amplitude. We set the reference temperature $T_0 = 0$ K in theoretical discussions for brevity. The envelope line of the actual temperature T can then be expressed as

$$\tilde{T} = \phi_0 e^{i(kx - \omega t)}. \tag{14.6}$$

The real part of Eq. (14.6) makes sense, which has been experimentally realized by periodically heating a material [24, 25]. The substitution of Eq. (14.6) into Eq. (14.4) yields

$$-i\omega\tilde{\rho} + ikC + k^2\tilde{\sigma} - \omega k S = 0. \tag{14.7}$$

Since thermal conduction features dissipation, the wave number k should be complex, i.e., $k = \mu + i\xi$ with μ and ξ being two real numbers. Equation (14.6) can then be rewritten as $\tilde{T} = \phi_0 e^{-\xi x} e^{i(\mu x - \omega t)}$. Therefore, the physical meaning of μ is the wave number and that of ξ is the spatial decay rate. With the complex k, Eq. (14.7) can be further reduced to

$$-i\omega\tilde{\rho} + i(\mu + i\xi)C + (\mu + i\xi)^2\tilde{\sigma} - \omega(\mu + i\xi)S = 0. \tag{14.8}$$

By independently considering the real and imaginary parts of Eq. (14.8), we can derive two equations,

$$-\xi C + (\mu^2 - \xi^2)\tilde{\sigma} - \omega\mu S = 0, \tag{14.9a}$$

$$\omega\tilde{\rho} - \mu C - 2\mu\xi\tilde{\sigma} + \omega\xi S = 0. \tag{14.9b}$$

The solution to Eq. (14.9) is

$$\mu = \frac{2S\omega + \sqrt{2}\varepsilon}{4\tilde{\sigma}}, \tag{14.10a}$$

$$\xi = \frac{-4C\omega(2\tilde{\sigma}\tilde{\rho} - CS) + 2\sqrt{2}(C^2 - S^2\omega^2)\varepsilon + \sqrt{2}\varepsilon^3}{8\tilde{\sigma}(2\tilde{\sigma}\tilde{\rho} - CS)\omega}, \tag{14.10b}$$

with $\varepsilon = \sqrt{-C^2 + S^2\omega^2 + \sqrt{\left(C^2 + S^2\omega^2\right)^2 + 16\omega^2\tilde{\sigma}\tilde{\rho}\left(\tilde{\sigma}\tilde{\rho} - CS\right)}}$.   Although
Eq. (14.10) is complicated, we can discuss some special conditions to have a
rough idea. For the forward case, we can know $\mu_f = \mu$ and $\xi_f = \xi$. For the
backward-1 case, we can derive $\mu_{b1} = \mu(-C, -S)$ and $\xi_{b1} = \xi(-C, -S)$. Due
to $\varepsilon(C, S) = \varepsilon(-C, -S)$, it does contribute to nonreciprocity, so the nonreciprocity origins of μ and ξ lie in S and C, respectively (Eq. (14.10)). We can then conclude
that nonreciprocal μ requires $S \neq 0$ (i.e., $\alpha \neq \operatorname{arccot}\Gamma \pm \pi/2$) and nonreciprocal ξ
requires $C \neq 0$ (i.e., $\alpha \neq \arctan\Gamma \pm \pi/2$). For the backward-2 case, we can derive
$\mu_{b2} = \mu[C(-u), S(-u)]$ and $\xi_{b2} = \xi[C(-u), S(-u)]$. When $\alpha = \pm\pi/2$, C, S,
and ε are all even functions of u, so nonreciprocity will disappear. Therefore, nonreciprocal μ (or ξ) requires $\alpha \neq \pm\pi/2$.

In general, it makes little sense to define a rectification ratio based on wave numbers. However, it is meaningful to define a rectification ratio (R_T) based on the
temperature amplitude $\left(\phi_0 e^{-\xi x}\right)$ or the reciprocal of spatial decay rate ($1/\xi$),

$$R_{T1} = \frac{1/\xi_f - 1/\xi_{b1}}{1/\xi_f + 1/\xi_{b1}} = \frac{\xi_{b1} - \xi_f}{\xi_{b1} + \xi_f}, \tag{14.11a}$$

$$R_{T2} = \frac{1/\xi_f - 1/\xi_{b2}}{1/\xi_f + 1/\xi_{b2}} = \frac{\xi_{b2} - \xi_f}{\xi_{b2} + \xi_f}, \tag{14.11b}$$

where R_{T1} and R_{T2} are defined for the backward-1 and backward-2 cases, respectively. We plot R_{T1} and R_{T2} as functions of α in Fig. 14.3. The results demonstrate
that a smaller Γ or a smaller ω yields larger nonreciprocity. Therefore, both R_{T1}
and R_{T2} can theoretically reach 1, and we can obtain a perfect thermal wave diode.
Especially when $\alpha = 0$, Eq. (14.11) can be reduce to

$$R_{T1} = R_{T2} = \frac{2\sqrt{2}C\omega\left(2\tilde{\sigma}\tilde{\rho} - CS\right)}{2\left(C^2 - S^2\omega^2\right)\varepsilon + \varepsilon^3}, \tag{14.12}$$

indicating that the two backward cases are equivalent when $\alpha = 0$.

Another possibility to define a rectification ratio (R_J) lies in nonreciprocal heat
fluxes J. For this purpose, we define the dynamic heat flux J according to Eq. (14.1),

$$J = -\sigma(x - ut)\frac{\partial T}{\partial x} = -\sigma(x - ut)\frac{\partial}{\partial x}\left[\phi(x - ut)e^{-\xi x}e^{i(\mu x - \omega t)}\right]$$
$$= -\sigma(x - ut)\left[\phi'(x - ut) + (-\xi + i\mu)\phi(x - ut)\right]e^{-\xi x}e^{i(\mu x - \omega t)}, \tag{14.13}$$

where $\phi'(x - ut) = \partial\phi(x - ut)/\partial x$. Since $\sigma(x - ut)$, $\phi(x - ut)$, and $\phi'(x - ut)$
are all periodic functions, the dynamic heat flux described by Eq. (14.13) varies with
temporal periodicity, but the heat flux amplitude decays along the x axis due to the
term of $e^{-\xi x}$. Therefore, we can also define R_J based on the reciprocal of spatial
decay rate ($1/\xi$) which should have the same form as Eq. (14.11), indicating that the
whole theoretical framework is self-consistent.

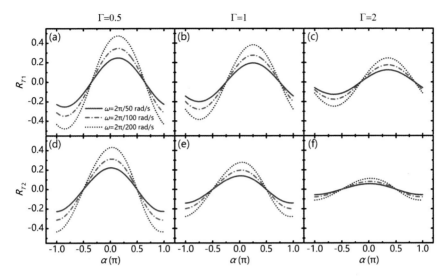

Fig. 14.3 R_{T1} and R_{T2} as functions of α. The parameters are the same as those for Fig. 14.2. Adapted from Ref. [35]

We can then draw a brief conclusion. Spatiotemporal modulation can generate two additional terms: the convective term associated with C and the Willis term related to S. C and S can be flexibly tuned by α. We also discuss two backward cases: (I) changing the source position and (II) changing the modulation direction, equivalent only when $\alpha = 0$. We further discuss their nonreciprocity conditions and define a rectification ratio (R_T or R_J) based on the reciprocal of spatial decay rate ($1/\xi$).

14.3 Finite-Element Simulation

We then perform simulations with COMSOL Multiphysics to confirm the theoretical analyses. For this purpose, we study the thermal conduction in a one-dimensional structure whose parameters are spatiotemporally modulated as described by Eq. (14.2) with $\Gamma = 1$. For accuracy, the mesh size is one-tenth of the modulation wavelength (γ), and the time tolerance is 10^{-6}.

We firstly discuss the backward-1 case, which requires changing the source position but keeping the modulation direction (Fig. 14.1b). As theoretically predicted (Eq. (14.11)), $R_{T1} = 0$ occurs when $\alpha = \pi/4 \pm \pi/2$. For brevity, we set $\alpha = -\pi/4$ to perform simulations. The temperature and heat flux evolutions are presented in Fig. 14.4a, c, respectively. We can observe that the forward and backward-1 propagations are the same, indicating reciprocal propagations. Moreover, R_{T1} reaches the maximum value when $\alpha = \pi/4$ as predicted. We also perform simulations with $\alpha = \pi/4$, and the results are presented in Fig. 14.4b, d. The temperature amplitudes

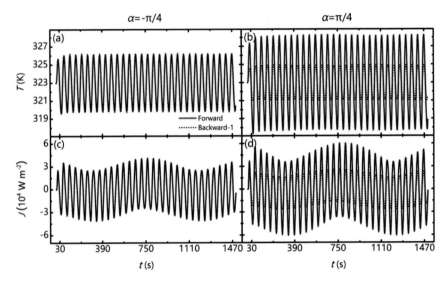

Fig. 14.4 Simulations of the backward-1 case. The parameters are the same as those for Fig. 14.2 with $\Gamma = 1$ and $L = 0.2$ m. The periodic temperature is set at $T_p = 40\cos(-2\pi t/50) + 323$ K. The detected position locates at the center of the structure. **a** and **b** Temperature evolution. **c** and **d** Heat flux evolution. Adapted from Ref. [35]

are different, indicating nonreciprocal propagations. The theoretical prediction of the forward and backward-1 temperature amplitudes are 5.13 and 1.87 K, respectively. The simulations show that the forward and backward-1 temperature amplitudes are 5.23 and 1.92 K, respectively. Therefore, the simulations agree well with the theoretical predictions.

We then discuss the backward-2 case, which requires changing the modulation direction but keeping the source position (Fig. 14.1c). Equation (14.11) tells that $R_{T2} = 0$ appears when $\alpha = \pm\pi/2$, and we set $\alpha = -\pi/2$ to perform simulations (Fig. 14.5a, c). The forward and backward-2 propagations have the same temperature (or heat flux) amplitudes, indicating reciprocal thermal waves. In addition, R_{T2} reaches the maximum value when $\alpha = 0$ as predicted, and the simulation results are presented in Fig. 14.5b, d. The theoretical prediction of the forward and backward-2 temperature amplitudes are 4.48 and 2.19 K, respectively. The simulations demonstrate that the forward and backward-1 temperature amplitudes are 4.53 and 2.24 K, respectively. Again, the simulations and theories have good agreement.

We finally provide some experimental suggestions to ensure the feasibility of practical implementations. The most crucial is to realize spatiotemporal modulations of σ (thermal conductivity) and ρ (the product of mass density and heat capacity). We firstly discuss the spatiotemporal modulation of σ. Many studies have shown that thermal conductivities can be flexibly controlled by external fields like electric fields [36, 37] and light fields [38]. The in-plane thermal conductivity can change two orders of magnitude with an out-of-plane electric field [36]. We then discuss the

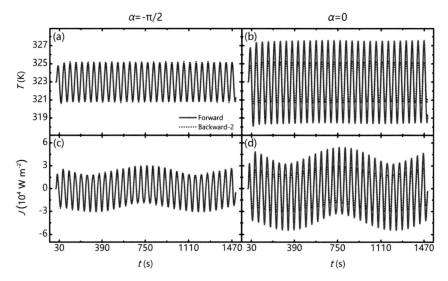

Fig. 14.5 Simulations of the backward-2 case. The parameters are the same as those for Fig. 14.4. The difference from Fig. 14.4 is that here we change the modulation speed instead of changing the source position. Adapted from Ref. [35]

spatiotemporal modulation of ρ by considering heat capacity. Many materials have a phase change [39] in the presence of an electric field, so heat capacities change with the phase change. Therefore, spatiotemporal modulations of σ and ρ can be realized with an electric field. Moreover, Ref. [19] also provides an insight into practical implementations, although the experiments were conducted in electrics. Since thermotics and electrics follow similar equations (thermal conductivity corresponds to electric conductivity and heat capacity corresponds to electric capacity), spatiotemporal modulations of σ and ρ might also be realized by rotating disks, as presented in Ref. [19]. A periodic temperature can be obtained by directly using a pulse heat source or alternately using a ceramic heater and a semiconductor cooler. Therefore, these results should be possible to be experimentally validated. Here, thermal waves are based on the Fourier law, and many other kinds of thermal waves remain further explored, i.e., those considering thermal relaxation [40–43].

14.4 Conclusion

We propose the mechanism of tunable thermal wave nonreciprocity with spatiotemporal modulation. The tunability lies in the phase difference (α) between two spatiotemporally modulated parameters. We reveal that the homogenized thermal conductivity ($\tilde{\sigma}$) and the homogenized product of mass density and heat capacity ($\tilde{\rho}$) are independent of the phase difference (α). Still, the convective term (C) and the

Willis term (S) are crucially dependent on the phase difference (α). We also discuss two backward cases: (I) changing the source position and (II) changing the modulation direction. The two cases are equivalent only when $\alpha = 0$. We further define a rectification ratio (R_{T1} or R_{T2}) based on the reciprocal of spatial decay rate ($1/\xi$) and discuss nonreciprocity conditions. These theoretical analyses are all confirmed by finite-element simulations, and experimental suggestions are also given to ensure feasibility. These results could provide distinct opportunities for nonreciprocal heat transfer.

14.5   Exercise and Solution

Exercise

1. Derive Eq. (14.5) by homogenizing spatiotemporal modulation.

Solution

1. We consider two variable substitutions of $n = x - ut$ and $\tau = t$, yielding $\partial/\partial x = \partial/\partial n$ and $\partial/\partial t = \partial/\partial \tau - u \partial/\partial n$. Equation (14.1) can then be reduced to

$$\rho(n) \frac{\partial T}{\partial \tau} - u\rho(n) \frac{\partial T}{\partial n} + \frac{\partial}{\partial n}\left[-\sigma(n) \frac{\partial T}{\partial n}\right] = 0. \qquad (14.14)$$

Similarly, Eq. (14.2) can also be simplified as

$$\sigma(n) = \sigma_A + \sigma_B \cos(Kn), \qquad (14.15a)$$

$$\rho(n) = \rho_A + \rho_B \cos(Kn + \alpha). \qquad (14.15b)$$

We rewrite Eq. (14.15) with the Fourier expansion,

$$\sigma(n) = \sum_{s=0,\pm 1} \sigma_s e^{iK_s n} = \sigma_0 e^{iK_0 n} + \sigma_{+1} e^{iK_{+1} n} + \sigma_{-1} e^{iK_{-1} n}, \qquad (14.16a)$$

$$\rho(n) = \sum_{s=0,\pm 1} \rho_s e^{iK_s n} = \rho_0 e^{iK_0 n} + \rho_{+1} e^{iK_{+1} n} + \rho_{-1} e^{iK_{-1} n}, \qquad (14.16b)$$

with $K_0 = 0$, $K_{\pm 1} = \pm K$, $\sigma_0 = \sigma_A$, $\sigma_{\pm 1} = \sigma_B/2$, $\rho_0 = \rho_A$, and $\rho_{\pm 1} = e^{\pm i\alpha} \rho_B/2$. With the Bloch theorem, we can express the temperature solution as

$$T(n, \tau) = \phi(n) e^{i(Gn - W\tau)} = \left(\sum_{s=0,\pm 1} \phi_s e^{iK_s n}\right) e^{i(Gn - W\tau)}$$

$$= \left(\phi_0 e^{iK_0 n} + \phi_{+1} e^{iK_{+1} n} + \phi_{-1} e^{iK_{-1} n}\right) e^{i(Gn - W\tau)}, \qquad (14.17)$$

where G and W are the wave number and circular frequency in the $n - \tau$ frame. $\phi(n)$ is the amplitude modulation function.

We can then express $\partial T/\partial \tau$ and $\partial T/\partial n$ as

$$\frac{\partial T}{\partial \tau} = -\mathrm{i}W\left(\phi_0 e^{\mathrm{i}K_0 n} + \phi_{+1} e^{\mathrm{i}K_{+1}n} + \phi_{-1} e^{\mathrm{i}K_{-1}n}\right) e^{\mathrm{i}(Gn-W\tau)}, \tag{14.18}$$

$$\frac{\partial T}{\partial n} = \mathrm{i}\left[(G + K_0)\,\phi_0 e^{\mathrm{i}K_0 n} + (G + K_{+1})\,\phi_{+1} e^{\mathrm{i}K_{+1}n} + (G + K_{-1})\,\phi_{-1} e^{\mathrm{i}K_{-1}n}\right] e^{\mathrm{i}(Gn-W\tau)}. \tag{14.19}$$

We can further write $\rho(n)\,\partial T/\partial \tau$ as

$$\begin{aligned}
\rho(n)\,\frac{\partial T}{\partial \tau} = &-\mathrm{i}W\left(\rho_0\phi_0 + \rho_{+1}\phi_{-1} + \rho_{-1}\phi_{+1}\right) e^{\mathrm{i}K_0 n} e^{\mathrm{i}(Gn-W\tau)} \\
&-\mathrm{i}W\left(\rho_0\phi_{+1} + \rho_{+1}\phi_0\right) e^{\mathrm{i}K_{+1}n} e^{\mathrm{i}(Gn-W\tau)} \\
&-\mathrm{i}W\left(\rho_0\phi_{-1} + \rho_{-1}\phi_0\right) e^{\mathrm{i}K_{-1}n} e^{\mathrm{i}(Gn-W\tau)} \\
&+ o\left(e^{\mathrm{i}K_{\pm 1}n}\right).
\end{aligned} \tag{14.20}$$

We can also express $-u\rho(n)\,\partial T/\partial n$ and $-\sigma(n)\,\partial T/\partial n$ as

$$\begin{aligned}
-u\rho(n)\,\frac{\partial T}{\partial n} = &-\mathrm{i}u\left[\rho_0(G + K_0)\,\phi_0 + \rho_{+1}(G + K_{-1})\,\phi_{-1} + \rho_{-1}(G + K_{+1})\,\phi_{+1}\right] e^{\mathrm{i}K_0 n} e^{\mathrm{i}(Gn-W\tau)} \\
&-\mathrm{i}u\left[\rho_0(G + K_{+1})\,\phi_{+1} + \rho_{+1}(G + K_0)\,\phi_0\right] e^{\mathrm{i}K_{+1}n} e^{\mathrm{i}(Gn-W\tau)} \\
&-\mathrm{i}u\left[\rho_0(G + K_{-1})\,\phi_{-1} + \rho_{-1}(G + K_0)\,\phi_0\right] e^{\mathrm{i}K_{-1}n} e^{\mathrm{i}(Gn-W\tau)} \\
&+ o\left(e^{\mathrm{i}K_{\pm 1}n}\right),
\end{aligned} \tag{14.21}$$

$$\begin{aligned}
-\sigma(n)\,\frac{\partial T}{\partial n} = &-\mathrm{i}\left[\sigma_0(G + K_0)\,\phi_0 + \sigma_{+1}(G + K_{-1})\,\phi_{-1} + \sigma_{-1}(G + K_{+1})\,\phi_{+1}\right] e^{\mathrm{i}K_0 n} e^{\mathrm{i}(Gn-W\tau)} \\
&-\mathrm{i}\left[\sigma_0(G + K_{+1})\,\phi_{+1} + \sigma_{+1}(G + K_0)\,\phi_0\right] e^{\mathrm{i}K_{+1}n} e^{\mathrm{i}(Gn-W\tau)} \\
&-\mathrm{i}\left[\sigma_0(G + K_{-1})\,\phi_{-1} + \sigma_{-1}(G + K_0)\,\phi_0\right] e^{\mathrm{i}K_{-1}n} e^{\mathrm{i}(Gn-W\tau)} \\
&+ o\left(e^{\mathrm{i}K_{\pm 1}n}\right).
\end{aligned} \tag{14.22}$$

With Eq. (14.22), we can further derive

$$\frac{\partial}{\partial n}\left[-\sigma\left(n\right)\frac{\partial T}{\partial n}\right]$$

$$= (G + K_0)\left[\sigma_0\left(G + K_0\right)\phi_0 + \sigma_{+1}\left(G + K_{-1}\right)\phi_{-1} + \sigma_{-1}\left(G + K_{+1}\right)\phi_{+1}\right]e^{iK_0 n}e^{i(Gn - W\tau)}$$

$$+ (G + K_{+1})\left[\sigma_0\left(G + K_{+1}\right)\phi_{+1} + \sigma_{+1}\left(G + K_0\right)\phi_0\right]e^{iK_{+1} n}e^{i(Gn - W\tau)}$$

$$+ (G + K_{-1})\left[\sigma_0\left(G + K_{-1}\right)\phi_{-1} + \sigma_{-1}\left(G + K_0\right)\phi_0\right]e^{iK_{-1} n}e^{i(Gn - W\tau)}$$

$$+ o\left(e^{iK_{\pm 1} n}\right). \tag{14.23}$$

By arranging the terms associated with $e^{iK_0 n}$, $e^{iK_{+1} n}$, and $e^{iK_{-1} n}$ in Eqs. (14.20), (14.21) and (14.23) together, we can obtain three equations,

$$-i\left[\rho_0\left(W + uG + uK_0\right)\phi_0 + \rho_{+1}\left(W + uG + uK_{-1}\right)\phi_{-1} + \rho_{-1}\left(W + uG + uK_{+1}\right)\phi_{+1}\right]$$
$$+ (G + K_0)\left[\sigma_0\left(G + K_0\right)\phi_0 + \sigma_{+1}\left(G + K_{-1}\right)\phi_{-1} + \sigma_{-1}\left(G + K_{+1}\right)\phi_{+1}\right] = 0, \tag{14.24a}$$

$$-i\left[\rho_0\left(W + uG + uK_{+1}\right)\phi_{+1} + \rho_{+1}\left(W + uG + uK_0\right)\phi_0\right]$$
$$+ (G + K_{+1})\left[\sigma_0\left(G + K_{+1}\right)\phi_{+1} + \sigma_{+1}\left(G + K_0\right)\phi_0\right] = 0, \tag{14.24b}$$
$$-i\left[\rho_0\left(W + uG + uK_{-1}\right)\phi_{-1} + \rho_{-1}\left(W + uG + uK_0\right)\phi_0\right]$$
$$+ (G + K_{-1})\left[\sigma_0\left(G + K_{-1}\right)\phi_{-1} + \sigma_{-1}\left(G + K_0\right)\phi_0\right] = 0. \tag{14.24c}$$

Equation (14.24) is written in the $n - \tau$ frame, and we can also express it in the $x - t$ frame by taking $k = G$ and $\omega = W + uG$ where k and ω are, respectively, the wave vector and circular frequency in the $x - t$ frame,

$$-i\left[\rho_0\left(\omega + uK_0\right)\phi_0 + \rho_{+1}\left(\omega + uK_{-1}\right)\phi_{-1} + \rho_{-1}\left(\omega + uK_{+1}\right)\phi_{+1}\right]$$
$$+ (k + K_0)\left[\sigma_0\left(k + K_0\right)\phi_0 + \sigma_{+1}\left(k + K_{-1}\right)\phi_{-1} + \sigma_{-1}\left(k + K_{+1}\right)\phi_{+1}\right] = 0, \tag{14.25a}$$

$$-i\left[\rho_0\left(\omega + uK_{+1}\right)\phi_{+1} + \rho_{+1}\left(\omega + uK_0\right)\phi_0\right]$$
$$+ (k + K_{+1})\left[\sigma_0\left(k + K_{+1}\right)\phi_{+1} + \sigma_{+1}\left(k + K_0\right)\phi_0\right] = 0, \tag{14.25b}$$
$$-i\left[\rho_0\left(\omega + uK_{-1}\right)\phi_{-1} + \rho_{-1}\left(\omega + uK_0\right)\phi_0\right]$$
$$+ (k + K_{-1})\left[\sigma_0\left(k + K_{-1}\right)\phi_{-1} + \sigma_{-1}\left(k + K_0\right)\phi_0\right] = 0. \tag{14.25c}$$

With Eqs. (14.25b) and (14.25c), we can derive the expressions of ϕ_{+1} and ϕ_{-1},

$$\phi_{+1} = -\frac{(k + K_{+1})\sigma_{+1}\left(k + K_0\right) - i\rho_{+1}\left(\omega + uK_0\right)}{(k + K_{+1})\sigma_0\left(k + K_{+1}\right) - i\rho_0\left(\omega + uK_{+1}\right)}\phi_0, \tag{14.26a}$$

$$\phi_{-1} = -\frac{(k + K_{-1})\sigma_{-1}\left(k + K_0\right) - i\rho_{-1}\left(\omega + uK_0\right)}{(k + K_{-1})\sigma_0\left(k + K_{-1}\right) - i\rho_0\left(\omega + uK_{-1}\right)}\phi_0. \tag{14.26b}$$

We then consider two approximations of $k \ll K$ and $\omega \ll uK$, so Eq. (14.26) can be reduced to

$$\phi_{+1} = -\frac{K_{+1}\sigma_{+1}k - i\rho_{+1}\omega}{K_{+1}\sigma_0 K_{+1} - i\rho_0 u K_{+1}}\phi_0, \tag{14.27a}$$

$$\phi_{-1} = -\frac{K_{-1}\sigma_{-1}k - i\rho_{-1}\omega}{K_{-1}\sigma_0 K_{-1} - i\rho_0 u K_{-1}}\phi_0. \tag{14.27b}$$

Similarly, Eq. (14.25a) can also be reduced to

$$k^2\sigma_0\phi_0 - i\omega\rho_0\phi_0 + (k\sigma_{-1}K_{+1} - i\rho_{-1}uK_{+1})\phi_{+1} + (k\sigma_{+1}K_{-1} - i\rho_{+1}uK_{-1})\phi_{-1} = 0. \tag{14.28}$$

The substitution of Eq. (14.27) into Eq. (14.28) yields

$$k^2\sigma_0\phi_0 - i\omega\rho_0\phi_0 - \frac{(k\sigma_{-1}K_{+1} - i\rho_{-1}uK_{+1})(K_{+1}\sigma_{+1}k - i\rho_{+1}\omega)}{K_{+1}\sigma_0 K_{+1} - i\rho_0 u K_{+1}}\phi_0$$
$$- \frac{(k\sigma_{+1}K_{-1} - i\rho_{+1}uK_{-1})(K_{-1}\sigma_{-1}k - i\rho_{-1}\omega)}{K_{-1}\sigma_0 K_{-1} - i\rho_0 u K_{-1}}\phi_0 = 0. \tag{14.29}$$

Equation (14.29) can be further arranged in a physical form,

$$-i\omega\left(\rho_0 + \frac{iKu\rho_{+1}\rho_{-1}}{\sigma_0 K^2 - i\rho_0 u K} + \frac{-iKu\rho_{+1}\rho_{-1}}{\sigma_0 K^2 + i\rho_0 u K}\right)\phi_0$$
$$+ ik\left(\frac{K^2 u\sigma_{+1}\rho_{-1}}{\sigma_0 K^2 - i\rho_0 u K} + \frac{K^2 u\sigma_{-1}\rho_{+1}}{\sigma_0 K^2 + i\rho_0 u K}\right)\phi_0$$
$$+ k^2\left(\sigma_0 - \frac{K^2\sigma_{+1}\sigma_{-1}}{\sigma_0 K^2 - i\rho_0 u K} - \frac{K^2\sigma_{+1}\sigma_{-1}}{\sigma_0 K^2 + i\rho_0 u K}\right)\phi_0$$
$$- \omega k\left(\frac{-iK\rho_{+1}\sigma_{-1}}{\sigma_0 K^2 - i\rho_0 u K} + \frac{iK\rho_{-1}\sigma_{+1}}{\sigma_0 K^2 + i\rho_0 u K}\right)\phi_0 = 0. \tag{14.30}$$

By taking $\partial/\partial t = -i\omega$, $\partial/\partial x = ik$, and $\tilde{T} = \phi_0 e^{i(kx-\omega t)}$, we can rewrite Eq. (14.30) as

$$\tilde{\rho}\frac{\partial\tilde{T}}{\partial t} + C\frac{\partial\tilde{T}}{\partial x} - \tilde{\sigma}\frac{\partial^2\tilde{T}}{\partial x^2} - S\frac{\partial^2\tilde{T}}{\partial x\partial t} = 0, \tag{14.31}$$

where the homogenized parameters take the form of

$$\tilde{\sigma} = \sigma_0 - \frac{K^2\sigma_{+1}\sigma_{-1}}{\sigma_0 K^2 - i\rho_0 u K} - \frac{K^2\sigma_{+1}\sigma_{-1}}{\sigma_0 K^2 + i\rho_0 u K}, \tag{14.32a}$$

$$\tilde{\rho} = \rho_0 + \frac{iKu\rho_{+1}\rho_{-1}}{\sigma_0 K^2 - i\rho_0 u K} + \frac{-iKu\rho_{+1}\rho_{-1}}{\sigma_0 K^2 + i\rho_0 u K}, \tag{14.32b}$$

$$C = \frac{K^2 u\sigma_{+1}\rho_{-1}}{\sigma_0 K^2 - i\rho_0 u K} + \frac{K^2 u\sigma_{-1}\rho_{+1}}{\sigma_0 K^2 + i\rho_0 u K}, \tag{14.32c}$$

$$S = \frac{-iK\sigma_{-1}\rho_{+1}}{\sigma_0 K^2 - i\rho_0 u K} + \frac{iK\sigma_{+1}\rho_{-1}}{\sigma_0 K^2 + i\rho_0 u K}. \tag{14.32d}$$

We can further reduce Eq. (14.32) to

$$\tilde{\sigma} = \sigma_A \left(1 - \frac{\sigma_B^2}{2\sigma_A^2}\frac{1}{1+\Gamma^2}\right), \tag{14.33a}$$

$$\tilde{\rho} = \rho_A \left(1 - \frac{\rho_B^2}{2\rho_A^2}\frac{\Gamma^2}{1+\Gamma^2}\right), \tag{14.33b}$$

$$C = u\frac{\sigma_B \rho_B}{2\sigma_A}\frac{1}{1+\Gamma^2}\left(\cos\alpha + \Gamma\sin\alpha\right), \tag{14.33c}$$

$$S = \frac{1}{u}\frac{\sigma_B \rho_B}{2\rho_A}\frac{\Gamma^2}{1+\Gamma^2}\left(\cos\alpha + \frac{1}{\Gamma}\sin\alpha\right), \tag{14.33d}$$

with $\Gamma = \rho_A u \gamma / (2\pi\sigma_A)$.

References

1. Yu, Z.F., Fan, S.H.: Complete optical isolation created by indirect interband photonic transitions. Nat. Photonics **3**, 91 (2009)
2. Sounas, D.L., Caloz, C., Alù, A.: Giant non-reciprocity at the subwavelength scale using angular momentum-biased metamaterials. Nat. Commun. **4**, 2407 (2013)
3. Sounas, D.L., Alù, A.: Angular-momentum-biased nanorings to realize magnetic-free integrated optical isolation. ACS Photonics **1**, 198 (2014)
4. Sounas, D.L., Alù, A.: Non-reciprocal photonics based on time modulation. Nat. Photonics **11**, 774 (2017)
5. Mock, A., Sounas, D., Alù, A.: Magnet-free circulator based on spatiotemporal modulation of photonic crystal defect cavities. ACS Photonics **6**, 2056 (2019)
6. Fleury, R., Sounas, D.L., Alù, A.: Subwavelength ultrasonic circulator based on spatiotemporal modulation. Phys. Rev. B **91**, 174306 (2015)
7. Shen, C., Li, J.F., Jia, Z.T., Xie, Y.B., Cummer, S.A.: Nonreciprocal acoustic transmission in cascaded resonators via spatiotemporal modulation. Phys. Rev. B **99**, 134306 (2019)
8. Shen, C., Zhu, X.H., Li, J.F., Cummer, S.A.: Nonreciprocal acoustic transmission in space-time modulated coupled resonators. Phys. Rev. B **100**, 054302 (2019)
9. Zhu, X.H., Li, J.F., Shen, C., Peng, X.Y., Song, A.L., Li, L.Q., Cummer, S.A.: Non-reciprocal acoustic transmission via space-time modulated membranes. Appl. Phys. Lett. **116**, 034101 (2020)
10. Zang, J.W., Correas-Serrano, D., Do, J.T.S., Liu, X., Alvarez-Melcon, A., Gomez-Diaz, J.S.: Nonreciprocal wavefront engineering with time-modulated gradient metasurfaces. Phys. Rev. Appl. **11**, 054054 (2019)
11. Guo, X.X., Ding, Y.M., Duan, Y., Ni, X.J.: Nonreciprocal metasurface with space-time phase modulation. Light-Sci. Appl. **8**, 123 (2019)
12. Wu, Q., Chen, H., Nassar, H., Huang, G.L.: Non-reciprocal Rayleigh wave propagation in space-time modulated surface. J. Mech. Phys. Solids **146**, 104196 (2021)
13. Wallen, S.P., Haberman, M.R.: Nonreciprocal wave phenomena in spring-mass chains with effective stiffness modulation induced by geometric nonlinearity. Phys. Rev. E **99**, 013001 (2019)

14. Chen, Y.Y., Li, X.P., Nassar, H., Norris, A.N., Daraio, C., Huang, G.L.: Nonreciprocal wave propagation in a continuum-based metamaterial with space-time modulated resonators. Phys. Rev. Appl. **11**, 064052 (2019)
15. Li, H.N., Moussa, H., Sounas, D., Alù, A.: Parity-time symmetry based on time modulation. Phys. Rev. Appl. **14**, 031002 (2020)
16. Ramakrishnan, V., Frazier, M.J.: Transition waves in multi-stable metamaterials with space-time modulated potentials. Appl. Phys. Lett. **117**, 151901 (2020)
17. Edwards, B., Engheta, N.: Asymmetrical diffusion through time-varying material parameters. In: Conference on Lasers and Electro-Optics JTu5A.34. Optical Society of America (2017)
18. Torrent, D., Poncelet, O., Batsale, J.-C.: Nonreciprocal thermal material by spatiotemporal modulation. Phys. Rev. Lett. **120**, 125501 (2018)
19. Camacho, M., Edwards, B., Engheta, N.: Achieving asymmetry and trapping in diffusion with spatiotemporal metamaterials. Nat. Commun. **11**, 3733 (2020)
20. Farhat, M., Chen, P.-Y., Bagci, H., Amra, C., Guenneau, S., Alù, A.: Thermal invisibility based on scattering cancellation and mantle cloaking. Sci. Rep. **5**, 9876 (2015)
21. Mulaveesala, R., Tuli, S.: Theory of frequency modulated thermal wave imaging for nondestructive subsurface defect detection. Appl. Phys. Lett. **89**, 191913 (2006)
22. Mulaveesala, R., Tuli, S.: Applications of frequency modulated thermal wave imaging for nondestructive characterization. AIP Conf. Proc. **1004**, 15 (2008)
23. Tuli, S., Chatterjee, K.: Frequency modulated thermal wave imaging. AIP Conf. Proc. **1430**, 523 (2012)
24. Li, Y., Peng, Y.-G., Han, L., Miri, M.-A., Li, W., Xiao, M., Zhu, X.-F., Zhao, J.L., Alù, A., Fan, S.H., Qiu, C.-W.: Anti-parity-time symmetry in diffusive systems. Science **364**, 170 (2019)
25. Xu, L.J., Wang, J., Dai, G.L., Yang, S., Yang, F.B., Wang, G., Huang, J.P.: Geometric phase, effective conductivity enhancement, and invisibility cloak in thermal convection-conduction. Int. J. Heat Mass Transf. **165**, 120659 (2021)
26. Cao, P.C., Li, Y., Peng, Y.G., Qiu, C.W., Zhu, X.F.: High-order exceptional points in diffusive systems: Robust APT symmetry against perturbation and phase oscillation at APT symmetry breaking. ES Energy Environ. **7**, 48 (2020)
27. Xu, L.J., Dai, G.L., Wang, G., Huang, J.P.: Geometric phase and bilayer cloak in macroscopic particle-diffusion systems. Phys. Rev. E **102**, 032140 (2020)
28. Xu, L.J., Huang, J.P.: Negative thermal transport in conduction and advection. Chin. Phys. Lett. **37**, 080502 (2020)
29. Farhat, M., Guenneau, S., Chen, P.-Y., Alù, A., Salama, K.N.: Scattering cancellation-based cloaking for the Maxwell-Cattaneo heat waves. Phys. Rev. Appl. **11**, 044089 (2019)
30. Xu, L.J., Huang, J.P.: Controlling thermal waves with transformation complex thermotics. Int. J. Heat Mass Transf. **159**, 120133 (2020)
31. Xu, L.J., Huang, J.P.: Active thermal wave cloak. Chin. Phys. Lett. **37**, 120501 (2020)
32. Chen, A.-L., Li, Z.-Y., Ma, T.-X., Li, X.-S., Wang, Y.-S.: Heat reduction by thermal wave crystals. Int. J. Heat Mass Transf. **121**, 215 (2018)
33. Xu, L.J., Huang, J.P.: Thermal convection-diffusion crystal for prohibition and modulation of wave-like temperature profiles. Appl. Phys. Lett. **117**, 011905 (2020)
34. Gandolfi, M., Giannetti, C., Banfi, F.: Temperonic crystal: a superlattice for temperature waves in graphene. Phys. Rev. Lett. **125**, 265901 (2020)
35. Xu, L.J., Huang, J.P., Ouyang, X.P.: Tunable thermal wave nonreciprocity by spatiotemporal modulation. Phys. Rev. E **103**, 032128 (2021)
36. Qin, G.Z., Qin, Z.Z., Yue, S.-Y., Yan, Q.-B., Hu, M.: External electric field driving the ultra-low thermal conductivity of silicene. Nanoscale **9**, 7227 (2017)
37. Deng, S.C., Yuan, J.L., Lin, Y.L., Yu, X.X., Ma, D.K., Huang, Y.W., Ji, R.C., Zhang, G.Z., Yang, N.: Electric-field-induced modulation of thermal conductivity in poly (vinylidene fluoride). Nano Energy **82**, 105749 (2021)
38. Shin, J., Sung, J., Kang, M., Xie, X., Lee, B., Lee, K.M., White, T.J., Leal, C., Sottos, N.R., Braun, P.V., Cahill, D.G.: Light-triggered thermal conductivity switching in azobenzene polymers. Proc. Natl. Acad. Sci. U. S. A. **116**, 5973 (2019)

39. Lu, N.P., Zhang, P.F., Zhang, Q.H., Qiao, R.M., He, Q., Li, H.-B., Wang, Y.J., Guo, J.W., Zhang, D., Duan, Z., Li, Z.L., Wang, M., Yang, S.Z., Yan, M.Z., Arenholz, E., Zhou, S.Y., Yang, W.L., Gu, L., Nan, C.-W., Wu, J., Tokura, Y., Yu, P.: Electric-field control of tri-state phase transformation with a selective dual-ion switch. Nature **546**, 124 (2017)
40. Joseph, D.D., Preziosi, L.: Heat waves. Rev. Mod. Phys. **61**, 41 (1989)
41. Nie, B.-D., Cao, B.-Y.: Three mathematical representations and an improved ADI method for hyperbolic heat conduction. Int. J. Heat Mass Transf. **135**, 974 (2019)
42. Gandolfi, M., Benetti, G., Glorieux, C., Giannetti, C., Banfi, F.: Accessing temperature waves: a dispersion relation perspective. Int. J. Heat Mass Transf. **143**, 118553 (2019)
43. Simoncelli, M., Marzari, N., Cepellotti, A.: Generalization of Fourier's Law into viscous heat equations. Phys. Rev. X **10**, 011019 (2020)

Chapter 15
Theory for Diffusive Fizeau Drag: Willis Coupling

Abstract In this chapter, we design a spatiotemporal thermal metamaterial based on heat transfer in porous media to demonstrate the diffusive analog to Fizeau drag. The space-related inhomogeneity and time-related advection enable the diffusive Fizeau drag effect. Thanks to the spatiotemporal coupling, different propagating speeds of temperature fields can be observed in two opposite directions, thus facilitating nonreciprocal thermal profiles. The phenomenon of diffusive Fizeau drag stands robustly even when the advection direction is perpendicular to the propagation of temperature fields. These results could pave an unexpected way toward realizing the nonreciprocal and directional transport of mass and energy.

Keywords Diffusive Fizeau drag · Willis coupling · Speed difference

15.1 Opening Remarks

Light travels at different speeds along and against the water flow, theoretically predicted by Fresnel [1] and experimentally verified by Fizeau [2]. This momentous discovery, generally referred to as Fizeau drag, has been well explained by relativistic kinematics. Similar effects have also been revealed in other moving [3, 4] or spatiotemporal [5, 6] media. Recently, two experimental studies have reported plasmonic Fizeau drag by the flow of electrons [7, 8], which results from the nonlinear kinematics of drifting Dirac electrons.

On the other hand, diffusion systems can also exhibit wavelike behaviors [9–14], which provides the possibility to realize diffusive Fizeau drag. However, unlike the dragging of photons and polaritons by the momentum interaction (Fig. 15.1a, b), it is intrinsically challenging to drag the macroscopic heat by the biased advection [16, 17] due to the absence of macroscopic heat momentum (Fig. 15.1c). Therefore, the forward and backward propagating speeds of temperature fields are always identical. Nevertheless, the amplitudes of temperature fields are different in opposite directions due to the dissipative property of heat transfer [18, 19]. Therefore, it is still an extremely challenging problem to realize diffusive Fizeau drag.

© The Author(s) 2023
L.-J. Xu and J.-P. Huang, *Transformation Thermotics and Extended Theories*,
https://doi.org/10.1007/978-981-19-5908-0_15

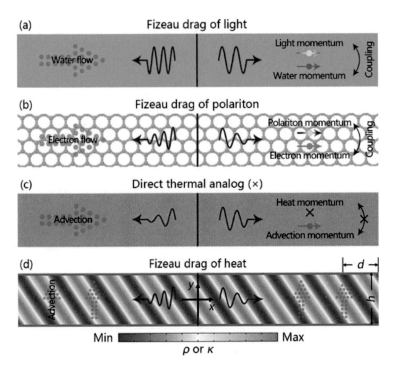

Fig. 15.1 Origin of diffusive Fizeau drag. Fizeau drag of **a** light and **b** polariton by the momentum interaction. **c** Failure of a direct thermal analog due to the lack of macroscopic heat momentum. **d** Fizeau drag of heat in a spatiotemporal thermal metamaterial by thermal Willis coupling. The red arrows contain the information on wave number and amplitude, indicating the forward and backward cases with (**a**), (**b**), (**d**) different wave numbers, and (**c**) different amplitudes. Adapted from Ref. [15]

15.2 Theoretical Foundation

We construct a spatiotemporal thermal metamaterial with space-related inhomogeneity and time-related advection to uncover diffusive Fizeau drag in heat transfer (Fig. 15.1d). Since the characteristic length of spatiotemporal modulation is much smaller than the wavelength of wavelike temperature fields, the proposed structure can be regarded as a metamaterial. Neither periodic inhomogeneity nor vertical advection alone contributes to the horizontal nonreciprocity, but their synergistic effect can give rise to diffusive Fizeau drag. The underlying mechanism lies in the coupling between heat flux and temperature change rate, which can be regarded as the thermal counterpart of Willis coupling in mechanical waves [20–24]. Therefore, the present nonreciprocity is distinctly different from the synthetic-motion-induced nonreciprocity [25, 26].

We first explain why the direct scheme presented in Fig. 15.1c fails. Heat transfer in porous media is described by $\rho_0 \partial_t T + \nabla \cdot (\phi \rho_a \boldsymbol{u} T - \kappa_0 \nabla T) = 0$, where ρ_0 (or ρ_a)

is the product of mass density and heat capacity of the porous medium (or fluid), κ_0 is the thermal conductivity of the porous medium, ϕ is the porosity, and $\boldsymbol{u}$ is the velocity of the fluid with the horizontal and vertical components of u_x and u_y, respectively. We consider a wavelike temperature field described by $T = Ae^{i(\beta x - \omega t)} + T_r$, where β and ω are the wave number and angular frequency, respectively. Here, we use "wavelike" because heat transfer is essentially governed by a diffusive equation rather than a wave equation. We set the temperature field amplitude of A as 1 and the balanced temperature of T_r as 0 for brevity. We apply a periodic source with a temperature of $T(x = 0) = e^{-i\omega t}$, thus leading to a real ω and a complex β. The imaginary part of β reflects the spatial decay rate of wavelike temperature fields. We focus on the real part of β because the propagating speed of wavelike temperature fields can be calculated by $v = \omega/\mathrm{Re}[\beta]$. The substitution of $T = e^{i(\beta x - \omega t)}$ with a preset real ω into the governing equation of heat transfer yields

$$\beta_{f,b} = \pm \frac{\sqrt{2}\gamma}{4\kappa_0} + i\frac{-8\phi\rho_a u_x \omega\rho_0\gamma_0 \pm \sqrt{2}\gamma\left(2\phi_a^{22}u_x^2 + \gamma^2\right)}{16\omega\rho_0\kappa_0^2}, \tag{15.1}$$

where β_f and β_b are, respectively, the forward and backward wavenumbers with a definition of $\gamma = \sqrt{-\phi^2\rho_a^2 u_x^2 + \sqrt{\phi^4\rho_a^4 u_x^4 + 16\omega^2\rho_0^2\kappa_0^2}}$. Since a nonzero u_x cannot generate different $|\mathrm{Re}[\beta]|$ in opposite directions, the forward and backward propagating speeds of temperature fields are identical, i.e., no diffusive Fizeau drag.

To achieve diffusive Fizeau drag, we introduce spatially-periodic inhomogeneity to the porous medium,

$$\rho(\xi) = \rho_0(1 + \Delta_\rho \cos(G\xi + \theta)), \tag{15.2a}$$
$$\kappa(\xi) = \kappa_0(1 + \Delta_\kappa \cos(G\xi)), \tag{15.2b}$$

where Δ_ρ and Δ_κ are the modulation amplitudes, $G = 2\pi/d$ is the modulation wave number, d is the horizontal modulation wavelength, $\xi = x + \zeta y$ is the generalized coordinate with a definition of $\zeta = d/h$, h is the vertical height, and θ is the modulation phase difference. To exclude the captivation that the horizontal advection can generate nonreciprocal amplitudes of temperature fields, as described by the imaginary part of Eq. (15.1), we consider the upward advection with a speed of u_y, which does not contribute to the horizontal nonreciprocity. The governing equation of heat transfer in spatiotemporal thermal metamaterials can be expressed as

$$\overline{\rho}(\xi)\frac{\partial T}{\partial t} + \phi u_y \frac{\partial T}{\partial y} + \frac{\partial}{\partial x}\left(-D_0\overline{\kappa}(\xi)\frac{\partial T}{\partial x}\right) + \frac{\partial}{\partial y}\left(-D_0\overline{\kappa}(\xi)\frac{\partial T}{\partial y}\right) = 0, \tag{15.3}$$

with definitions of $\overline{\rho}(\xi) = \rho(\xi)/\rho_0$, $\overline{\kappa}(\xi) = \kappa(\xi)/\kappa_0$, $\epsilon = \rho_a/\rho_0$, and $D_0 = \kappa_0/\rho_0$.

We further consider a wavelike temperature field with a spatially-periodic modulation,

$$T = F(\xi)e^{i(\beta x - \omega t)} = \left(\sum_s F_s e^{isG\xi}\right) e^{i(\beta x - \omega t)}, \qquad (15.4)$$

where $F(\xi)$ is a Bloch modulation function with parameters of $s = 0, \pm 1, \pm 2, \ldots,$ $\pm\infty$ and $F_0 = 1$. We can treat $e^{i(\beta x - \omega t)}$ as the temperature field envelope and $F(\xi)$ as local inhomogeneity. The substitution of Eq. (15.4) into Eq. (15.3) yields a series of component equations related to the order of s. For accuracy, we consider $s = 0, \pm 1, \pm 2, \cdots, \pm 10$ and $F_{|s|>10} = 0$ to obtain twenty-one equations with twenty-one unknown numbers including β and $F_{|s|\leq 10}$, so β can be numerically calculated.

The properties of spatiotemporal modulation are reflected in three crucial dimensionless parameters of $2\pi\Gamma = \phi u_y d/D_0$, $\Lambda = \Delta_\rho \cos\theta/\Delta_\kappa$, and $\zeta = d/h$. The parameter of $2\pi\Gamma$ is similar to the Peclet number, which can describe the ratio of advection to diffusion. The parameters of Λ and ζ reflect the influences of modulation amplitude and wavelength, respectively. We define the speed ratio as $\eta = |v_f/v_b| = |\text{Re}[\beta_b]/\text{Re}[\beta_f]|$ to discuss the degree of nonreciprocity, where v_f and v_b are the forward and backward propagating speeds of temperature fields, respectively.

We first discuss Λ when $\zeta = 0.2$ (Fig. 15.2a). Since $2\pi\Gamma = 0$ and $2\pi\Gamma \to \infty$ always yield $\eta = 1$, it is necessary to introduce the vertical advection, but not the larger, the better. Meanwhile, a speed difference still exists when $\Lambda = 0$ (i.e., $\Delta_\rho = 0$), so it is unnecessary to modulate ρ and κ simultaneously. We find two types of curves in Fig. 15.2a. Type I features that η is always larger than 1 (the top three curves). Type II features that η is first larger and then smaller than 1 (the bottom three curves). The transition between types I and II is at the critical point of $\Lambda = 1$ (the third curve from the top), where the modulations in Eqs. (15.2a) and (15.2b) do not affect the effective thermal diffusivity in the vertical direction. When we change ζ from 0.2 to 1 (Fig. 15.2b) and 2 (Fig. 15.2c), type III curves appear, with η always smaller than 1. These three types indicate that nonreciprocal speeds can be flexibly manipulated.

We further discuss θ when $\zeta = 1$ (Fig. 15.2d), so $\Lambda = \Delta_\rho \cos\theta/\Delta_\kappa$ can be both positive and negative. The critical point of $\Lambda = 1$ still determines the transition between types I and II. Moreover, since $\theta = \pi/2$ always leads to $\Lambda = 0$, the curves in Fig. 15.2e are almost overlapped. We also discuss the thermal diffusivity of $D = \kappa/\rho$ (Fig. 15.2f), where κ is the balanced value of the periodic thermal conductivity and ρ is the balanced value of the periodic product of mass density and heat capacity. The peaks of η appear at almost the same value of $2\pi\Gamma$. Meanwhile, the peak of η gets larger as the thermal diffusivity decreases, which does not mean that the smaller the thermal diffusivity is, the better. We do not discuss the small thermal diffusivity because the system becomes insulated.

We further plot the thermal dispersion in Fig. 15.3a. The thermal dispersion curve is symmetric when $2\pi\Gamma = 0$, but becomes asymmetric when $2\pi\Gamma = 8$, which is the proof of diffusive Fizeau drag. We also plot the wavenumber difference $\Delta\text{Re}[\beta] = \text{Re}[\beta_f] + \text{Re}[\beta_b]$ in Fig. 15.3b, demonstrating linear responses to ω. More intuitively, a speed difference leads to a time difference of temperature field evolution at two

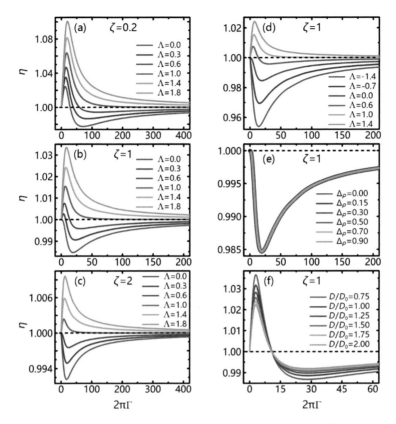

Fig. 15.2 Numerical results of the speed ratio of $\eta = |v_f/v_b|$ as a function of $2\pi\Gamma = \phi u_y d/D_0$. $\Lambda = \Delta_\rho \cos\theta/\Delta_\kappa$ is tuned by **a**–**c** Δ_ρ or (**d**) θ. Except the parameters presented in **a**–**f**, the others are $\phi = 0.1$, $\epsilon = 1$, $D_0 = 5 \times 10^{-5}$ m^2/s, $d = 0.02$ m, and $\omega = \pi/10$ rad/s for **a**–**f**; $\Delta_\rho = 0.7$ for (**d**); $\Delta_\rho = 0.6$ for (**f**); $\Delta_\kappa = 0.5$ for (a)-(e); $\Delta_\kappa = 0.9$ for (**f**); $\theta = 0$ for (**a**)–(**d**) and (**f**); and $\theta = \pi/2$ for (**e**). Adapted from Ref. [15]

symmetric positions of x and $C - x$ to reach the same phases. The forward phase at x is Re$[\beta_f]x - \omega t_f$, and the backward phase at $-x$ is $-$Re$[\beta_b]x - \omega t_b$. The same phases correspond to a time difference of $\Delta t = t_f - t_b$, which can be calculated by

$$\Delta t = \Delta\text{Re}[\beta]|x|/\omega. \tag{15.5}$$

Since Δt increases linearly with $|x|$, we focus on the parameter of $\Delta t/|x| = \Delta\text{Re}[\beta]/\omega$ in Fig. 15.3c, which is almost invariant as ω changes.

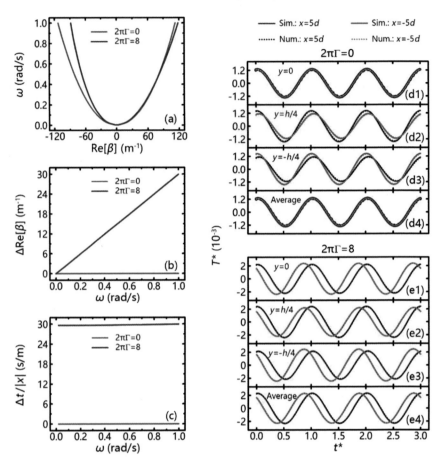

Fig. 15.3 Simulation results of diffusive Fizeau drag. **a** Thermal dispersion. **b** Wave number difference $\Delta \mathrm{Re}[\beta] = \mathrm{Re}[\beta_f] + \mathrm{Re}[\beta_b]$ as a function of ω. **c** Time difference per unit of distance $\Delta t/|x| = \Delta \mathrm{Re}[\beta]/\omega$ as a function of ω. Evolution of T^* when (d1)–(d4) $2\pi\Gamma = 0$ or (e1)–(e4) $2\pi\Gamma = 8$, corresponding to $u_y = 0$ or $u_y = 0.2$ m/s, respectively. Parameters: $\phi = 0.1$, $\epsilon = 1$, $D_0 = 5 \times 10^{-5}$ m²/s, $\Delta_\rho = 0.9$, $\Delta_\kappa = 0.9$, $\theta = \pi$, $d = 0.02$ m, $h = 0.02$ m, and $t_0 = 20$ s. The simulation length is $30d = 0.6$ m. The left and right boundaries are insulated. The upper and lower boundaries are set with periodic conditions. Sim.: Simulation; and Num.: Numerical. Adapted from Ref. [15]

15.3   Finite-Element Simulation

Finite-element simulations are also performed with COMSOL Multiphysics. For brevity, we define a dimensionless temperature of $T^* = (T - T_r)/A$ and a dimensionless time of $t^* = t/t_0$, where t_0 is the time periodicity of the temperature source. When $2\pi\Gamma = 0$ (Fig. 15.3d1), the forward and backward cases are identical at $y = 0$, but a slight difference appears at $y = \pm h/4$ (Fig. 15.3d2, d3) due to the

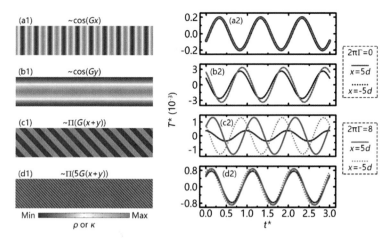

Fig. 15.4 Influences of inhomogeneity on thermal Willis coupling. The left column shows different kinds of inhomogeneity. The right column shows the evolution of T^*. The parameters and boundary conditions are the same as those in Fig. 15.3. Adapted from Ref. [15]

local inhomogeneity described by the $F(\xi)$ in Eq. (15.4). As long as we discuss the average temperature in the vertical direction, the effect of local inhomogeneity can be excluded, so the forward and backward cases become identical again (Fig. 15.3d4). We further set $2\pi\Gamma = 8$, and the simulation results demonstrate a time difference of $\Delta t^* = 0.14$, which can be observed locally (Fig. 15.3e1–e3) and globally (Fig. 15.3e4). The numerical results predict a time difference of $\Delta t^* = 0.15$, indicating that the numerical calculations are convincing. Meanwhile, we plot the numerical results with dotted curves, which agree well with the simulation results.

We analytically homogenize the governing equation to reveal the underlying mechanism of diffusive Fizeau drag. We find two high-order terms of ∂_t^2 and $\partial_t\partial_x$ in the homogenized equation. This situation is similar to the properties of Willis metamaterials that result from the homogenization of inhomogeneous media [20–24]. The modified constitutive relation describing the heat flux of J can be approximately expressed as $\tau\partial_t J + J = -\kappa_e\partial_x T_0 + \sigma_2\partial_t T_0$, where τ, κ_e, σ_2, and T_0 are the homogenized parameters. Besides the temperature gradient of $\partial_x T_0$, the horizontal heat flux is also coupled with the temperature change rate of $\partial_t T_0$, which can be referred to as the thermal Willis term. Moreover, the thermal Willis term can lead to nonreciprocal $|\mathrm{Re}[\beta]|$, but cannot generate nonreciprocal $|\mathrm{Im}[\beta]|$. This property indicates an obvious speed difference but no amplitude difference in opposite directions, which agrees with the simulation results in Fig. 15.3e1–e4.

Inhomogeneity is crucial to thermal Willis coupling. It will disappear when we consider only the horizontal inhomogeneity (Fig. 15.4a1, a2) or only the vertical inhomogeneity (Fig. 15.4b1, b2). We further change the modulations from cosine functions to square wave functions denoted by Π. When the periodicity of inhomogeneity is the same as that in Fig. 15.3e and $2\pi\Gamma = 8$, a time difference of $\Delta t^* = 0.32$

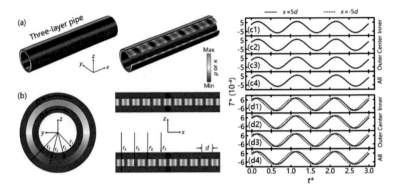

Fig. 15.5 Experimental suggestions. **a** Three-dimensional and **b** two-dimensional diagrams of a three-layer pipe. Temperature evolution when (c1)–(c4) $\Omega = 0$ or (d1)–(d4) $\Omega = 2\pi$ rad/s. The parameters of the center layer are $\rho_0 = 2 \times 10^6$ J m^{-3} K^{-1}, $\kappa_0 = 200$ W m^{-1} K^{-1}, $\Delta_\rho = 0.9$, $\Delta_\kappa = 0.9$, and $\theta = \pi$. Those of the inner and outer layers are $\rho = 2 \times 10^6$ J m^{-3} K^{-1} and $\kappa = 10$ W m^{-1} K^{-1}. The other parameters are $d = 20$ mm, $r_1 = 2.43$ mm, $r_2 = 2.93$ mm, $r_3 = 3.43$ mm, $r_4 = 3.93$ mm, and $t_0 = 20$ s. The simulation length is $15d = 300$ mm. The left and right boundaries are insulated. Adapted from Ref. [15]

can be observed (Fig. 15.4c1, c2). Therefore, the square wave modulation is more efficient than the cosine modulation ($\Delta t^* = 0.14$). We further reduce the modulation wavelength by a factor of five (Fig. 15.4d1). When $2\pi \Gamma = 8$, a time difference of $\Delta t^* = 0.04$ appears, but it is far smaller than that of $\Delta t^* = 0.32$ in Fig. 15.4c2. Therefore, the more homogeneous parameters yield weaker thermal Willis coupling, which is consistent with the current understanding in mechanical waves [20–24].

15.4 Experimental Suggestion

For experimental suggestions, we design a three-dimensional structure without fluids, i.e., a three-layer solid pipe (Fig. 15.5a, b). The inner and outer layers are homogeneous with the same angular velocities of Ω. The center layer is stationary with spatially-periodic parameters of $\rho(\xi') = \rho_0(1 + \Delta_\rho \cos(G\xi' + \theta))$ and $\kappa(\xi') = \kappa_0(1 + \Delta_\kappa \cos(G\xi'))$, where $\xi' = x + \alpha/G$ is the generalized coordinate in three dimensions with definitions of $\cos\alpha = z/\sqrt{y^2 + z^2}$ and $\sin\alpha = y/\sqrt{y^2 + z^2}$. The two rotating layers can provide the surface advection to the center layer, which has a similar effect as the bulk advection. The simulation results without and with angular rotation are presented in Fig. 15.5c1–c4, d1–d4, respectively. The detecting locations are in the inner, center, outer, and all layers, respectively. With proper angular rotation, a time difference between the evolution of the forward and backward temperature field indicates diffusive Fizeau drag.

15.5 Conclusion

We conclude the distinctive features of diffusive Fizeau drag. (I) As described by Eq. (15.1), only the biased advection cannot realize diffusive Fizeau drag. (II) Diffusive Fizeau drag in spatiotemporal thermal metamaterials results from thermal Willis coupling between heat flux and temperature change rate. (III) Diffusive Fizeau drag is unexpected because the vertical advection can generally not induce horizontal nonreciprocity. (IV) Three curves in Fig. 15.2 indicate that diffusive Fizeau drag can be flexibly controlled.

We have revealed diffusive Fizeau drag in a spatiotemporal thermal metamaterial, featuring a speed difference of temperature field propagation in opposite directions. Spatial or temporal modulation alone cannot realize the horizontal nonreciprocity, so spatiotemporal modulation necessarily introduces the high-order coupling, referred to as thermal Willis coupling, between heat flux and temperature change rate. Diffusive Fizeau drag has also been visualized by observing the time difference of temperature field evolution at two symmetric positions. These results suggest a distinct mechanism to achieve nonreciprocal diffusion [25–28] by thermal Willis coupling and also have potential applications for controlling nonequilibrium heat and mass transfer [30, 31].

15.6 Exercise and Solution

Exercise

1. Derive Eq. (15.1).

Solution

1. Heat transfer in a two-dimensional homogeneous porous medium is governed by

$$\rho_0 \frac{\partial T}{\partial t} + \nabla \cdot (\phi \rho_a u T - \kappa_0 \nabla T) = 0, \tag{15.6}$$

with definitions of $\rho_0 = \phi \rho_a + (1 - \phi)\rho_{s0}$ and $\kappa_0 = \phi \kappa_a + (1 - \phi)\kappa_{s0}$. ρ_a (or ρ_{s0}) is the product of mass density and heat capacity of the fluid (or solid). κ_a (or κ_{s0}) is the thermal conductivity of the fluid (or solid). The substitution of a wavelike temperature field described by $T = e^{i(\beta x - \omega t)}$ into Eq. (15.6) yields

$$-i\omega\rho_0 + i\beta\phi\rho_a u_x + \beta^2 \kappa_0 = 0. \tag{15.7}$$

Since we apply a periodic source with a temperature of $T(x = 0) = e^{(-i\omega t)}$, ω is real and β is complex. We can take $\beta = p + iq$, with p and q being two real numbers, so Eq. (15.7) can be rewritten as

$$-i\omega\rho_0 + i(p + iq)\phi\rho_a u_x + (p + iq)^2 \kappa_0 = 0, \tag{15.8}$$

which can be further decomposed into two equations according to its real and imaginary parts,

$$-q\phi\rho_a u_x + (p^2 - q^2)\kappa_0 = 0, \tag{15.9a}$$

$$-\omega\rho_0 + p\phi\rho_a u_x + 2pq\kappa_0 = 0. \tag{15.9b}$$

The solution to Eqs. (15.9a) and (15.9b) is just Eq. (15.1).

References

1. Fresnel, A.: Lettre d'Augustin Fresnel à François Arago sur l'influence du mouvement terrestre dans quelques phénomènes d'optique. Ann. Chem. Phys. **9**, 57 (1818)
2. Fizeau, H.: Sur les hypothèses relatives àl'éther lumineux. C. R. Acad. Sci. **33**, 349 (1851)
3. Kuan, P.-C., Huang, C., Chan, W.S., Kosen, S., Lan, S.-Y.: Large Fizeau's light-dragging effect in a moving electromagnetically induced transparent medium. Nat. Commun. **7**, 13030 (2016)
4. Qin, T., Yang, J.F., Zhang, F.X., Chen, Y., Shen, D.Y., Liu, W., Chen, L., Jiang, X.S., Chen, X.F., Wan, W.J.: Fast- and slow-light-enhanced light drag in a moving microcavity. Commun. Phys. **3**, 118 (2020)
5. Huidobro, P.A., Galiffi, E., Guenneau, S., Craster, R.V., Pendry, J.B.: Fresnel drag in space Ctime-modulated metamaterials. Proc. Natl. Acad. Sci. U. S. A. **116**, 24943 (2019)
6. Huidobro, P.A., Silveirinha, M.G., Galiffi, E., Pendry, J.B.: Homogenization theory of space-time metamaterials. Phys. Rev. Appl. **16**, 014044 (2021)
7. Dong, Y., Xiong, L., Phinney, I.Y., Sun, Z., Jing, R., McLeod, A.S., Zhang, S., Liu, S., Ruta, F.L., Gao, H., Dong, Z., Pan, R., Edgar, J.H., Jarillo-Herrero, P., Levitov, L.S., Millis, A.J., Fogler, M.M., Bandurin, D.A., Basov, D.N.: Fizeau drag in graphene plasmonics. Nature **594**, 513 (2021)
8. Zhao, W.Y., Zhao, S.H., Li, H.Y., Wang, S., Wang, S.X., Iqbal Bakti Utama, M., Kahn, S., Jiang, Y., Xiao, X., Yoo, S., Watanabe, K., Taniguchi, T., Zettl, A., Wang, F.: Efficient Fizeau drag from dirac electrons in monolayer graphene. Nature **594**, 517 (2021)
9. Farhat, M., Guenneau, S., Chen, P.-Y., Alù, A., Salama, K.N.: Scattering cancellation-based cloaking for the Maxwell-Cattaneo heat waves. Phys. Rev. Appl. **11**, 044089 (2019)
10. Gandolfi, M., Giannetti, C., Banfi, F.: Temperonic crystal: a superlattice for temperature waves in graphene. Phys. Rev. Lett. **125**, 265901 (2020)
11. Li, Y., Peng, Y.-G., Han, L., Miri, M.-A., Li, W., Xiao, M., Zhu, X.-F., Zhao, J.L., Al, A., Fan, S.H., Qiu, C.-W.: Anti-parity-time symmetry in diffusive systems. Science **364**, 170 (2019)
12. Xu, L.J., Wang, J., Dai, G.L., Yang, S., Yang, F.B., Wang, G., Huang, J.P.: Geometric phase, effective conductivity enhancement, and invisibility cloak in thermal convection-conduction. Int. J. Heat Mass Transf. **165**, 120659 (2021)
13. Voti, R.L., Bertolotti, M.: Thermal waves emitted by moving sources and the Doppler effect. Int. J. Heat Mass Transf. **176**, 121098 (2021)
14. Xu, G.Q., Li, Y., Li, W., Fan, S.H., Qiu, C.-W.: Configurable phase transitions in a topological thermal material. Phys. Rev. Lett. **127**, 105901 (2021)
15. Xu, L.J., Xu, G.Q., Huang, J.P., Qiu, C.-W.: Diffusive Fizeau drag in spatiotemporal thermal metamaterials. Phys. Rev. Lett. **128**, 145901 (2022)
16. Li, Y., Zhu, K.-J., Peng, Y.-G., Li, W., Yang, T.Z., Xu, H.-X., Chen, H., Zhu, X.-F., Fan, S.H., Qiu, C.-W.: Thermal metadevice in analog of zero-index photonics. Nat. Mater. **18**, 48 (2019)
17. Xu, G.Q., Dong, K.C., Li, Y., Li, H.G., Liu, K.P., Li, L.Q., Wu, J.Q., Qiu, C.-W.: Tunable analog thermal material. Nat. Commun. **11**, 6028 (2020)

18. Xu, L.J., Huang, J.P., Ouyang, X.P.: Tunable thermal wave nonreciprocity by spatiotemporal modulation. Phys. Rev. E **103**, 032128 (2021)
19. Xu, L.J., Huang, J.P., Ouyang, X.P.: Nonreciprocity and isolation induced by an angular momentum bias in convection-diffusion systems. Appl. Phys. Lett. **118**, 221902 (2021)
20. Willis, J.R.: Variational principles for dynamic problems for inhomogeneous elastic media. Wave Motion **3**, 1 (1981)
21. Milton, G.W., Willis, J.R.: On modifications of Newton's second law and linear continuum elastodynamics. Proc. R. Soc. A **463**, 855 (2007)
22. Sieck, C.F., Alù, A., Haberman, M.R.: Origins of Willis coupling and acoustic bianisotropy in acoustic metamaterials through source-driven homogenization. Phys. Rev. B **96**, 104303 (2017)
23. Pernas-Salomón, R., Shmuel, G.: Fundamental principles for generalized Willis metamaterials. Phys. Rev. Appl. **14**, 064005 (2020)
24. Meng, Y., Hao, Y.R., Guenneau, S., Wang, S.B., Li, J.: Willis coupling in water waves. New J. Phys. **23**, 073004 (2021)
25. Torrent, D., Poncelet, O., Batsale, J.-C.: Nonreciprocal thermal material by spatiotemporal modulation. Phys. Rev. Lett. **120**, 125501 (2018)
26. Camacho, M., Edwards, B., Engheta, N.: Achieving asymmetry and trapping in diffusion with spatiotemporal metamaterials. Nat. Commun. **11**, 3733 (2020)
27. Li, Y., Shen, X.Y., Wu, Z.H., Huang, J.Y., Chen, Y.X., Ni, Y.S., Huang, J.P.: Temperature-dependent transformation thermotics: From switchable thermal cloaks to macroscopic thermal diodes. Phys. Rev. Lett. **115**, 195503 (2015)
28. Li, Y., Li, J.X., Qi, M.H., Qiu, C.-W., Chen, H.S.: Diffusive nonreciprocity and thermal diode. Phys. Rev. B **103**, 014307 (2021)
29. Wong, M.Y., Tso, C.Y., Ho, T.C., Lee, H.H.: A review of state of the art thermal diodes and their potential applications. Int. J. Heat Mass Transf. **164**, 120607 (2021)
30. Yang, S., Wang, J., Dai, G.L., Yang, F.B., Huang, J.P.: Controlling macroscopic heat transfer with thermal metamaterials: Theory, experiment and application. Phys. Rep. **908**, 1 (2021)
31. Li, Y., Li, W., Han, T.C., Zheng, X., Li, J.X., Li, B.W., Fan, S.H., Qiu, C.-W.: Transforming heat transfer with thermal metamaterials and devices. Nat. Rev. Mater. **6**, 488 (2021)

Chapter 16
Theory for Thermal Wave Refraction: Advection Regulation

Abstract In this chapter, we study thermal waves of conduction and advection and further design advection-assisted metamaterials to realize the positive, vertical, and negative refraction of thermal waves. These results have a phenomenological analog of electromagnetic wave refraction despite different mechanisms. The negative refraction of thermal waves means that the incident and refractive thermal waves are on the same side of the normal, but the wave vector and energy flow are still in the same direction. As a model application, we apply the refractive behavior to design a thermal wave concentrator that can increase wave numbers and energy flows. This work provides insights into thermal wave manipulation, which may have potential thermal imaging applications.

Keywords Thermal wave refraction · Conduction and advection · Anisotropic permeability

16.1 Opening Remarks

Electromagnetic waves are dominated by the Maxwell equations, which have the hyperbolic feature. Due to the generality of hyperbolic equations, electromagnetic phenomena can be extended to other physical fields like acoustics without much difficulty. However, it is crucially different from the Fourier conduction because the Fourier equation is parabolic [1–3]. By modulating the Fourier equation with thermal relaxation [4–9], the parabolic equation can become hyperbolic, which can support the propagation of thermal waves and avoid the infinite speed of thermal propagation. Thermal waves refer to wave-like temperature profiles. The refractive behaviors can thus be studied [10–15].

Besides thermal relaxation, by combining the Fourier equation with convection, the dominant equation can also have the hyperbolic feature [16], which leads to novel thermal phenomena such as nonreciprocity [17, 18], anti-parity-time symmetry [19–21], negative transport [22], cloaks [23–25], and crystals [26]. However, the refractive behaviors have been rarely touched, let alone the negative refraction of thermal waves. Although it is difficult to discuss the refractive behaviors of diffusion waves [1–3],

© The Author(s) 2023
L.-J. Xu and J.-P. Huang, *Transformation Thermotics and Extended Theories*,
https://doi.org/10.1007/978-981-19-5908-0_16

we introduce advection to provide the hyperbolic property for heat transfer [16]. Therefore, it is possible to discuss the refractive behaviors of thermal waves based on conduction and convection. Despite the hyperbolic property, the entire investigation is not related to thermal relaxation [4–9]. In addition to the hyperbolic property, advection is also ubiquitous, so taking it into account is necessary and meaningful. Moreover, advection also clarifies the concept of thermal wave vectors, so different refractive behaviors can be visualized. We do not consider radiation (another basic mode of heat transfer) because it essentially belongs to electromagnetic waves.

We then consider conduction and advection with a time-harmonic temperature source to generate thermal waves. Since the heat transfer efficiency of advection is much higher than that of conduction, we can generally conclude that conduction leads to the decay of thermal waves, and advection contributes to the propagation of thermal waves [22]. Therefore, controlling thermal waves is essentially controlling advection velocities. We then focus our discussions on advection velocities which are preset (ideal model), determined by the Darcy equation (practical model), and controlled by layered structures (experimental suggestion).

16.2 Theoretical Foundation

We firstly consider an ideal model base on conduction and convection in a porous medium (composed of fluid and solid) whose dominant equation is

$$\rho_m C_m \frac{\partial T}{\partial t} + \nabla \cdot (-\kappa_m \nabla T) + \rho_f C_f \boldsymbol{v} \cdot \nabla T = 0, \tag{16.1}$$

where T and t are temperature and time, respectively. $\rho_m C_m$ is the product of the density and heat capacity of the porous medium, and κ_m is the thermal conductivity of the porous medium, which can be both calculated by the weighted average of the fluid and solid [28]. $\rho_f C_f$ is the product of the density and heat capacity of the fluid, and $\boldsymbol{v}$ is the convective velocity of the fluid. When we discuss the pure fluid shown in Fig. 16.1a, the solid does not exist, so we can derive $\rho_m C_m = \rho_f C_f$ and $\kappa_m = \kappa_f$ with κ_f being the thermal conductivity of the fluid. The convective term $\rho_f C_f \boldsymbol{v} \cdot \nabla T$ introduces the hyperbolic property to heat transfer. A heat source with time-harmonic temperature (THT) is located on the upper-left boundary. The time-harmonic temperature can be expressed as $T = A \cos(\omega t) + B$, where A, ω, and B are the temperature amplitude, circular frequency, and reference temperature of the thermal wave, respectively. The other boundaries are set with open boundary condition (OBC), indicating no reflection of heat energy (i.e., with semi-infinite length). The ideal model is composed of pure fluid with incident velocity $\boldsymbol{v}_i$ and refractive velocity $\boldsymbol{v}_r$, whose angles to the normal are θ_i and θ_r, respectively. The heat source with time-harmonic temperature generates thermal waves with the assistance of convection. Therefore, the direction of thermal waves follows that of convective velocities, which can be concluded as

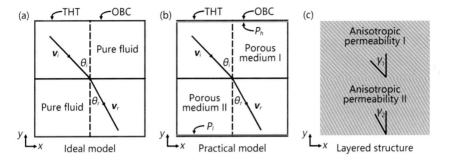

Fig. 16.1 Schematic diagrams of **a** ideal model, **b** practical model, and **c** layered structure. THT, time-harmonic temperature; OBC, open boundary condition. Adapted from Ref. [27]

$$\tan \theta_i = \frac{v_{ix}}{-v_{iy}}, \tag{16.2a}$$

$$\tan \theta_r = \frac{v_{rx}}{-v_{ry}}, \tag{16.2b}$$

where minus signs ensure $-v_{iy} > 0$ and $-v_{ry} > 0$. The subscript x (or y) denotes the x-component (or y-component) of convective velocities. The continuity of convective velocities along the y axis gives

$$v_{iy} = v_{ry}. \tag{16.3}$$

Equation (16.2) can then be simplified as

$$\frac{\tan \theta_i}{\tan \theta_r} = \frac{v_{ix}}{v_{rx}}, \tag{16.4}$$

which is similar to the tangent law for conductive refractions [29–31], but also has a different physical connotation. Compared with conductive refractions where thermal conductivities are critical, thermal wave refractions are mainly determined by convective velocities.

We then further discuss convective velocities with a practical model, as presented in Fig. 16.1b. Compared with the ideal model where convective velocities are preset, we explain the origination of convective velocities in the practical model. Although many fluid models [32–37] are applicable, we use the Darcy equation in porous media [32–35] for brevity, i.e., $\mathbf{v} = -(\sigma/\mu) \cdot \nabla P$, where $\mathbf{v}$ is convective velocity, σ is permeability, μ is dynamic viscosity, and P is pressure. The upper and lower boundaries are additionally set with high pressure P_h and low pressure P_l. The anisotropic permeabilities of the incident region σ_i and refractive region σ_r are, respectively, expressed as

$$\boldsymbol{\sigma}_i = \begin{pmatrix} \sigma_{ixx} & \sigma_{ixy} \\ \sigma_{iyx} & \sigma_{iyy} \end{pmatrix},$$ (16.5a)

$$\boldsymbol{\sigma}_r = \begin{pmatrix} \sigma_{rxx} & \sigma_{rxy} \\ \sigma_{ryx} & \sigma_{ryy} \end{pmatrix}.$$ (16.5b)

We can then express the incident velocity $\boldsymbol{v}_i$ and refractive velocity $\boldsymbol{v}_r$ with the Darcy equation as

$$\boldsymbol{v}_i = -\frac{\boldsymbol{\sigma}_i}{\mu} \cdot \nabla P_i = \frac{-1}{\mu} \begin{pmatrix} \sigma_{ixx} & \sigma_{ixy} \\ \sigma_{iyx} & \sigma_{iyy} \end{pmatrix} \begin{pmatrix} \nabla P_{ix} \\ \nabla P_{iy} \end{pmatrix}$$
$$= \frac{-1}{\mu} \begin{pmatrix} \sigma_{ixx} \nabla P_{ix} + \sigma_{ixy} \nabla P_{iy} \\ \sigma_{iyx} \nabla P_{ix} + \sigma_{iyy} \nabla P_{iy} \end{pmatrix} = \begin{pmatrix} v_{ix} \\ v_{iy} \end{pmatrix},$$ (16.6a)

$$\boldsymbol{v}_r = -\frac{\boldsymbol{\sigma}_r}{\mu} \cdot \nabla P_r = \frac{-1}{\mu} \begin{pmatrix} \sigma_{rxx} & \sigma_{rxy} \\ \sigma_{ryx} & \sigma_{ryy} \end{pmatrix} \begin{pmatrix} \nabla P_{rx} \\ \nabla P_{ry} \end{pmatrix}$$
$$= \frac{-1}{\mu} \begin{pmatrix} \sigma_{rxx} \nabla P_{rx} + \sigma_{rxy} \nabla P_{ry} \\ \sigma_{ryx} \nabla P_{rx} + \sigma_{ryy} \nabla P_{ry} \end{pmatrix} = \begin{pmatrix} v_{rx} \\ v_{ry} \end{pmatrix}.$$ (16.6b)

With Eq. (16.6), we can obtain

$$\tan \theta_i = \frac{v_{ix}}{-v_{iy}} = -\frac{\sigma_{ixx} \nabla P_{ix} + \sigma_{ixy} \nabla P_{iy}}{\sigma_{iyx} \nabla P_{ix} + \sigma_{iyy} \nabla P_{iy}},$$ (16.7a)

$$\tan \theta_r = \frac{v_{rx}}{-v_{ry}} = -\frac{\sigma_{rxx} \nabla P_{rx} + \sigma_{rxy} \nabla P_{ry}}{\sigma_{ryx} \nabla P_{rx} + \sigma_{ryy} \nabla P_{ry}},$$ (16.7b)

which further yields

$$\frac{\nabla P_{iy}}{\nabla P_{ix}} = -\frac{\sigma_{ixx} + \sigma_{iyx} \tan \theta_i}{\sigma_{ixy} + \sigma_{iyy} \tan \theta_i},$$ (16.8a)

$$\frac{\nabla P_{ry}}{\nabla P_{rx}} = -\frac{\sigma_{rxx} + \sigma_{ryx} \tan \theta_r}{\sigma_{rxy} + \sigma_{ryy} \tan \theta_r}.$$ (16.8b)

The boundary conditions on the interface of incident and refractive regions are

$$v_{iy} = v_{ry},$$ (16.9a)
$$\nabla P_{ix} = \nabla P_{rx},$$ (16.9b)

which indicate the continuities of convective velocities along the y axis (Eq. (16.9a)) and pressure gradients along the x axis (Eq. (16.9b)). With Eq. (16.9), we can derive

$$\sigma_{iyx} + \sigma_{iyy} \frac{\nabla P_{iy}}{\nabla P_{ix}} = \sigma_{ryx} + \sigma_{ryy} \frac{\nabla P_{ry}}{\nabla P_{rx}}.$$ (16.10)

The substitution of Eq. (16.8) into Eq. (16.10) yields

$$\frac{\sigma_{ixx}\sigma_{iyy} - \sigma_{ixy}\sigma_{iyx}}{\sigma_{ixy} + \sigma_{iyy}\tan\theta_i} = \frac{\sigma_{rxx}\sigma_{ryy} - \sigma_{rxy}\sigma_{ryx}}{\sigma_{rxy} + \sigma_{ryy}\tan\theta_r}. \tag{16.11}$$

Equation (16.11) is an extension of Eq. (16.4), revealing the refractive behaviors of thermal waves with a practical model determined by permeabilities. In other words, thermal wave refractions can be controlled by designing specific permeabilities.

We then suppose that the anisotropic permeabilities of porous media I and II have the same eigenvalues, i.e., σ_s and σ_p. Therefore, $\boldsymbol{\sigma}_i$ and $\boldsymbol{\sigma}_r$ can be obtained by anticlockwise rotating the eigenvalues with angles of γ_1 and γ_2, respectively. The permeabilities can then be expressed as [38–42]

$$\boldsymbol{\sigma}_i = \begin{pmatrix} \sigma_s \cos^2\gamma_1 + \sigma_p \sin^2\gamma_1 & (\sigma_s - \sigma_p)\cos\gamma_1\sin\gamma_1 \\ (\sigma_s - \sigma_p)\cos\gamma_1\sin\gamma_1 & \sigma_s \sin^2\gamma_1 + \sigma_p \cos^2\gamma_1 \end{pmatrix}, \tag{16.12a}$$

$$\boldsymbol{\sigma}_r = \begin{pmatrix} \sigma_s \cos^2\gamma_2 + \sigma_p \sin^2\gamma_2 & (\sigma_s - \sigma_p)\cos\gamma_2\sin\gamma_2 \\ (\sigma_s - \sigma_p)\cos\gamma_2\sin\gamma_2 & \sigma_s \sin^2\gamma_2 + \sigma_p \cos^2\gamma_2 \end{pmatrix}. \tag{16.12b}$$

The substitution of Eq. (16.12) into Eq. (16.11) yields

$$\frac{\sigma_s\sigma_p}{\sigma_{ixy} + \sigma_{iyy}\tan\theta_i} = \frac{\sigma_s\sigma_p}{\sigma_{rxy} + \sigma_{ryy}\tan\theta_r}, \tag{16.13}$$

which can be further reduced to

$$\frac{\sigma_s}{(\sigma_s/\sigma_p - 1)\cos\gamma_1\sin\gamma_1 + (\sigma_s/\sigma_p \sin^2\gamma_1 + \cos^2\gamma_1)\tan\theta_i} = \frac{\sigma_s}{(\sigma_s/\sigma_p - 1)\cos\gamma_2\sin\gamma_2 + (\sigma_s/\sigma_p \sin^2\gamma_2 + \cos^2\gamma_2)\tan\theta_r}. \tag{16.14}$$

With the further assumption that the eigenvalues are highly anisotropic,

$$\sigma_p \gg \sigma_s \approx 0, \tag{16.15}$$

we can obtain $\sigma_s/\sigma_p \approx 0$ and reduce Eq. (16.14) to

$$\theta_i \approx \gamma_1, \tag{16.16a}$$

$$\theta_r \approx \gamma_2. \tag{16.16b}$$

Equation (16.16) is a further extension of Eq. (16.11), which indicates that the direction of thermal waves approximately follows the orientation of anisotropic permeabilities. Therefore, by designing γ_1 and γ_2, we can effectively control the angles of incidence and refraction.

16.3 Finite-Element Simulation

We further perform finite-element simulations to confirm the theory. We use three templates: heat transfer in fluids, porous media, and the Darcy law in COMSOL Multiphysics. We choose a free triangular mesh with a maximum element size of 2×10^{-4} m, a minimum element size of 10^{-6} m, a maximum element growth rate of 1.1, a curvature factor of 0.2, and a resolution of narrow regions of 1. The transient simulations are conducted with times from 0 s to 200 s with a step of 0.1 s, and the relative tolerance is 0.0001. We also introduce a dimensionless characteristic parameter to understand the two modes of heat transfer, i.e., the Peclet number $Pe = |v|L/D$, where v is advection velocity, L is characteristic length, and $D = \kappa_m / (\rho_m C_m)$ is heat diffusion coefficient. In our investigation, the Peclet number $Pe = 10^4$ is large, so our system is dominated by advection and featured by the hyperbolic property.

The results are presented in Figs. 16.2, 16.3 and 16.4. We also consider three typical cases. The first one is positive refraction, indicating that the incident and

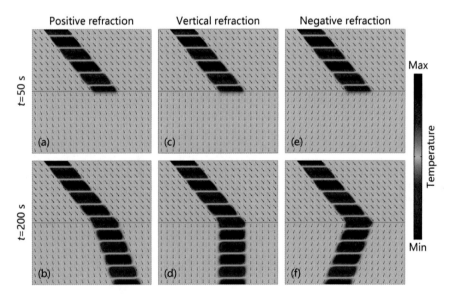

Fig. 16.2 Simulations of **a** and **b** positive refraction, **c** and **d** vertical refraction, and **e** and **f** negative refraction. Colors denote temperatures, and solid arrows denote convective velocities. The simulation size is 10×10 cm². The time-harmonic temperature is set at $T = 40 \cos (\pi t/10) + 323$ K with length 2 cm, whose left endpoint has a 1 cm distance from the left-upper corner. The initial temperature is set at 323 K. The advection velocities of the incident regions are $(v_0 \tan \theta_0, -v_0)^\tau$ where τ denotes transpose. The advection velocities of the refractive regions are $(v_0 \tan \theta_p, -v_0)^\tau$ for **(a)** and **(b)**, $(v_0 \tan \theta_v, -v_0)^\tau$ for **(c)** and **(d)**, and $(v_0 \tan \theta_n, -v_0)^\tau$ for **(e)** and **(f)**. Concrete parameters: $v_0 = 1$ mm/s, $\theta_0 = 4\pi/18$ rad, $\theta_p = 2\pi/18$ rad, $\theta_v = 0$ rad, $\theta_n = -2\pi/18$ rad, $\kappa_f = 0.01$ W m⁻¹ K⁻¹, $C_f = 1000$ J kg⁻¹ K⁻¹, and $\rho_f = 1000$ kg/m³, respectively. Adapted from Ref. [27]

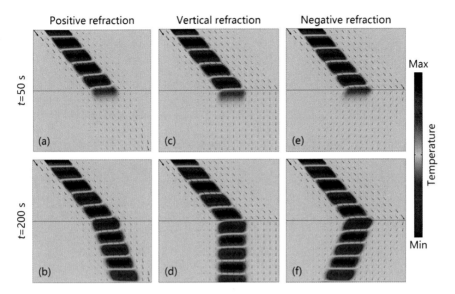

Fig. 16.3 Simulations based on the practical model. The upper and lower boundaries are additionally set with $P_h = 15000$ Pa and $P_l = 0$ Pa, respectively. The parameters of thermal conductivity, heat capacity, and density of porous media I and II are 0.01 W m^{-1} K^{-1}, 1000 J kg^{-1} K^{-1}, and 1000 kg/m^3, respectively. The anisotropic permeabilities of porous media I and II have the same eigenvalues, i.e., $\sigma_s = 10^{-13}$ m^2 and $\sigma_p = 10^{-11}$ m^2. The anisotropic permeabilities in the incident regions are described by Eq. (16.12a) with $\gamma_1 = 4\pi/18$ rad, and those in the refractive regions are described by Eq. (16.12b) with $\gamma_2 = 2\pi/18$ rad for (**a**) and (**b**), $\gamma_2 = 0$ rad for (**c**) and (**d**), and $\gamma_2 = -2\pi/18$ rad for (**e**) and (**f**). The dynamic viscosity of the fluid is 0.001 Pa·s. Adapted from Ref. [27]

refractive thermal waves are at both sides of the normal. The second is vertical refraction with the refractive thermal wave along the normal. The third one is negative refraction, indicating that the incident and refractive thermal waves are on the same side of the normal.

The simulations based on the ideal model are shown in Fig. 16.2. We take the positive refraction as an example. The thermal wave takes about 50 s to reach the interface (Fig. 16.2a). Since the convective velocities of the incident and refractive regions are different, the thermal wave changes its propagation direction (Fig. 16.2b). Since v_{ix} and v_{rx} are both positive, the result is positive refraction. We then maintain the incident velocity and change the refractive velocity. Suppose the direction of the refractive velocity is vertical to the interface. In that case, that of the refractive thermal wave is also vertical to the interface, yielding vertical refraction (Fig. 16.2c, d). We finally set the refractive velocity to have a negative x-component, so the refractive and incident thermal waves are at the same side of the normal, yielding negative refraction (Fig. 16.2e, f). Intuitively speaking, where thermal convection flows, where thermal waves propagate. The temperature evolutions in more detail are animated in the supplemental media.

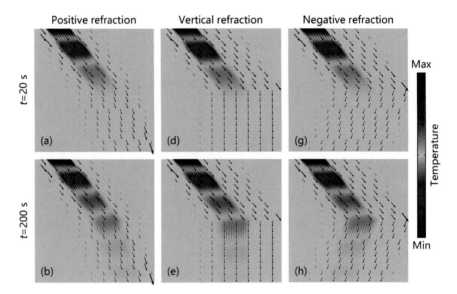

Fig. 16.4 Simulations based on layered structures. Anisotropic permeabilities are obtained with two isotropic porous media A and B (with porosity 0.1), whose permeabilities are $\sigma_a = 5 \times 10^{-14}$ m^2 and $\sigma_b = 2 \times 10^{-11}$ m^2, respectively. The fluid is water whose thermal conductivity, heat capacity, density, and dynamic viscosity are 0.6 W m^{-1} K^{-1}, 4200 J kg^{-1} K^{-1}, 1000 kg/m^3, and 0.001 Pa·s, respectively. The thermal conductivity, heat capacity, and density of the solid is 0.1 W m^{-1} K^{-1}, 800 J kg^{-1} K^{-1}, and 3000 kg/m^3, respectively. Porous media A and B are supposed to have only a permeability difference. Adapted from Ref. [27]

The simulations in Fig. 16.2 are based on pure fluid. We then perform simulations based on the Darcy equation in porous media where anisotropic permeabilities can guide advection velocities. The positive, vertical, and negative refractions results are shown in Fig. 16.3, which are the same as Fig. 16.2. Convective velocities (denoted by arrows) are no longer ideally distributed, which, however, does not affect phenomenon observations. The propagation of thermal waves at 50 s is similar in Fig. 16.3a, c, e because of the same permeabilities in the incident regions. Positive, vertical, and negative refractions occur in Fig. 16.3b, d, f, respectively. Therefore, the simulations are consistent with the theoretical predictions.

Despite the practical model, the parameters are still too ideal. Therefore, we further use a layered structure to obtain the desired anisotropic permeability (Fig. 16.1c). Porous medium A with permeability σ_a and width a and porous medium B with permeability σ_b and width b are arranged alternately. The effective permeabilities with series connection σ_s and parallel connection σ_p can be expressed as [38–42]

$$\sigma_s = \frac{a+b}{a/\sigma_a + b/\sigma_b},$$ (16.17a)

$$\sigma_p = \frac{a\sigma_a + b\sigma_b}{a+b}.$$ (16.17b)

In this way, we can use two isotropic porous media to realize the anisotropic permeability described by Eq. (16.12).

The simulations of positive, vertical, and negative refractions are shown in Fig. 16.4a–c, d–f, g–i, respectively. Since we keep the eigenvalues of the anisotropic permeabilities the same, only two porous media with isotropic permeabilities are required. We can obtain the expected permeabilities by alternately arranging porous media A and B and anticlockwise rotating the structure with different angles. Compared with the advection velocities in Figs. 16.2 and 16.3, those in Fig. 16.4 are more complicated, but the general directions are still as expected. Therefore, the expected control of thermal waves can be obtained. Meanwhile, the simulations in Fig. 16.4 can be regarded as experimental suggestions because we choose practical parameters like water. For this reason, the wavelength and decay rate of thermal waves are very different from those in Figs. 16.2 and 16.3.

Although thermal wave refractions have a phenomenal analog to electromagnetic ones, the underlying mechanisms are very different. The former requires anisotropy, but the latter does not. Therefore, anisotropy guides the direction of convective velocities and further affects the propagation of thermal waves. In other words, convection helps generate thermal waves. From this perspective, we further understand the negative refraction of thermal waves. Although we have observed that the incident and refractive thermal waves are on the same side of the normal, it is phenomenological negative refraction. Generally, negative refractions feature the opposite directions of energy flows and wave vectors. In our system, energy flows follow convective velocities, and wave vectors follow thermal waves. Therefore, energy flows and wave vectors are along the same direction because convective velocities yield thermal waves (i.e., the casualty in thermotics [22]). We can then conclude that the negative refraction of thermal waves is phenomenally observed. However, the wave vector and energy flow are still in the same direction, not violating the casualty.

16.4 Model Application

The refractive behaviors of thermal waves can also be applied in practice, such as in designing a thermal wave concentrator. Since an anisotropic permeability can guide the direction of thermal waves, we are allowed to design the orientation of the anisotropic permeability to point towards the center, as schematically shown in Fig. 16.5a. Hence, thermal waves propagate along with the anisotropic permeability orientation, thus being guided and concentrated (see the solid lines with arrows in Fig. 16.5a). The simulations at 100 and 200 s are presented in Fig. 16.5b, c, respectively. We also plot the temperature and heat flux distributions at 200 s on the central

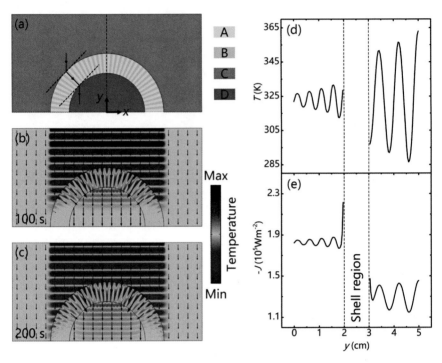

Fig. 16.5 Thermal wave concentrator. **a** Schematic diagram with simulation size $10 \times 5\,\mathrm{cm}^2$, inner radius 2 cm, and outer radius 3 cm. Simulations at **b** 100 s and **c** 200 s. **d** Temperature distribution and **e** energy flow distribution along $x = 0$ cm. The upper and lower boundaries are set with $P_h = 3750$ Pa and $P_l = 0$ Pa, respectively. The time-harmonic temperature is set at $T = 40\cos(\pi t/10) + 323$ K with length 6 cm in the center of the upper boundary. Other boundaries are set with open boundary condition. Four porous media (A, B, C, and D) are applied. Parameters: The permeabilities of A and B are, respectively, 5×10^{-14} and $2 \times 10^{-11}\,\mathrm{m}^2$, and those of C and D are the same, i.e. $\times 10^{-12}\,\mathrm{m}^2$. The thermal conductivity, heat capacity, density, and dynamic viscosity of the fluid in A-D are $0.01\,\mathrm{W\,m^{-1}\,K^{-1}}$, $1000\,\mathrm{J\,kg^{-1}\,K^{-1}}$, $1000\,\mathrm{kg/m^3}$, and $0.001\,\mathrm{Pa\cdot s}$, respectively. The thermal conductivity and density of the solid in A-D are $0.01\,\mathrm{W\,m^{-1}\,K^{-1}}$ and $1000\,\mathrm{kg/m^3}$, respectively. The heat capacity of the solid in A-C is $1000\,\mathrm{J\,kg^{-1}\,K^{-1}}$, and that for D is $2500\,\mathrm{J\,kg^{-1}\,K^{-1}}$. Adapted from Ref. [27]

dashed line, which are shown in Fig. 16.5d, e, respectively. The wave number and heat flux in the center are larger than those in the background. Therefore, this scheme provides guidance to control thermal waves beyond scattering cancellation [23, 24] and coordinate transformation [25].

16.5 Conclusion

We reveal the refractive behaviors of thermal waves between different media and present the thermal wave counterpart of electromagnetic wave refractions, including the positive, vertical, and negative refractions. We also design convection-assisted metamaterials to control thermal wave refractions and provide experimental suggestions to observe the desired phenomena. We further propose a potential application of thermal wave concentrators, which can be used for energy collection. These results are helpful in understanding and controlling the refractive behaviors of thermal waves and may have potential applications in thermal wave imaging [46–50] and intelligent thermal management.

16.6 Exercise and Solution

Exercise

1. Prove Eq. (16.16) in detail.

Solution

1. With Eqs. (16.15) and (16.14) can be reduced to

$$\frac{\sigma_s}{-\cos \gamma_1 \sin \gamma_1 + \cos^2 \gamma_1 \tan \theta_i} = \frac{\sigma_s}{-\cos \gamma_2 \sin \gamma_2 + \cos^2 \gamma_2 \tan \theta_r}. \tag{16.18}$$

Since the numerators of Eq. (16.18) are approximately zero (i.e., $\sigma_s \approx 0$), the denominators of Eq. (16.18) should also be approximately zero to ensure that Eq. (16.18) is nonzero,

$$-\cos \gamma_1 \sin \gamma_1 + \cos^2 \gamma_1 \tan \theta_i =$$
$$-\cos \gamma_2 \sin \gamma_2 + \cos^2 \gamma_2 \tan \theta_r \approx 0. \tag{16.19}$$

The physical meaning is that the convective velocities along the y axis are nonzero. Solving Eq. (16.19), we can then derive

$$\tan \theta_i \approx \frac{\cos \gamma_1 \sin \gamma_1}{\cos^2 \gamma_1} = \tan \gamma_1, \tag{16.20a}$$

$$\tan \theta_r \approx \frac{\cos \gamma_2 \sin \gamma_2}{\cos^2 \gamma_2} = \tan \gamma_2. \tag{16.20b}$$

References

1. Mandelis, A.: Diffusion waves and their uses. Phys. Today **53**, 29 (2000)
2. Mandelis, A., Nicolaides, L., Chen, Y.: Structure and the reflectionless/refractionless nature of parabolic diffusion-wave fields. Phys. Rev. Lett. **87**, 020801 (2001)
3. Salazar, A.: On thermal diffusivity. Eur. J. Phys. **24**, 351 (2003)
4. Joseph, D.D., Preziosi, L.: Heat waves. Rev. Mod. Phys. **61**, 41 (1989)
5. Mongiovi, M.S., Jou, D., Sciacca, M.: Non-equilibrium thermodynamics, heat transport and thermal waves in laminar and turbulent superfluid helium. Phys. Rep. **726**, 1 (2018)
6. Nie, B.-D., Cao, B.-Y.: Three mathematical representations and an improved ADI method for hyperbolic heat conduction. Int. J. Heat Mass Transf. **135**, 974 (2019)
7. Gandolfi, M., Benetti, G., Glorieux, C., Giannetti, C., Banfi, F.: Accessing temperature waves: a dispersion relation perspective. Int. J. Heat Mass Transf. **143**, 118553 (2019)
8. Domenico, M.D., Jou, D., Sellitto, A.: Nonlinear heat waves and some analogies with nonlinear optics. Int. J. Heat Mass Transf. **156**, 119888 (2020)
9. Simoncelli, M., Marzari, N., Cepellotti, A.: Generalization of Fourier's law into viscous heat equations. Phys. Rev. X **10**, 011019 (2020)
10. Shendeleva, M.L.: Thermal wave reflection and refraction at a plane interface: two-dimensional geometry. Phys. Rev. B **65**, 134209 (2002)
11. Tsai, C.-S., Hung, C.-I.: Thermal wave propagation in a bi-layered composite sphere due to a sudden temperature change on the outer surface. Int. J. Heat Mass Transf. **46**, 5137 (2003)
12. Ramadan, K., Al-Nimr, M.D.A.: Thermal wave reflection and transmission in a multilayer slab with imperfect contact using the dual-phase-lag model. Heat Transf. Eng. **30**, 677 (2009)
13. Guo, Z.-Y., Hou, Q.-W.: Thermal wave based on the thermomass model. J. Heat Transf. **132**, 072403 (2010)
14. Xu, M.T., Guo, J.F., Wang, L.Q., Cheng, L.: Thermal wave interference as the origin of the overshooting phenomenon in dual-phase-lagging heat conduction. Int. J. Therm. Sci. **50**, 825 (2011)
15. Kang, Z.X., Zhu, P.A., Gui, D.Y., Wang, L.Q.: A method for predicting thermal waves in dual-phase-lag heat conduction. Int. J. Heat Mass Transf. **115**, 250 (2017)
16. Sobolev, S.L.: On hyperbolic heat-mass transfer equation. Int. J. Heat Mass Transf. **122**, 629 (2018)
17. Torrent, D., Poncelet, O., Batsale, J.-C.: Nonreciprocal thermal material by spatiotemporal modulation. Phys. Rev. Lett. **120**, 125501 (2018)
18. Camacho, M., Edwards, B., Engheta, N.: Achieving asymmetry and trapping in diffusion with spatiotemporal metamaterials. Nat. Commun. **11**, 3733 (2020)
19. Li, Y., Peng, Y.-G., Han, L., Miri, M.-A., Li, W., Xiao, M., Zhu, X.-F., Zhao, J.L., Alù, A., Fan, S.H., Qiu, C.-W.: Anti-parity-time symmetry in diffusive systems. Science **364**, 170 (2019)
20. Cao, P.C., Li, Y., Peng, Y.G., Qiu, C.W., Zhu, X.F.: High-order exceptional points in diffusive systems: robust APT symmetry against perturbation and phase oscillation at APT symmetry breaking. ES Energy Environ. **7**, 48 (2020)
21. Xu, L.J., Wang, J., Dai, G.L., Yang, S., Yang, F., Wang, G., Huang, J.P.: Geometric phase, effective conductivity enhancement, and invisibility cloak in thermal convection-conduction. Int. J. Heat Mass Transf. **165**, 120659 (2021)
22. Xu, L.J., Huang, J.P.: Negative thermal transport in conduction and advection. Chin. Phys. Lett. **37**, 080502 (2020)
23. Farhat, M., Chen, P.-Y., Bagci, H., Amra, C., Guenneau, S., Alù, A.: Thermal invisibility based on scattering cancellation and mantle cloaking. Sci. Rep. **5**, 9876 (2015)
24. Farhat, M., Guenneau, S., Chen, P.-Y., Alù, A., Salama, K.N.: Scattering cancellation-based cloaking for the Maxwell-Cattaneo heat waves. Phys. Rev. Appl. **11**, 044089 (2019)
25. Xu, L.J., Huang, J.P.: Controlling thermal waves with transformation complex thermotics. Int. J. Heat Mass Transf. **159**, 120133 (2020)

26. Xu, L.J., Huang, J.P.: Thermal convection-diffusion crystal for prohibition and modulation of wave-like temperature profiles. Appl. Phys. Lett. **117**, 011905 (2020)
27. Xu, L.J., Yang, S., Huang, J.P.: Controlling thermal waves of conduction and convection. EPL **133**, 20006 (2021)
28. Bear, J., Corapcioglu, M.Y.: Fundamentals of Transport Phenomena in Porous Media. Springer, Netherlands (1984)
29. Tan, A., Holland, L.R.: Tangent law of refraction for heat conduction through an interface and underlying variational principle. Am. J. Phys. **58**, 988 (1990)
30. Vemuri, K.P., Bandaru, P.R.: Anomalous refraction of heat flux in thermal metamaterials. Appl. Phys. Lett. **104**, 083901 (2014)
31. Hu, R., Xie, B., Hu, J.Y., Chen, Q., Luo, X.B.: Carpet thermal cloak realization based on the refraction law of heat flux. EPL **111**, 54003 (2015)
32. Urzhumov, Y.A., Smith, D.R.: Fluid flow control with transformation media. Phys. Rev. Lett. **107**, 074501 (2011)
33. Dai, G.L., Shang, J., Huang, J.P.: Theory of transformation thermal convection for creeping flow in porous media: cloaking, concentrating, and camouflage. Phys. Rev. E **97**, 022129 (2018)
34. Yeung, W.-S., Mai, V.-P., Yang, R.-J.: Cloaking: controlling thermal and hydrodynamic fields simultaneously. Phys. Rev. Appl. **13**, 064030 (2020)
35. Xu, L.J., Huang, J.P.: Chameleonlike metashells in microfluidics: a passive approach to adaptive responses. Sci. China-Phys. Mech. Astron. **63**, 228711 (2020)
36. Park, J., Youn, J.R., Song, Y.S.: Hydrodynamic metamaterial cloak for drag-free flow. Phys. Rev. Lett. **123**, 074502 (2019)
37. Park, J., Youn, J.R., Song, Y.S.: Fluid-flow rotator based on hydrodynamic metamaterial. Phys. Rev. Appl. **12**, 061002 (2019)
38. Vemuri, K.P., Bandaru, P.R.: Geometrical considerations in the control and manipulation of conductive heat flux in multilayered thermal metamaterials. Appl. Phys. Lett. **103**, 133111 (2013)
39. Vemuri, K.P., Canbazoglu, F.M., Bandaru, P.R.: Guiding conductive heat flux through thermal metamaterials. Appl. Phys. Lett. **105**, 193904 (2014)
40. Yang, T.Z., Vemuri, K.P., Bandaru, P.R.: Experimental evidence for the bending of heat flux in a thermal metamaterial. Appl. Phys. Lett. **105**, 083908 (2014)
41. Xu, L.J., Yang, S., Huang, J.P.: Thermal theory for heterogeneously architected structure: fundamentals and application. Phys. Rev. E **98**, 052128 (2018)
42. Xu, L.J., Yang, S., Huang, J.P.: Designing the effective thermal conductivity of materials of core-shell structure: theory and simulation. Phys. Rev. E **99**, 022107 (2019)
43. Huang, J.P.: Theoretical Thermotics: Transformation Thermotics and Extended Theories for Thermal Metamaterials. Springer, Singapore (2020)
44. Li, Y., Zhu, K.-J., Peng, Y.-G., Li, W., Yang, T.Z., Xu, H.X., Chen, H., Zhu, X.-F., Fan, S.H., Qiu, C.-W.: Thermal meta-device in analogue of zero-index photonics. Nat. Mater. **18**, 48 (2019)
45. Xu, L.J., Yang, S., Huang, J.P.: Effectively infinite thermal conductivity and zero-index thermal cloak. EPL **131**, 24002 (2020)
46. Orth, T., Netzelmann, U., Pelzl, J.: Thermal wave imaging by photothermally modulated ferromagnetic resonance. Appl. Phys. Lett. **53**, 1979 (1988)
47. Busse, G., Wu, D., Karpen, W.: Thermal wave imaging with phase sensitive modulated thermography. J. Appl. Phys. **71**, 3962 (1992)
48. Mulaveesala, R., Tuli, S.: Theory of frequency modulated thermal wave imaging for nondestructive subsurface defect detection. Appl. Phys. Lett. **89**, 191913 (2006)
49. Mulaveesala, R., Tuli, S.: Applications of frequency modulated thermal wave imaging for nondestructive characterization. AIP Conf. Proc. **1004**, 15 (2008)
50. Tuli, S., Chatterjee, K.: Frequency modulated thermal wave imaging. AIP Conf. Proc. **1430**, 523 (2012)

Part II
Outside Metamaterials

Chapter 17
Theory for Active Thermal Control: Thermal Dipole Effect

Abstract In this chapter, we establish a theory for thermal-dipole-based thermotics. Tailoring the thermal dipole moment allows thermal invisibility without the requirements of singular and uncommon thermal conductivities. Furthermore, finite-element simulations and laboratory experiments both validate the theoretical analyses. Thermal-dipole-based thermotics offers a distinct mechanism to achieve thermal invisibility and provides guidance to other physical fields, such as electrostatics, magnetostatics, and particle diffusion. These results also pave the way for heat regulation with thermal dipoles, and potential applications can be expected in thermal protection, infrared detection, etc.

Keywords Active thermal control · Thermal dipole · Thermal invisibility

17.1 Opening Remarks

With growing concerns about energy issues, many researchers have turned their research focus to heat management. The emerging field of thermal metamaterials mainly drove this trend in the last decade. The most representative example is thermal invisibility [1–16], which has almost run through the development of thermal metamaterials. Thermal invisibility is characterized by the uniform thermal field of the matrix. Many schemes have been proposed for this realization, but they have shortcomings. The initial exploration is based on transformation thermotics [1–6], which is the thermal counterpart of transformation optics [17]. However, transformation thermotics leads to four severe problems, thus limiting practical applications. The first is anisotropy which requires different radial and tangential components of the tensorial thermal conductivity. The second is inhomogeneity which means a spatially-distributed thermal conductivity. The third is singularity which requires zero and/or infinite thermal conductivities. The fourth is extremely large thermal conductivities which are uncommon. Thermal conductivities of natural materials range only from $0.026\,\mathrm{W\,m^{-1}\,K^{-1}}$ (air) to $430\,\mathrm{W\,m^{-1}\,K^{-1}}$ (silver). Thermal conductivities out of this range are uncommon [16].

© The Author(s) 2023

L.-J. Xu and J.-P. Huang, *Transformation Thermotics and Extended Theories*,

https://doi.org/10.1007/978-981-19-5908-0_17

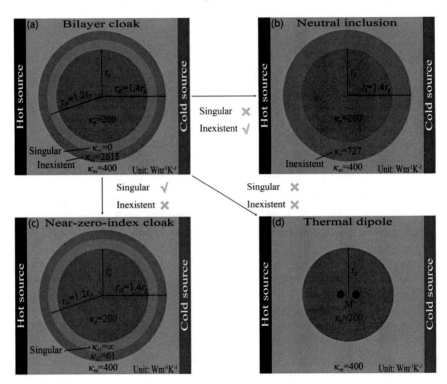

Fig. 17.1 Approaches to thermal invisibility with **a** a bilayer cloak, **b** the concept of neutral inclusion, **c** a near-zero-index cloak, and **d** a thermal dipole. None of these approaches can simultaneously remove the problem of singular and uncommon thermal conductivities except for our thermal-dipole-based scheme. Adapted from Ref. [19]

Although these problems restrict practical applications, they also promote the development of thermal metamaterials by solving them. Fortunately, the issues of anisotropy and inhomogeneity were solved soon [7–16]. However, the issue of singular and uncommon thermal conductivity still cannot be solved simultaneously. For example, we discuss a matrix with a very high thermal conductivity, such as copper ($400\,\mathrm{W\,m^{-1}\,K^{-1}}$) because high thermal conductivities correspond to the high efficiency of heat transfer. When a bilayer cloak [7–11] is designed, the thermal conductivities of the inner and outer shells are, respectively, 0 and $2615\,\mathrm{W\,m^{-1}\,K^{-1}}$, which are singular and inexistent; see Fig. 17.1a. When the concept of neutral inclusion [12–15] is used, the thermal conductivity of the shell should be $727\,\mathrm{W\,m^{-1}\,K^{-1}}$, which is also uncommon; see Fig. 17.1b. When a near-zero-index cloak [16] is designed, the thermal conductivity of the inner shell should tend to infinity, which is singular; see Fig. 17.1c. These two problems (singular and uncommon) largely restrict the further development of thermal metamaterials because uncommon thermal conductivity is difficult to achieve, and the realization of singularity (mainly the infinite thermal conductivity) depends on very complex devices, such as thermal convection [16].

To completely solve these two problems, we propose a theory for thermal-dipole-based thermotics, which can simultaneously remove the requirements of singular and uncommon thermal conductivities. We even do not require to design any shell (or metamaterial), and a thermal dipole is enough; see Fig. 17.1d. This advantage originates from the particularity of the thermal field of a thermal dipole, which can just offset the influence of a particle by designing the thermal dipole moment (M). In what follows, we establish the theory for thermal-dipole-based thermotics in two dimensions. Finite-element simulations and laboratory experiments further validate the approach. The thermal dipole effect provides a distinct mechanism for controlling heat with heat, inspiring the thermal counterpart of coherent perfect absorbtion [18].

17.2 Thermal-Dipole-Based Thermotics

Thermal invisibility aims to keep the thermal field of the matrix undistorted. Therefore, we focus on the thermal field of the matrix in what follows. In the presence of an external uniform thermal field G_0, when there is a particle (with thermal conductivity κ_p and radius r_p) embedded in the matrix (with thermal conductivity κ_m), it will distort the uniform thermal field of the matrix. The thermal field of the matrix (generated by the external uniform thermal field), G_{me}, can be expressed as

$$G_{me} = -\nabla T_{me}. \tag{17.1}$$

T_{me} is the temperature distribution given by [20]

$$T_{me} = -G_0 r \cos\theta - \frac{\kappa_m - \kappa_p}{\kappa_m + \kappa_p} r_p^2 G_0 r^{-1} \cos\theta + T_0, \tag{17.2}$$

where (r, θ) denotes cylindrical coordinates whose origin is at the center of the particle. $G_0 = |G_0|$, and T_0 is the temperature at $\theta = \pm\pi/2$.

When there is a thermal dipole (with hot source power Q and distance l) at the center of the particle, it will generate a thermal field in the matrix. The thermal field of the matrix (generated by the thermal dipole), G_{md}, can be expressed as

$$G_{md} = -\nabla T_{md}. \tag{17.3}$$

T_{md} is the temperature distribution given by

$$T_{md} = \frac{M}{\pi \left(\kappa_m + \kappa_p\right)} r^{-1} \cos\theta + T_0, \tag{17.4}$$

where M is the thermal dipole moment given by $M = Ql$. Equation (17.4) is valid only when $r \gg l$, and details will be shown in the discussion part.

Because of the superposition principle of vector fields, the thermal field of the matrix (generated by the external uniform thermal field and the thermal dipole), $\boldsymbol{G}_s$, can be expressed as

$$\boldsymbol{G}_s = \boldsymbol{G}_{me} + \boldsymbol{G}_{md} = -\nabla T_s. \tag{17.5}$$

T_s is the temperature distribution given by

$$T_s = -G_0 r \cos\theta - \left[\frac{\kappa_m - \kappa_p}{\kappa_m + \kappa_p} r_p^2 G_0 - \frac{M}{\pi \left(\kappa_m + \kappa_p\right)} \right] r^{-1} \cos\theta + T_0. \tag{17.6}$$

As mentioned at the very beginning, thermal invisibility is characterized by the uniform thermal field of the matrix, and thus the second term on the right side of Eq. (17.6) should be zero,

$$\frac{\kappa_m - \kappa_p}{\kappa_m + \kappa_p} r_p^2 G_0 - \frac{M}{\pi \left(\kappa_m + \kappa_p\right)} = 0. \tag{17.7}$$

Solving Eq. (17.7), we can derive the thermal dipole moment,

$$M = \left(\kappa_m - \kappa_p\right) f G_0, \tag{17.8}$$

where $f = \pi r_p^2$ is the acreage of the particle. When the thermal dipole moment is set as required by Eq. (17.8), thermal invisibility can be achieved.

17.3   Finite-Element Simulation

We further perform finite-element simulations with COMSOL Multiphysics to validate the theoretical analyses. In Fig. 17.2a, d, the temperatures of the left and right boundaries are set at 323 and 283 K, and the top and bottom boundaries are insulated. If there is a particle with different thermal conductivity from the matrix in the center, isotherms are contracted due to the smaller thermal conductivity of the particle; see Fig. 17.2d. The distorted temperature profile makes the particle visible with infrared detection. Then, we explore the thermal profile of a thermal dipole; see Fig. 17.2b, e. All boundaries are insulated, and we set the temperature at $\theta = \pm\pi/2$ to 303 K as the reference temperature. The temperature profile is presented in Fig. 17.2e. Finally, we combine the structures shown in Fig. 17.2a, b and obtain the structure presented in Fig. 17.2c. As predicted by Eq. (17.8), the distorted temperature profile is restored; see Fig. 17.2f. Therefore, the particle becomes invisible with infrared detection, and thermal invisibility is achieved.

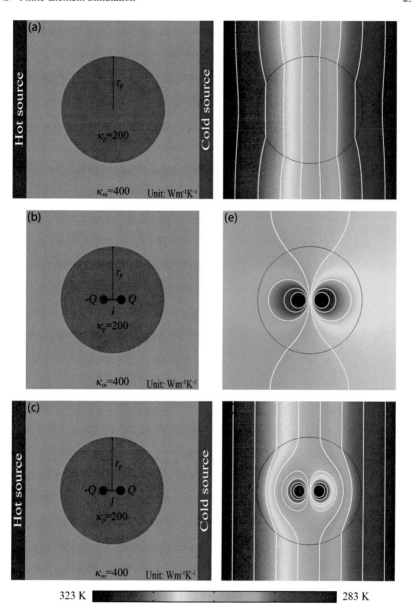

Fig. 17.2 Finite-element simulations in the presence of **a** and **d** an external uniform thermal field, **b** and **e** a thermal dipole, and **c** and **f** both an external uniform thermal field and a thermal dipole. The simulation box is 20×20 cm^2, $r_p = 6$ cm, and $l = 2$ cm. The thermal conductivities of the particle and the matrix are 200 and 400 W m^{-1} K^{-1}, respectively. The thermal dipole moment should be 452.4 W m as required by Eq. (17.8), which leads to $Q = 22620$ W. Each source of the thermal dipole has a radius of 0.5 cm. White lines represent isotherms. Temperatures higher than 323 K are shown as 323 K, and temperatures lower than 283 K are shown as 283 K. Adapted from Ref. [19]

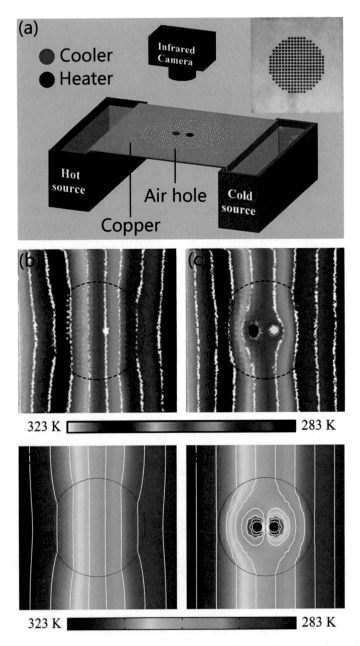

Fig. 17.3 Laboratory experiments. **a** Schematic diagrams of the sample and experimental devices. **b** and **c** are the measured results without and with a thermal dipole, respectively. **d** and **e** are the corresponding finite-element simulations based on the structure in (**a**). Copper: thermal conductivity 400 W m^{-1} K^{-1}, density 8960 kg m^{-3}, and heat capacity 385 J kg^{-1} K^{-1}; air: thermal conductivity 0.026 W m^{-1} K^{-1}, density 1.29 × 10^{-3} kg m^{-3}, and heat capacity 1005 J kg^{-1}K^{-1}. The radius of the 256 air holes is 0.22 cm, and the distance between air holes is 2/3 cm. Adapted from Ref. [19]

17.4 Laboratory Experiment

We also perform laboratory experiments to validate the theoretical analyses and finite-element simulations. We fabricate the sample based on a copper plate ($400\,\text{W m}^{-1}\,\text{K}^{-1}$); see Fig. 17.3a. Air holes ($0.026\,\text{W m}^{-1}\,\text{K}^{-1}$) are engraved on the copper plate by laser cut, which makes the effective thermal conductivity of the corresponding region to be $200\,\text{W m}^{-1}\,\text{K}^{-1}$. The upper and lower surfaces are covered with transparent and foamed plastic (insulated) to reduce infrared reflection and thermal convection.

The thermal dipole is realized by a ceramic heater and a semiconductor cooler. The designed power of a heater (or cooler) is 22620 (or -22620) W, which is an extremely large (or small) value. On the one hand, it maintains the uniform field of the matrix. On the other hand, it generates a higher (or lower) temperature inside the heater (or cooler) than the hot (or cold) source. However, the higher (or lower) temperature inside the heater (or cooler) does not contribute to the effect of thermal invisibility because only the edge temperature of the heater (or cooler) makes sense. Therefore, we only need to keep the temperature of the heater (or cooler) at 325 (or 281) K, as ensured by the uniqueness theorem in thermotics. The two temperatures can be directly obtained from finite-element simulations, depending on the heater's size (or cooler).

We measure the temperature profile with an infrared camera (FLIR E60) between the hot source (323 K) and the cold source (283 K). The measured results without and with a thermal dipole are presented in Fig. 17.3b, c. We also perform finite-element simulations based on the structure presented in Fig. 17.3a; see Fig. 17.3d, e. The experimental results (Fig. 17.3b, c) and finite-element simulations (Fig. 17.2d, f, d, e) both validate that the thermal dipole is reliable and flexible to achieve thermal invisibility.

17.5 Discussion

There is only one approximation (say, $r \gg l$) in the whole process to ensure the validity of Eq. (17.4). Therefore, we discuss the effect of this approximation on thermal invisibility. First, we compare our thermal-dipole-based result (Fig. 17.4a) with a reference (Fig. 17.4b). The temperature distributions of the matrix are the same. We also plot the temperature-difference distribution ($\Delta T = T_1 - T_2$) of the matrix (Fig. 17.4c) to perform quantitative analyses. The maximum value of the temperature difference (ΔT_{max}) is 0.04 K. Compared with the temperature difference between the hot and cold source (40 K), the relative error is only 0.1%, which shows the excellent performance of the thermal-dipole-based scheme.

We also find that the maximum value of the temperature difference (ΔT_{max}) can reflect the effect of the thermal dipole on thermal invisibility. Therefore, we calculate ΔT_{max} with different r_d (the radius of the source) and l (the distance between

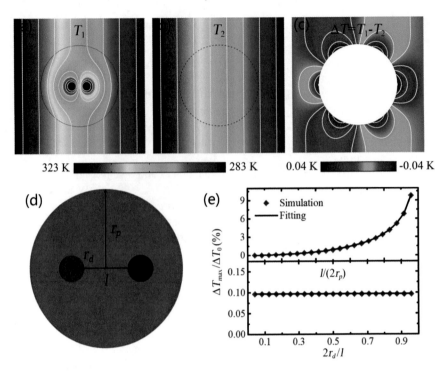

Fig. 17.4 Effects of the thermal dipole on thermal invisibility. **a** shows the thermal-dipole-based temperature distribution. **b** presents the temperature distribution when the thermal conductivities of the matrix and particle are the same (say, 400 W m^{-1} K^{-1}). **c** exhibits the temperature-difference distribution of the matrix. In **d** and **e**, we explore the effects of two parameters (l and r_d) on thermal invisibility. The upper panel in (**e**) is with $r_d = 0$ cm, say two point sources of the dipole. The lower panel in (**e**) is with $l = 2$ cm. Adapted from Ref. [19]

sources); see Fig. 17.4d, and plot two curves, showing ΔT_{max} changing with $l/2r_p$ and $2r_d/l$; see Fig. 17.4e. The top curve in Fig. 17.4e shows that the performance of the thermal dipole decreases with the increment of $l/2r_p$. When $l/2r_p = 0$ (say, $l = 0$), $\Delta T_{max} = 0$, which indicates the perfect performance. However, the bottom curve in Fig. 17.4e shows that the performance of the thermal dipole keeps unchanged with the increment of $2r_d/l$. Therefore, only one parameter (say, the distance l) mainly influences the effect of the thermal dipole on thermal invisibility, and the shorter, the better.

Further explorations on thermal dipoles could be surely expected. For example, thermal dipoles might be used to realize other thermal phenomena beyond thermal invisibility, such as thermal camouflage. Thermal dipoles may also exhibit novel properties in different systems, such as thermal Janus structures [21] and many-particle systems [22]. The properties of thermal quadrupoles may contain other interesting points.

Although the concept of a dipole originates from electromagnetism, its development in thermotics may, in turn, promote the further development of electrostatics [23] and magnetostatics [24, 25]. Indeed, the concept of a dipole may also be extended to other physical fields such as heat and mass diffusive fields [26, 27].

17.6 Conclusion

We have established a theory for thermal-dipole-based thermotics, which helps realize thermal invisibility by tailoring thermal dipole moments. The thermal-dipole-based scheme removes the requirements of singular and uncommon thermal conductivities, contributing to practical applications and further developments in thermal management. Both finite-element simulations and laboratory experiments validate the theoretical analyses. The potential applications of thermal dipoles are to mislead infrared detections, simplify the fabrication of thermal metamaterials, enhance the efficiency of heat management, etc.

17.7 Exercise and Solution

Exercise

1. Derive the three-dimensional dipole moment for thermal invisibility.

Solution

1. The thermal field of the matrix (generated by the external uniform thermal field), G'_{me}, can be expressed as

$$G'_{me} = -\nabla T'_{me}. \tag{17.9}$$

T'_{me} is the temperature distribution given by

$$T'_{me} = -G'_0 r \cos\theta - \frac{\kappa'_m - \kappa'_p}{2\kappa'_m + \kappa'_p} r'^3_p G'_0 r^{-2} \cos\theta + T'_0. \tag{17.10}$$

The thermal field of the matrix (generated by the thermal dipole), G'_{md}, can be expressed as

$$G'_{md} = -\nabla T'_{md}. \tag{17.11}$$

T'_{md} is the temperature distribution given by

$$T'_{md} = \frac{3M'}{4\pi \left(2\kappa'_m + \kappa'_p\right)} r^{-2} \cos\theta + T'_0. \tag{17.12}$$

Detailed derivations of Eq. (17.12) are as follows. The general solution to the heat conduction equation in three dimensions is

$$T = \sum_{i=0}^{\infty} \left(A_i r^{-1/2+\sqrt{1/4+i(i+1)}} + B_i r^{-1/2-\sqrt{1/4+i(i+1)}} \right) P_i (\cos\theta), \quad (17.13)$$

where P_i is Legendre polynomial. Then, we perform similar limit analyses to determine the forms of T'_{pd} and T'_{md}. We suppose $r'_p \to \infty$, and then the temperature distribution of the particle generated by the thermal dipole in three dimensions can be expressed as

$$T'_{pd}\left(r'_p \to \infty\right) = \frac{Q'}{4\pi\kappa'_p} r'^{-1}_+ + \frac{-Q'}{4\pi\kappa'_p} r'^{-1}_- = \frac{Q'l'}{4\pi\kappa'_p} r^{-2} \cos\theta = \frac{M'}{4\pi\kappa'_p} r^{-2} \cos\theta.$$

$$(17.14)$$

Equation (17.14) is valid only when $r \gg l'$ (or $l' \to 0$). The temperature distribution of a thermal dipole in three dimensions is characterized by $r^{-2} \cos\theta$. Further, we consider a finite r_p. Similar to the analyses in two dimensions, T'_{pd} and T'_{md} can be concluded as

$$T'_{pd} = \frac{M'}{4\pi\kappa'_p} r^{-2} \cos\theta + \alpha' r \cos\theta + T'_0, \quad (17.15)$$

$$T'_{md} = \beta' r^{-2} \cos\theta + T'_0. \quad (17.16)$$

The boundary conditions are given by the continuous temperatures and heat fluxes,

$$T'_{pd}\left(r_p\right) = T'_{md}\left(r_p\right), \quad (17.17)$$

$$\left(-\kappa_p \partial T'_{pd}/\partial r\right)_{r_p} = \left(-\kappa_m \partial T'_{md}/\partial r\right)_{r_p}, \quad (17.18)$$

Therefore, the undetermined coefficients can be calculated,

$$\alpha' = \frac{-M'\left(\kappa'_m - \kappa'_p\right)}{2\pi r'^3_p \kappa'_p \left(2\kappa'_m + \kappa'_p\right)}, \quad (17.19)$$

$$\beta' = \frac{3M'}{4\pi \left(2\kappa'_m + \kappa'_p\right)}. \quad (17.20)$$

Then, Eq. (17.16) turns to

$$T'_{md} = \frac{3M'}{4\pi \left(2\kappa'_m + \kappa'_p\right)} r^{-2} \cos \theta + T'_0, \tag{17.21}$$

which is just Eq. (17.12).

Because of the superposition principle, the thermal field of the matrix (generated by the external uniform thermal field and the thermal dipole), G'_s, can be expressed as

$$G'_s = G'_{me} + G'_{md} = -\nabla T'_s. \tag{17.22}$$

T'_s is the temperature distribution given by

$$T'_s = -G'_0 r \cos \theta - \left[\frac{\kappa'_m - \kappa'_p}{2\kappa'_m + \kappa'_p} r'^3_p G'_0 - \frac{3M'}{4\pi \left(2\kappa'_m + \kappa'_p\right)} \right] r^{-2} \cos \theta + T'_0. \tag{17.23}$$

Thermal invisibility requires the second term on the right side of Eq. (17.23) to be zero,

$$\frac{\kappa'_m - \kappa'_p}{2\kappa'_m + \kappa'_p} r'^3_p G'_0 - \frac{3M'}{4\pi \left(2\kappa'_m + \kappa'_p\right)} = 0, \tag{17.24}$$

Solving Eq. (17.24), we can derive the thermal dipole moment,

$$M' = \left(\kappa'_m - \kappa'_p\right) f' G'_0, \tag{17.25}$$

where $f' = 4\pi r'^3_p/3$ is the volume of the particle.

References

1. Fan, C.Z., Gao, Y., Huang, J.P.: Shaped graded materials with an apparent negative thermal conductivity. Appl. Phys. Lett. **92**, 251907 (2008)
2. Chen, T.Y., Weng, C.N., Chen, J.S.: Cloak for curvilinearly anisotropic media in conduction. Appl. Phys. Lett. **93**, 114103 (2008)
3. Narayana, S., Sato, Y.: Heat flux manipulation with engineered thermal materials. Phys. Rev. Lett. **108**, 214303 (2012)
4. Guenneau, S., Amra, C., Veynante, D.: Transformation thermodynamics: cloaking and concentrating heat flux. Opt. Express **20**, 8207 (2012)
5. Schittny, R., Kadic, M., Guenneau, S., Wegener, M.: Experiments on transformation thermodynamics: molding the flow of heat. Phys. Rev. Lett. **110**, 195901 (2013)
6. Li, Y., Bai, X., Yang, T.Z., Luo, H., Qiu, C.W.: Structured thermal surface for radiative camouflage. Nat. Commun. **9**, 273 (2018)
7. Xu, H.Y., Shi, X.H., Gao, F., Sun, H.D., Zhang, B.L.: Ultrathin three-dimensional thermal cloak. Phys. Rev. Lett. **112**, 054301 (2014)
8. Han, T.C., Bai, X., Gao, D.L., Thong, J.T.L., Li, B.W., Qiu, C.W.: Experimental demonstration of a bilayer thermal cloak. Phys. Rev. Lett. **112**, 054302 (2014)

9. Ma, Y.G., Liu, Y.C., Raza, M., Wang, Y.D., He, S.L.: Experimental demonstration of a multiphysics cloak: manipulating heat flux and electric current simultaneously. Phys. Rev. Lett. **113**, 205501 (2014)
10. Han, T.C., Bai, X., Thong, J.T.L., Li, B.W., Qiu, C.W.: Full control and manipulation of heat signatures: cloaking, camouflage and thermal metamaterials. Adv. Mater. **26**, 1731 (2014)
11. Han, T.C., Yang, P., Li, Y., Lei, D.Y., Li, B.W., Hippalgaonkar, K., Qiu, C.W.: Full-parameter omnidirectional thermal metadevices of anisotropic geometry. Adv. Mater. **30**, 1804019 (2018)
12. He, X., Wu, L.Z.: Thermal transparency with the concept of neutral inclusion. Phys. Rev. E **88**, 033201 (2013)
13. Zeng, L.W., Song, R.X.: Experimental observation of heat transparency. Appl. Phys. Lett. **104**, 201905 (2014)
14. Yang, T.Z., Bai, X., Gao, D.L., Wu, L.Z., Li, B.W., Thong, J.T.L., Qiu, C.W.: Invisible sensors: simultaneous sensing and camouflaging in multiphysical fields. Adv. Mater. **27**, 7752 (2015)
15. Yang, T.Z., Su, Y., Xu, W., Yang, X.D.: Transient thermal camouflage and heat signature control. Appl. Phys. Lett. **109**, 121905 (2016)
16. Li, Y., Zhu, K.J., Peng, Y.G., Li, W., Yang, T.Z., Xu, H.X., Chen, H., Zhu, X.F., Fan, S.H., Qiu, C.W.: Thermal meta-device in analogue of zero-index photonics. Nat. Mater. **18**, 48 (2019)
17. Pendry, J.B., Schurig, D., Smith, D.R.: Controlling electromagnetic fields. Science **312**, 1780 (2006)
18. Li, Y., Qi, M., Li, J., Cao, P.-C., Wang, D., Zhu, X.-F., Qiu, C.-W., Chen, H.: Heat transfer control using a thermal analogue of coherent perfect absorption. Nat. Commun. **13**, 2683 (2022)
19. Xu, L.J., Yang, S., Huang, J.P.: Dipole-assisted thermotics: experimental demonstration of dipole-driven thermal invisibility. Phys. Rev. E **100**, 062108 (2019)
20. Xu, L.J., Yang, S., Huang, J.P.: Designing effective thermal conductivity of materials of core-shell structure: theory and simulation. Phys. Rev. E **99**, 022107 (2019)
21. Xu, L.J., Yang, S., Huang, J.P.: Thermal theory for heterogeneously architected structure: fundamentals and application. Phys. Rev. E **98**, 052128 (2018)
22. Xu, L.J., Yang, S., Huang, J.P.: Thermal transparency induced by periodic interparticle interaction. Phys. Rev. Appl. **11**, 034056 (2019)
23. Xu, L.J., Huang, J.P.: Electrostatic chameleons: theory of intelligent metashells with adaptive response to inside objects. Eur. Phys. J. B **92**, 53 (2019)
24. Batlle, R.M., Parra, A., Laut, S., Valle, N.D., Navau, C., Sanchez, A.: Magnetic illusion: transforming a magnetic object into another object by negative permeability. Phys. Rev. Appl. **9**, 034007 (2018)
25. Xu, L.J., Huang, J.P.: Magnetostatic chameleonlike metashells with negative permeabilities. EPL **125**, 64001 (2019)
26. Guenneau, S., Puvirajesinghe, T.M.: Fick's second law transformed: one path to cloaking in mass diffusion. J. R. Soc. Interface **10**, 20130106 (2013)
27. Guenneau, S., Petiteau, D., Zerrad, M., Amra, C., Puvirajesinghe, T.: Transformed Fourier and Fick equations for the control of heat and mass diffusion. AIP Adv. **5**, 053404 (2015)

Chapter 18
Theory for Thermal Bi/Multistability: Nonlinear Thermal Conductivity

Abstract In this chapter, we theoretically design diffusive bistability (and even multistability) in the macroscopic scale, which has a similar phenomenon but a different mechanism from its microscopic counterpart (Wang et al., Phys. Rev. Lett. 101, 267203 (2008)); the latter has been extensively investigated in the literature, e.g., for building nanometer-scale memory components. By introducing second- and third-order nonlinear terms (opposite in sign) into diffusion coefficient matrices, bistable energy or mass diffusion occurs with two different steady states, identified as "0" and "1". In particular, we study heat conduction in a two-terminal three-body system. This bistable system exhibits a macro-scale thermal memory effect with tailored nonlinear thermal conductivities. Finite-element simulations confirm the theoretical analysis. Also, we suggest experiments with metamaterials based on shape memory alloys. This framework blazes a trail in constructing intrinsic bistability or multistability in diffusive systems for macroscopic energy or mass management.

Keywords Thermal bi/multistability · Nonlinear thermal conductivity · Macroscopic heat transfer

18.1 Opening Remarks

Modern electronic techniques face increasingly prominent heat dissipation problems due to shrinking chip sizes and increasing integration levels [1]. Fortunately, the past decade has witnessed the possibility of manipulating heat transport at nanoscale [2–7], which provides a promising method for evolving electron-based computation. Phononics, a microscale interpretation of controlling heat flow to carry and process information, has flourished since then [8]. To date, indispensable elements of phononic computers, including thermal diodes [2], thermal logical gates [3], and thermal memories [4], have been proposed theoretically and experimentally. The thermal memory requests a nonlinear bistable thermal circuit for basic phononic information storage. Two different steady states can be demonstrated as "0" and "1" beyond the same boundary condition, just like the electronic counterpart. Although this concept was proposed in 2008 [4], the studies of thermal bistability (TBIS) devices are still far

© The Author(s) 2023

L.-J. Xu and J.-P. Huang, *Transformation Thermotics and Extended Theories*,
https://doi.org/10.1007/978-981-19-5908-0_18

from being satisfactory (say, compared with existing research on optical bistability), which prohibits its practical applications. This situation is because most studies are executed at a microscopic scale, but the nanofabrication capacity is limited.

Recent progress in TBIS focuses on achieving bistability by introducing nonlinear thermal radiation for forming the negative differential thermal resistance (NDTR) [9–15], in which the Steffan-Boltzmann's law deviates. With success in optical bistability [16–20], it is natural to migrate its methods into thermal radiation for TBIS because both optical and thermal-radiation processes can be classified as wave physics. Comprehensively, TBIS is realized mainly in two ways: the radiative phase transition at a specific temperature region [9–12] and the anomalous radiative phenomenon such as near-field radiation or nonlinear optical resonances [13–15]. The switching time between two states has been improved to several hundred μs in the laboratory. The nonlinear thermal radiation can be a potential theoretical scheme for achieving TBIS.

However, in a macroscopic diffusive system such as heat conduction, TBIS has never been touched because of the absence of a theoretical framework analogous to its counterpart in wave systems. Nevertheless, heat conduction, a sort of major heat transfer mode described by the diffusion equation [21], can not apply to the method in wave processes because of the distinction of governing equations between diffusive and wave systems [22]. Hence, it is necessary to consider the conduction TBIS due to its ubiquity. On the one hand, thermal conduction still plays a primary role of heat dissipation in traditional electron-based computation. Conduction TBIS devices may well couple thermal and electronic memory. On the other hand, great progress has been made in manipulating macroscopic thermal conduction at will, especially in recent decades, by using the theory of transformation thermotics and thermal metamaterials [23–32], which may facilitate the design and manufacture of conduction TBIS devices. Here, we establish a bistability theory for treating diffusive systems. We take heat conduction as a classical diffusive system and deduce the non-linear heat-conduction parameters by adopting two theoretical methods. Finite-element simulations confirm it and further demonstrate a practical thermal memory process. We also give a proof-of-principle experimental design by adopting the temperature-trapping theory [32]. The theoretical framework applies to tailoring diffusion coefficient matrices for bistability (and multistability) in diffusion.

18.2 Theoretical Foundation

A diffusive system is usually described by force causing a flux. For example, Fourier's law $\boldsymbol{J} = -\kappa \nabla T$ implies that the heat flux is induced by a temperature gradient, similar to Ohm's law $\boldsymbol{I} = -\epsilon \nabla U$ and Fick's law $\boldsymbol{q} = -D \nabla n$. Generally, the relation between fluxes and forces of a diffusive system can be written as

$$Y_i = \sum_{j=1}^{n} K_{ij} X_j, \tag{18.1}$$

Fig. 18.1 a A two-terminal model for thermal bistability. Heat transfers along the x-axis. A and B are two different heat-conduction materials. C is a region for reading out and writing in. T_h and T_c are temperatures of heat baths. T_0 is the temperature of region C. **b** Schematic diagram of heat flow in region A (dotted red line)and B (dashed blue line), and the net flow of region C (solid black line). A and B have different nonlinear thermal conductivities, resulting in three intersections. Adapted from Ref. [34]

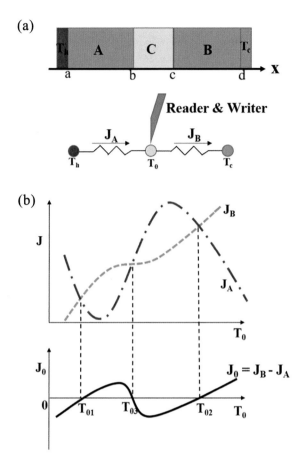

where i represents the variety of fluxes and j symbolizes different kinds of forces. Considering a simple single-field diffusion, $i = j$, Eq. (18.1) comes into no-coupled transport in the system. If the elements of transport coefficient tensor K_{ii} are constant, the relation between a flux Y_i and a force X_i is linear. However, bistability requests that the system deviates from a linear relation between Y_i and X_i. The nonlinearity of elements in the coefficient matrices becomes necessary for getting two or more steady-state solutions in the diffusion equation. Let us take Fourier's law as an example. Here, nonlinearity in macroscopic heat conduction can owe to the temperature-dependence of thermal conductivity, $\kappa(T)$ [33]. Thus, by engineering $\kappa(T)$ of a thermal circuit, an NDTR [4, 5] will work, which induces anomalous thermal diffusion. This property is essentially important for obtaining macroscopic TBIS.

Inspired by the model proposed in Ref. [4], we consider a two-terminal three-body heat transport model presented in Fig. 18.1a without loss of generality. In this case, heat flows along the x axis. A and B are two heat-conduction materials with the same sizes denoted as L (length) and S (cross-sectional area). The middle small region C shows a uniform temperature distribution due to a relatively high thermal conductivity. So C is set for extracting state information of the system. We aim to observe two divergent steady temperatures within C under the same boundary condition to achieve TBIS. Both extremes are connected to heat baths. We fix T_h and T_c as the temperature of two heat baths respectively and set T_0 as the temperature of region C. According to the continuity of heat flow, T_0 has a unique solution under the steady state if A and B are linear heat-conduction materials (namely, their thermal conductivities κ_A and κ_B are temperature-independent constants). The heat flows J_A and J_B running through A and B are linear monotonic functions of T_0, which can be verified by $J_A = \kappa_A(T_h - T_0)/L$ and $J_B = \kappa_B(T_0 - T_c)/L$. Their changes concerning T_0 are two straight lines with one intersection point [$J_A(T_0) = J_B(T_0)$], which refers to the unique heat-conduction steady state. However, steady states can increase if A and B are two nonlinear heat-conduction materials (their thermal conductivities depend on the temperature). Here we denote $\kappa_A(T)$ and $\kappa_B(T)$ as their thermal conductivities, respectively

$$\kappa_A(T) = \kappa_{A0} + \sum_m \chi_{Am} T^m, \tag{18.2}$$

$$\kappa_B(T) = \kappa_{B0} + \sum_n \chi_{Bn} T^n, \tag{18.3}$$

where m and n are positive integers. The linear relation between heat flow J_A (or J_B) and T_0 deviates. As illustrated in Fig. 18.1b, more than one intersection point of J_A and J_B. That is to say, thermal bistability or multistability phenomena can appear due to nonlinear heat conduction.

A. Calculations of Net Heat Flow

We define $J_0 = J_B - J_A$ as the net heat flow from the region C. $J_0 = 0$ is the necessary condition that a steady-state system should satisfy. In a TBIS system, $J_0(T_0) = 0$ has three real solutions. These three points are candidates of steady points. But the point where $\partial J_0/\partial T_0 < 0$ should be excluded because it is an unstable equilibrium point. Then, which steady state will the system come into? This depends on the initial conditions. As shown in Fig. 18.1b, a cubic function (rather than a quadratic function) can construct a bistable system perfectly. Thus, we can speculate that the index terms in Eqs. (18.2) and (18.3) should be kept up to the second terms. This means $m = n = 2$. Accordingly, Eqs. (18.2) and (18.3) can be reduced as

$$\kappa_A(T) = \kappa_{A0} + \chi_{A1} T + \chi_{A2} T^2, \tag{18.4}$$

$$\kappa_B(T) = \kappa_{B0} + \chi_{B1} T + \chi_{B2} T^2. \tag{18.5}$$

Distinctly, two factors influence such a TBIS system. When one factor dominates, the system will become state I (on) and vice versa into state II (off). For TBIS, the dominating factors depend on the temperature-evolution direction. For example, if relaxing from a low-temperature state, factor I dominates, and the system will enter state I. On the contrary, an initial high-temperature state will conclude in another final state. So χ_{A1} and χ_{A2} (or χ_{B1} and χ_{B2}) are inferred to have opposite signs. Based on the above analysis, we are in a position to calculate the thermal conductivity parameters for a TBIS system.

The nonlinear thermal conductivity values show position-dependent (one-to-one mapping to position x) in a steady state. But the heat flows J_A and J_B are independent of x due to the heat flow conservation. J_A can be written as $J_A = \kappa_{eA} S \langle \nabla T_A \rangle$, where κ_{eA} is the effective thermal conductivity and $\langle \nabla T_A \rangle$ is the corresponding average temperature gradient of A. So is J_B. Then we can derive J_A and J_B as

$$J_A = \kappa_{eA} S \langle \nabla T_A \rangle = \frac{\kappa_{eA}(T_h - T_0)S}{L}, \tag{18.6}$$

$$J_B = \kappa_{eB} S \langle \nabla T_B \rangle = \frac{\kappa_{eB}(T_0 - T_c)S}{L}. \tag{18.7}$$

To conclude J_A and J_B, the effective thermal conductivities should be deduced. For simplicity, we assume B is a linear heat-conduction material ($\chi_{Bn} = 0$), and only hold A's nonlinearity. This simplification will not affect the cubic relation between net heat flow J_0 and T_0. Then κ_{eA} and κ_{eB} can be written as

$$\kappa_{eA} = \frac{\int_{T_0}^{T_h} \kappa_A(T)}{T_h - T_0} = \frac{\kappa_{A0}T_h + \frac{1}{2}\chi_{A1}T_h^2 + \frac{1}{3}\chi_{A2}T_h^3 - \left(\kappa_{A0}T_0 h + \frac{1}{2}\chi_{A1}T_0^2 + \frac{1}{3}\chi_{A2}T_0^3\right)}{T_h - T_0}, \tag{18.8}$$

$$\kappa_{eB} = \kappa_{B0}. \tag{18.9}$$

Substituting Eqs. (18.8) and (18.9) into Eqs. (18.6) and (18.7), we get

$$J_A = -\frac{S}{L}\left[\frac{1}{3}\chi_{A2}T_0^3 + \frac{1}{2}\chi_{A1}T_0^2 + \kappa_{A0}T_0 - \left(\frac{1}{3}\chi_{A2}T_h^3 + \frac{1}{2}\chi_{A1}T_h^2 + \kappa_{A0}T_h\right)\right], \tag{18.10}$$

$$J_B = \frac{S}{L}(\kappa_{B0}T_0 - \kappa_{B0}T_c). \tag{18.11}$$

Defining shape factor $\Gamma = S/L$, then J_0 can be expressed as

$$J_0 = J_B - J_A$$
$$= \Gamma\left[\frac{1}{3}\chi_{A2}T_0^3 + \frac{1}{2}\chi_{A1}T_0^2 + (\kappa_{A0} + \kappa_{B0})T_0 - \left(\frac{1}{3}\chi_{A2}T_h^3 + \frac{1}{2}\chi_{A1}T_h^2 + \kappa_{A0}T_h + \kappa_{B0}T_c\right)\right].$$

$$(18.12)$$

Equations (18.4) and (18.5) describe the nonlinear heat conduction. Generally, it is hard to solve the nonlinear heat conduction differential equation. So we may adopt an effective-thermal-conductivity approximation to avoid the nonlinear terms above. In addition, Kirchhoff's transformation provides another way to make the nonlinear equation linearization [35]. As it works well in one-dimensional heat conduction problems, we can get exact solutions to temperature distributions in our model. Then comparing the two results, we can verify the above approximation results.

Let us still consider the nonlinear heat conduction in region A and assume region B has a linear thermal conductivity. Under a steady state, heat conduction in region A can be described as

$$\frac{\partial}{\partial x}\left[\kappa_A(T)\frac{\partial T}{\partial x}\right] = 0. \qquad (18.13)$$

Here, we define a new variable U, which has the same unit as a temperature,

$$U = U(T) = \int_{T_{ref}}^{T} \frac{\kappa_A(T')}{\kappa_A(T_{ref})}dT', \qquad (18.14)$$

where T_{ref} is an arbitrary reference temperature. And Eq. (18.13) can be transformed as

$$\frac{\partial}{\partial x}\left[\kappa_A(T)\frac{\partial T}{\partial U}\frac{\partial U}{\partial x}\right] = 0. \qquad (18.15)$$

Combing Eqs. (18.14) and (18.15), we can get a heat-conduction equation with U,

$$\frac{\partial^2 U}{\partial x^2} = 0. \qquad (18.16)$$

If we take $T_{ref} = 0$ K, the variable U and corresponding upper and lower bounds can be deduced as

$$U(T) = \frac{\int_0^T \left(\kappa_{A0} + \chi_{A1}T' + \chi_{A2}T'^2\right)dT'}{\kappa_{A0}} = \frac{\kappa_{A0}T + \frac{1}{2}\chi_{A1}T^2 + \frac{1}{3}\chi_{A2}T^3}{\kappa_{A0}},$$

$$(18.17)$$

and

$$\begin{cases} U_1 = \dfrac{\kappa_{A0}T_h + \frac{1}{2}\chi_{A1}T_h^2 + \frac{1}{3}\chi_{A2}T_h^3}{\kappa_{A0}} \quad (x = a), \\[3mm] U_2 = \dfrac{\kappa_{A0}T_0 + \frac{1}{2}\chi_{A1}T_0^2 + \frac{1}{3}\chi_{A2}T_0^3}{\kappa_{A0}} \quad (x = b). \end{cases} \qquad (18.18)$$

Combing Eqs. (18.16) and (18.18) together, we can solve the expression of U as

$$U(x) = \frac{U_2 - U_1}{L}x + U_1, \tag{18.19}$$

which indicates the value of U at each position. It is easy to migrate $U(x)$ back to $T(x)$. Thus, by means of the intermediate variable U, we can find the relation between T and x as

$$U = \frac{\kappa_{A0}T + \frac{1}{2}\chi_{A1}T^2 + \frac{1}{3}\chi_{A2}T^3}{\kappa_{A0}} = \frac{U_2 - U_1}{L}x + U_1. \tag{18.20}$$

Taking the derivative of Eq. (18.20) with respect to x in region A, we get $\left.\frac{\partial T}{\partial x}\right|_A$,

$$\left.\frac{\partial T}{\partial x}\right|_A = \frac{\kappa_{A0}(U_2 - U_1)}{\kappa_A(T)L}. \tag{18.21}$$

Then, the net outflow of heat from C can be written as

$$J_0^* = J_B^* - J_A^* = \kappa_{B0}\frac{T_0 - T_c}{L}S + \kappa_A(T)\left.\frac{\partial T}{\partial x}\right|_A S$$

$$= \Gamma\left[\frac{1}{3}\chi_{A2}T_0^3 + \frac{1}{2}\chi_{A1}T_0^2 + (\kappa_{A0} + \kappa_{B0})T_0 - \left(\frac{1}{3}\chi_{A2}T_h^3 + \frac{1}{2}\chi_{A1}T_h^2 + \kappa_{A0}T_h + \kappa_{B0}T_c\right)\right], \tag{18.22}$$

which echoes with Eq. (18.12). It is definite that a nonlinear one-dimensional heat conduction process can be simplified by executing the space averaging of $\kappa(T)$, which makes a detour around the nonlinear terms. This will facilitate the disposal of nonlinear-heat-conduction case.

B. Tailoring Nonlinear-thermal-Conductivities Coefficients

We can see J_0 satisfies a cubic relation with T_0. Now we construct another cubic function $J_0'(T_0)$ with three zero points T_{01}, T_{02}, T_{03} (suppose $T_c < T_{01} < T_{03} < T_{02} < T_h$). J_0' can be written as

$$J_0' = \alpha[(T_0 - T_{01})(T_0 - T_{02})(T_0 - T_{03})]$$
$$= \alpha[T_0^3 - (T_{01} + T_{02} + T_{03})T_0^2 + (T_{01}T_{02} + T_{01}T_{03} + T_{02}T_{03})T_0 - T_{01}T_{02}T_{03}]. \tag{18.23}$$

α is the pre-coefficient with a unit J/K. T_{01} and T_{02} are the two designed stable temperatures of region C. By comparing the coefficient and constant terms of Eqs. (18.12) and (18.23), we acquire a set of equations

$$\begin{cases} \frac{1}{3}\Gamma\chi_{A2} = \alpha, \\[2mm] \frac{1}{2}\Gamma\chi_{A1} = -\alpha(T_{01} + T_{02} + T_{03}), \\[2mm] \Gamma(\kappa_{A0} + \kappa_{B0}) = \alpha(T_{01}T_{02} + T_{01}T_{03} + T_{02}T_{03}), \\[2mm] -\Gamma\left(\frac{1}{3}\chi_{A2}T_h^3 + \frac{1}{2}\chi_{A1}T_h^2 + \kappa_{A0}T_h + \kappa_{B0}T_c\right) = -\alpha T_{01}T_{02}T_{03}. \end{cases} \tag{18.24}$$

Then we achieve

$$\kappa_{A0} = \frac{\alpha}{\Gamma}\left[\frac{-T_h^3 + (T_{01} + T_{02} + T_{03})T_h^2 - (T_{01}T_{02} + T_{01}T_{03} + T_{02}T_{03})T_c + T_{01}T_{02}T_{03}}{T_h - T_c}\right],$$

$$\kappa_{B0} = \frac{\alpha}{\Gamma}\left[\frac{T_h^3 - (T_{01} + T_{02} + T_{03})T_h^2 + (T_{01}T_{02} + T_{01}T_{03} + T_{02}T_{03})T_h - T_{01}T_{02}T_{03}}{T_h - T_c}\right],$$

$$\chi_{A1} = -\frac{2\alpha}{\Gamma}(T_{01} + T_{02} + T_{03}),$$

$$\chi_{A2} = \frac{3\alpha}{\Gamma}. \tag{18.25}$$

We can see χ_{A1} and χ_{A2} have opposite signs definitely, which echo the inference above. Therefore, a bistable system features that two kinds of factors compete in evolution from a non-equilibrium state to an equilibrium state. T_{01} and T_{02} are representations of two different states, while T_{03} can not exist in a steady state. Equation (18.25) provides guidance in designing nonlinear parameters of heat-conduction objects to realize TBIS. We can calculate the coefficients according to the pre-set zero-point temperatures (T_{01}, T_{02}, and T_{03}), the temperatures of heat baths, and two factors Γ and α.

This method allows a diffusive system to exhibit bistable states by engineering nonlinear transport coefficients. This intrinsic bistability depends on two competitive factors, reflected by two nonlinear terms with opposite signs. We prove that the second- and third-order nonlinearity of transport coefficients makes bistability effects valid. If the nonlinearity orders are higher, multistability can come to appear. And the switching time depends on the diffusion velocity of heat or mass.

18.3   Numerical Analysis and Simulation

We draw the graphs to illustrate our methods for tailoring nonlinear thermal conductivities. On basis of the model shown in Fig. 18.1a, we set $T_{01} = 330$ K, $T_{02} = 370$ K, and $T_{03} = 350$ K. The heat and cold baths are fixed at 400 K and 300 K, respectively. Two factors are set as $\alpha = 0.001$ J/K and $\Gamma = 1$ m. The substitution of these parameters into Eq. (18.25) yields $\kappa_{A0} = 366.05$ J/(m K), $\kappa_{B0} = 1.05$ J/(m K),

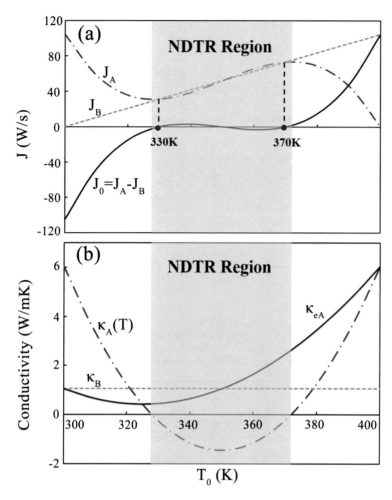

Fig. 18.2 Analysis of the bistability and NDTR based on the analytical model discussed in the text. **a** Heat flow in region A (dotted red line), B (dashed blue line), and net flow in region C (solid black line) versus T_0 of the system. Here B is a linear heat-conduction material, and the J_B curve is a straight line. J_A and J_B have three intersections. **b** thermal conductivities of A (dotted red line) and B (dashed blue line) versus T_0. The effective thermal conductivity of A is also shown with a solid red line by an integral average of T_0. The NDTR region is shadowed in yellow, containing two stable temperature points. Adapted from Ref. [34]

$\chi_{A1} = -2.1$ J/(m K^2), and $\chi_{A2} = 0.003$ J/(m K^3). The curves of J_A, J_B, and J_0 versus T_0 are shown in Fig. 18.2a. Three intersections emerge, corresponding to the pre-set parameters T_{01}, T_{02}, and T_{03}. In Fig. 18.2b, the thermal conductivities of A and B versus temperature are depicted. We can see that $\kappa_A(T)$ has negative values in a certain temperature region. This value is calculated as (328.02 K, 371.98 K), which refers to the NDTR region (see the yellow-shadowed region in Fig. 18.2). The region

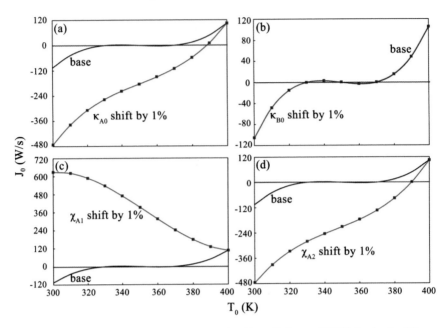

Fig. 18.3 Net flow J_0 versus T_0 for different small-shift coefficients. **a** TBIS behavior for different linear coefficients κ_{A0}. **b** TBIS behavior for different linear coefficients κ_{B0}. **c** TBIS behavior for different second-order coefficients χ_{A1}. **d** TBIS behavior for different third-order coefficients χ_{A2}. Adapted from Ref. [34]

contains two stable temperatures, confirming that the NDTR induces the desired TBIS. These two graphs accord with our expected results as sketched in Fig. 18.1b.

When the coefficients of nonlinear thermal conductivities have slight variations, will TBIS be broken? Here, we give a 1% value shift to four parameters respectively (κ_{A0}, κ_{B0}, χ_{A1}, and χ_{A2} are increased by 1%, respectively). According to the comparisons in Fig. 18.3, the small shift of κ_{B0} cannot affect the TBIS, which can be interpreted by the steady heat flow in the system kept almost unchanged. While the thermal conductivity of A varies slightly, TBIS will not exist anymore. So we can conclude that the TBIS of heat conduction is parameter-sensitive. This strict limitation makes it hard to observe the TBIS phenomenon in practical heat-conduction materials. But we can carefully tailor an intrinsic TBIS with pre-designed zero-point temperatures.

We perform finite-element simulations based on the commercial software COMSOL Multiphysics. We build a model with 9 cm in length and 1 cm in width. Heat conducts along the x-axis. The thermostat region is placed in the center with $\kappa_C = 1000$ J/(m K). We give 400 K, 500 K, and 600 K as three pre-designed zero points for the thermal conductivities of the left and right parts. Γ is 1/4 m according to the model's geometry. α is arbitrary, and here we take it as 0.0001 J/K. Thus, we can calculate that $\kappa_{A0} = 290$ J/(m K), $\kappa_{B0} = 6$ J/(m K), $\chi_{A1} = -1.2$ J/(m K^2),

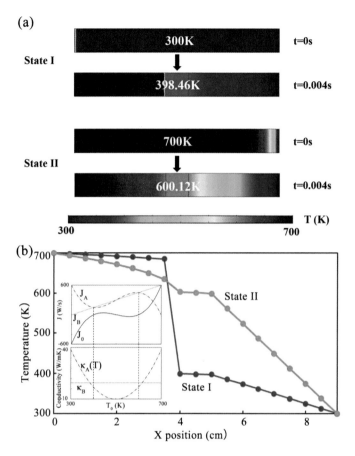

Fig. 18.4 Finite-element simulations of TBIS. **a** Transient simulation results beyond fixed heat baths' temperatures. After 0.004 s, the system becomes stable with two different T_0 due to the different initial temperatures. **b** Temperature distribution along x-axis. The left end of the model is set as the origin point ($x = 0$). Theoretical heat flows and thermal conductivities of the model are compared with the simulation results in vignettes. Adapted from Ref. [34]

and $\chi_{A2} = 0.0012$ J/(m K^3). The density and specific heat of all materials are set as 10 kg/m^3 and 10 J/(kg K). Boundary conditions are fixed at 700 K (left) and 300 K (right). Then, we give 300 K and 700 K as initial surface temperatures, see Fig. 18.4a. After the temperature evolution within 0.004 s, the system comes into stable states. However, the final temperatures of C are different according to different initial temperatures, representing two different stable states. The initial temperature of 300 K induces 398.46 K (stable state I) in C, while 700 K leads to 600.12 K (stable state II). The states of C depend on the initial surface temperatures. We fetch the final-state-temperature data of the model along the x-axis and curve it in Fig. 18.4b. State I and II have two different platform temperatures in region C (4~5 cm). In addition, we plot the theoretical results of heat flow and thermal conductivities versus T_0 as inset

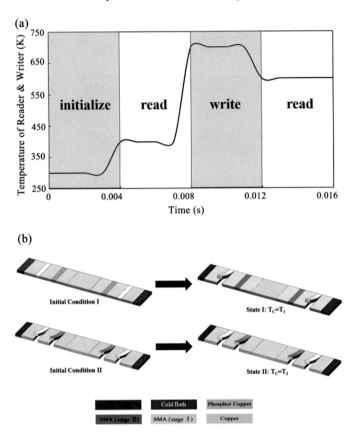

Fig. 18.5 **a** A demonstration of the thermal memory process with the model we design. Four stages are displayed initialization, reading-out, writing-in, and reading-out. **b** An experimental design based on the temperature-trapping theory. Two stages with different types of SMA are arrayed. The central temperatures depend on the SMA stages' critical temperatures. Adapted from Ref. [34]

diagrams in Fig. 18.4b. Both 400 K and 600 K are the pre-set stable values for designing the thermal conductivity parameters. And the simulation results well confirm the theoretical values.

Then, we demonstrate an overall thermal memory process with the designed conduction TBIS in Fig. 18.5a, which is based on the simulation results above. Firstly, we initialize the model by a temperature-writer in 300 K as an initial temperature. After 0.004 s, the system will become steady, and we read out T_0 in region C by a temperature-reader. It is 398.46 K now. And then we write in another temperature as 700 K. After 0.004 s, a steady temperature of 600.12 K can be read out. This model's switching time is 0.004 s, which depends on each part's density and specific heat. So these two parameters should be considered and optimized when devices are in practical application. This memory process makes the conduction TBIS practicable in fabricating macroscopic thermal memory components.

18.4 Experimental Suggestion

The temperature-trapping theory [32] inspires us to design a proof-of-principle experiment. This theory implies a thermostat region in the center of a spatially symmetric structure within shape memory alloys (SMAs). The thermostat's temperature depends entirely on the critical temperature of SMAs. Here, we improve this structure and design a two-stage SMAs device to achieve TBIS, as shown in Fig. 18.5b. Two pairs of SMAs are arrayed on both sides, in white and gray, forming two-stage thermal switches. Different types of SMAs are applied in each pair. In particular, these two stages have different critical temperatures T_1 and T_2, where $T_1 < T_2$. In detail, the white stage on the left levels below T_1 and bends above T_1, while the right one shows the same T_1 but inverse deformation. A similar rule works on the gray stage. Heat and cold baths are fixed on both sides with $T_h > T_2$ and $T_c < T_1$. When the whole device is initialized under a low temperature on the left and a high temperature on the right, all the SMAs get straight. The outer stage bends, and the heat flow cannot run into the inner layer when coming to the steady state. The thermostat's temperature approaches T_1. When the initial condition reverses, all stages bend. They will not be level at steady state as $T_h > T_2$ and $T_c < T_1$. This process induces another steady state that T_2 is the final temperature of the thermostat. As the SMAs are commercially available, assembling such a two-stage structure is feasible. But the thermal contact resistances may affect the experimental results, which should be considered further.

18.5 Discussion

For the temperature-dependent thermal conductivity of A depicted in Fig. 18.1a, the third-order nonlinearity is just a necessary condition. We can find that $|\kappa_{A0}| \sim |\chi_{A1}T| \sim |\chi_{A2}T^2|$ is another parameter requirement. Fortunately, these extraordinary thermal properties were proved to emerge in some bulk nonmetallic solids [36]. For example, the thermal conductivity of bulk ZrO_2 is $4.00 - 8.72 \times 10^{-3}T + 1.28 \times 10^{-5}T^2 - 5.82 \times 10^{-9}T^3$[W/(m K)], which agrees qualitatively with the conduction TBIS requires at 10^3 K level. It can be applied as material A in our model combing with a common material B. By solving the inverse solutions of Eq. (18.24), namely working out α, T_{01}, T_{02}, and T_{03}, one can estimate the experimental bistable temperature for such a structure composed of a nonlinear bulk heat-conduction material plus a common material. Thus, the observation of conduction TBIS in natural materials is practically probable. Besides, using the composite effect of nonlinear heat transfer [37], the fabrication of a conduction TBIS device with composite materials is possible. In this case, the nonlinear thermal conductivities can be well-tailored if adjusting the fraction or configuration of composite bulk components [38]. For example, a core-shell structure [33] and a particle-embedded-in-host structure [39] may be candidates. So we also suggest the composite manufacture method as material A in fabricating the device for application scenarios.

We have established a theoretical framework for achieving bistability in diffusive heat systems. We prove that the TBIS phenomenon exists not only in wave processes (say, nonlinear thermal radiation) but also can be realized in heat-conduction systems. Second- and third-order nonlinearity of thermal conductivity can induce a bistable thermal circuit. When the nonlinearity orders go higher, multistability can be observed as well. We have also given numerical calculation results and show the parameter-sensitive TBIS in heat conduction. Besides, a completed thermal memory process is demonstrated with four stages as an evident consequence. Except for thermal memories, a thermal switch is another possible application. As the designed experiment implies, the switch is initial-temperature-forced and can barrier or allow heat flows due to distinguishable thermal conductivities. As waste heat is dissipated mainly by the diffusive process in traditional computers, conduction TBIS devices can thus be coupled with electronic devices, facilitating thermal calculation based on existing electric calculation.

18.6 Conclusion

We have introduced an approach to designing macroscopic bistability by taking the heat conduction process as a typical case. Due to the form-similarity of governing equations, this method is applicable in other diffusive systems, such as direct current or particle diffusion systems. Bistability or multistability can be realized by carefully tailoring spatial asymmetry and nonlinearity of diffusive parameters. This method helps generate a significant physical phenomenon in the macroscopic diffusive process and is a potential tool in macroscopic energy or mass management.

18.7 Exercise and Solution

Exercise

1. Discuss the differences between a negative differential thermal resistance and a negative thermal conductivity.

Solution

1. A negative differential thermal resistance means that the heat flux decreases when the temperature difference increases, which naturally does not violate the second law of thermodynamics.
   A negative thermal conductivity means heat can spontaneously transport from low to high temperatures, which is usually impossible due to the second law of thermodynamics. However, it can still be effectively realized if an external source is applied.

References

1. Maldovan, M.: Sound and heat revolutions in phononics. Nature **503**, 209 (2013)
2. Li, B., Wang, L., Casati, G.: Thermal diode: rectification of heat flux. Phys. Rev. Lett. **93**, 184301 (2004)
3. Wang, L., Li, B.: Thermal logic gates: computation with phonons. Phys. Rev. Lett. **99**, 177208 (2007)
4. Wang, L., Li, B.: Thermal memory: a storage of phononic information. Phys. Rev. Lett. **101**, 267203 (2008)
5. Lee, W., Kim, K., Jeong, W., Zotti, L.A., Pauly, F., Cuevas, J.C., Reddy, P.: Heat dissipation in atomic-scale junctions. Nature **498**, 209 (2013)
6. Balram, K.C., Davanco, M.I., Song, J.D., Srinivasan, K.: Coherent coupling between radiofrequency, optical and acoustic waves in piezo-optomechanical circuits. Nat. Photonics **10**, 346 (2016)
7. Lee, J., Lee, W., Wehmeyer, G., Dhuey, S., Olynick, D.L., Cabrini, S., Dames, C., Urban, J.J.: and P, Yang, Investigation of phonon coherence and backscattering using silicon nanomeshes. Nat. Commun. **8**, 14054 (2017)
8. Li, N., Ren, J., Wang, L., Zhang, G., Hanggi, P., Li, B.: Colloquium: Phononics: manipulating heat flow with electronic analogs and beyond. Rev. Mod. Phys. **84**, 1045 (2012)
9. Xie, R., Bui, C.T., Varghese, B., Zhang, Q., Sow, C.H., Li, B., Thong, J.T.L.: An electrically tuned solid-state thermal memory based on metal-insulator transition of single-crystalline VO_2 nanobeams. Adv. Funct. Mater. **21**, 1602 (2011)
10. Wu, C., Feng, F., Xie, Y.: Design of vanadium oxide structures with controllable electrical properties for energy applications. Chem. Soc. Rev. **42**, 5157 (2013)
11. Ben-Abdallah, P., Biehs, S.: Near-field thermal transistor. Phys. Rev. Lett. **112**, 044301 (2014)
12. Kubytskyi, V., Ben-Abdallah, P., Biehs, S.: Radiative bistability and thermal memory. Phys. Rev. Lett. **113**, 074301 (2014)
13. Elzouka, M., Ndao, S.: Near-field nanothermomechanical memory. Appl. Phys. Lett. **105**, 243510 (2014)
14. Khandekar, C., Rodriguez, A.W.: Thermal bistability through coupled photonic resonances. Appl. Phys. Lett. **111**, 083104 (2017)
15. Morsy, A.M., Biswas, R., Povinelli, M.L.: High temperature, experimental thermal memory based on optical resonances in photonic crystal slabs. APL Photonics **4**, 010804 (2019)
16. Bergman, D.J., Levy, O., Stroud, D.: Theory of optical bistability in a weakly nonlinear composite medium. Phys. Rev. B **49**, 129 (1994)
17. Gao, L., Gu, L., Li, Z.: Optical bistability and tristability in nonlinear metal/dielectric composite media of nonspherical particles. Phys. Rev. E **68**, 066601 (2003)
18. Litchinitser, N.M., Gabitov, I.R., Maimistov, A.I.: Optical bistability in a nonlinear optical coupler with a negative index channel. Phys. Rev. Lett. **99**, 113902 (2007)
19. Chen, P.Y., Farhat, M., Alù, A.: Bistable and self-tunable negative-index metamaterial at optical frequencies. Phys. Rev. Lett. **106**, 105503 (2011)
20. Huang, Y., Gao, L.: Tunable Fano resonances and enhanced optical bistability in composites of coated cylinders due to nonlocality. Phys. Rev. B **93**, 235439 (2016)
21. Narasimhan, T.N.: Fourier's heat conduction equation: history, influence, and connections. Rev. Geophys. **37**, 151 (1999)
22. Zhukovsky, K.V., Srivastava, H.M.: Analytical solutions for heat diffusion beyond Fourier law. Appl. Math. Compt. **293**, 423 (2017)
23. Fan, C.Z., Gao, Y., Huang, J.P.: Shaped graded materials with an apparent negative thermal conductivity. Appl. Phys. Lett. **92**, 251907 (2008)
24. Chen, T.Y., Weng, C.-N., Chen, J.-S.: Cloak for curvilinearly anisotropic media in conduction. Appl. Phys. Lett. **93**, 114103 (2008)

25. Guenneau, S., Amra, C., Veynante, D.: Transformation thermodynamics: cloaking and concentrating heat flux. Opt. Express **20**, 8207 (2012)
26. Narayana, S., Sato, Y.: Heat flux manipulation with engineered thermal materials. Phys. Rev. Lett. **108**, 214303 (2012)
27. Schittny, R., Kadic, M., Guenneau, S., Wegener, M.: Experiments on transformation thermodynamics: molding the flow of heat. Phys. Rev. Lett. **110**, 195901 (2013)
28. Xu, H.Y., Shi, X.H., Gao, F., Sun, H.D., Zhang, B.L.: Ultrathin three-dimensional thermal cloak. Phys. Rev. Lett. **112**, 054301 (2014)
29. Han, T.C., Bai, X., Gao, D.L., Thong, J.T.L., Li, B.W., Qiu, C.-W.: Experimental demonstration of a bilayer thermal cloak. Phys. Rev. Lett. **112**, 054302 (2014)
30. Ma, Y.G., Liu, Y.C., Raza, M., Wang, Y.D., He, S.L.: Experimental demonstration of a multiphysics cloak: manipulating heat flux and electric current simultaneously. Phys. Rev. Lett. **113**, 205501 (2014)
31. Li, Y., Shen, X.Y., Wu, Z.H., Huang, J.Y., Chen, Y.X., Ni, Y.S., Huang, J.P.: Temperature-dependent transformation thermotics: from switchable thermal cloaks to macroscopic thermal diodes. Phys. Rev. Lett. **115**, 195503 (2015)
32. Shen, X.Y., Li, Y., Jiang, C.R., Huang, J.P.: Temperature Trapping: energy-free maintenance of constant temperatures as ambient temperature gradients change. Phys. Rev. Lett. **117**, 055501 (2016)
33. Yang, S., Xu, L.J., Huang, J.P.: Metathermotics: nonlinear thermal responses of core-shell metamaterials. Phys. Rev. E **99**, 042144 (2019)
34. Wang, J., Dai, G.L., Yang, F.B., Huang, J.P.: Designing bistability or multistability in macroscopic diffusive systems. Phys. Rev. E **101**, 022119 (2020)
35. Nakwaski, W.: An application of Kirchhoff transformation to solving the non-linear thermal conduction equation for a laser diode. Opt. Appl. **10**, 281 (1980)
36. Touloukian, Y.S., Livey, P.E., Saxena, S.C.: Thermal Conductivity: Nonmetallic Solids. IFI/Plenum Press, New York (1970)
37. Lu, G., Wang, X.D., Duan, Y.Y., Li, X.W.: Effects of non-ideal structures and high temperatures on the insulation properties of aerogel-based composite materials. J. Non-Cryst. Solids **357**, 3822 (2011)
38. Dai, Y.J., Tang, Y.Q., Fang, W.Z., Zhang, H., Tao, W.Q.: A theoretical model for the effective thermal conductivity of silica aerogel composites. Appl. Therm. Eng. **128**, 1634 (2018)
39. Dai, G.L., Shang, J., Wang, R.Z., Huang, J.P.: Nonlinear thermotics: nonlinearity enhancement and harmonic generation in thermal metasurfaces. Eur. Phys. J. B **91**, 59 (2018)

Chapter 19
Theory for Negative Thermal Transport: Complex Thermal Conductivity

Abstract In this chapter, we coin a complex thermal conductivity whose imaginary part corresponds to the real part of a complex refractive index. Therefore, the thermal counterpart of a negative refractive index is just a negative imaginary thermal conductivity, featuring the opposite directions of energy flow and wave vector in thermal conduction and advection, thus called negative thermal transport herein. We design an open system with energy exchange and explore three different cases to reveal negative thermal transport to avoid violating causality. We further provide experimental suggestions with a solid ring structure. All finite-element simulations agree with the theoretical analyses, indicating that negative thermal transport is physically feasible. These results have potential applications such as designing the inverse Doppler effect in thermal conduction and advection.

Keywords Negative thermal transport · Complex thermal conductivity · Conduction and advection

19.1 Opening Remarks

Negative refraction is one of the most attractive phenomena in wave systems, which was first revealed with negative permeability and permittivity [1]. With the proposal of metal wire arrays [2] and split rings [3], negative refractive index was soon designed and fabricated [4–6], which gave birth to broad applications like breaking diffraction limit [7–9]. One representative property of a negative refractive index is the opposite directions of energy flow (or Poynting vector) and wave vector [Fig. 19.1(a)]. Based on this property, related phenomena were revealed intensively, such as the inverse Doppler effect [10], the inverse Cerenkov radiation [11], and the abnormal Goos-Hanschen shift [12].

Refractive phenomena were also studied with a thermal wave with a time-periodic heat source [13]. Moreover, multilayered structures were also proposed to guide heat flow [14–17], yielding practical applications such as thermal lens [18] and thermal cloaks [19]. These studies attempted to connect thermal phenomena and

© The Author(s) 2023

L.-J. Xu and J.-P. Huang, *Transformation Thermotics and Extended Theories*,

https://doi.org/10.1007/978-981-19-5908-0_19

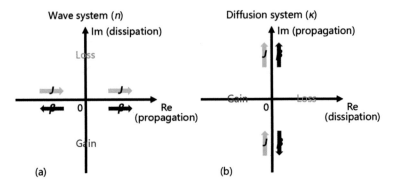

Fig. 19.1 Comparison between **a** wave system and **b** diffusion system. n and κ denote complex refractive index and complex thermal conductivity, respectively. J and β denote energy flow and wave vector, respectively. Adapted from Ref. [20]

electromagnetic phenomena. However, some basic concepts are still ambiguous, especially the correspondence between thermal conductivity and refractive index.

To promote the related physics in thermotics with a clear physical picture, we manage to coin a complex thermal conductivity κ as the thermal counterpart of a complex refractive index n (Fig. 19.1). The imaginary part of a complex thermal conductivity is analogous to the real part of a complex refractive index. Therefore, the thermal counterpart of a negative refractive index is just a negative imaginary thermal conductivity, which is characterized by the opposite directions of energy flux J and wave vector β, thus called negative thermal transport. We design an open system with energy exchange to observe negative thermal transport and provide experimental suggestions with a three-dimensional solid ring structure. All theoretical analyses and finite-element simulations indicate that negative thermal transport is physical.

19.2   Complex Thermal Conductivity

Thermal conduction-advection process is dominated by the famous equation

$$\rho C \partial T / \partial t + \nabla \cdot (-\sigma \nabla T + \rho C \boldsymbol{v} T) = 0, \tag{19.1}$$

where $\rho, C, \sigma, \boldsymbol{v}, T$, and t are density, heat capacity, thermal conductivity, convective velocity, temperature, and time, respectively. Equation (19.1) indicates the energy conservation of thermal conduction and advection. We assume that the convective term $(\rho C \boldsymbol{v} T)$ results from the motion of solid, so density and heat capacity can be seen as two constants which do not depend on time or temperature [21, 22]. Therefore, the mass and momentum conservations of thermal advection are naturally satisfied.

To proceed, we apply a plane-wave solution for temperature [21–23],

$$T = A_0 e^{i(\boldsymbol{\beta}\cdot\boldsymbol{r}-\omega t)} + T_0, \tag{19.2}$$

where A_0, $\boldsymbol{\beta}$, $\boldsymbol{r}$, ω, and T_0 are the amplitude, wave vector, position vector, frequency, and reference temperature of wave-like temperature profile, respectively. $i = \sqrt{-1}$ is imaginary unit. Only the real part of Eq. (19.2) makes sense. We substitute Eq. (19.2) into Eq. (19.1), and derive a dispersion relation,

$$\omega = \boldsymbol{v}\cdot\boldsymbol{\beta} - i\frac{\sigma\beta^2}{\rho C}. \tag{19.3}$$

With the wave-like temperature profile described by Eq. (19.2), we can derive $\nabla T = i\boldsymbol{\beta}T$ (T_0 is neglected for brevity). Then, Eq. (19.1) can be rewritten as

$$\rho C \partial T/\partial t + \nabla \cdot (-i\sigma\boldsymbol{\beta}T + \rho C\boldsymbol{v}T) = 0. \tag{19.4}$$

With the mass conservation of thermal advection, we can obtain $\nabla \cdot (\rho\boldsymbol{v}) = 0$ or $\nabla \cdot \boldsymbol{v} = 0$ (for ρ is a constant). Meanwhile, $\boldsymbol{\beta}$ is a constant vector, so Eq. (19.4) can be reduced to

$$\rho C \partial T/\partial t - i\sigma\boldsymbol{\beta}\cdot\nabla T + \rho C\boldsymbol{v}\cdot\nabla T = 0, \tag{19.5}$$

which can be further reduced with $\nabla T = i\boldsymbol{\beta}T$ to

$$\rho C \partial T/\partial t + \sigma\boldsymbol{\beta}^2 T + i\rho C\boldsymbol{v}\cdot\boldsymbol{\beta}T = 0. \tag{19.6}$$

Now, it is natural to coin a complex thermal conductivity κ as

$$\kappa = \sigma + i\frac{\rho C\boldsymbol{v}\cdot\boldsymbol{\beta}}{\beta^2}, \tag{19.7}$$

with which Eq. (19.6) can be simplified as

$$\rho C \partial T/\partial t + \kappa\boldsymbol{\beta}^2 T = 0. \tag{19.8}$$

With $\nabla T = i\boldsymbol{\beta}T$, Eq. (19.8) is equivalent to the familiar equation of thermal conduction,

$$\rho C \partial T/\partial t + \nabla \cdot (-\kappa\nabla T) = 0. \tag{19.9}$$

Clearly, the thermal conduction-advection equation (Eq. (19.1)) is converted to the complex thermal conduction equation (Eq. (19.9)) with a complex thermal conductivity (Eq. (19.7)). Although thermal conductivity is generally defined by fixing moving parts (advection) as zero, advection can be mathematically regarded as a complex form of conduction. Conduction and advection are mathematically unified within a conductive framework (despite different physical mechanisms). Therefore, coining complex thermal conductivity makes sense by treating advection as a complex form of conduction.

With the substitution of Eq. (19.2) into Eq. (19.9), we can derive a dispersion relation,

$$\omega = -i\frac{\kappa\beta^2}{\rho C} = v \cdot \beta - i\frac{\sigma\beta^2}{\rho C}, \tag{19.10}$$

which is in accordance with the result of Eq. (19.3), indicating that complex thermal conductivity is physical.

To understand the complex frequency ω, we substitute Eq. (19.10) into Eq. (19.2), and the wave-like temperature profile becomes

$$T = A_0 e^{\text{Im}(\omega)t} e^{i[\beta \cdot r - \text{Re}(\omega)t]} + T_0. \tag{19.11}$$

Obviously, $\text{Re}(\omega)$ and $\text{Im}(\omega)$ determine propagation and dissipation, respectively. Meanwhile, $\text{Re}(\omega)$ and $\text{Im}(\omega)$ are related to Im (κ) and Re (κ), respectively. In other words, Re (κ) and Im (κ) are related to dissipation and propagation, respectively. The physical connotation can be clearly understood with Fig. 19.1b. Positive (or negative) Re (κ) means loss (or gain), indicating the amplitude decrement (or increment) of wave-like temperature profile. Im (κ) is of our interest, which is discussed later.

We further confirm complex thermal conductivity in a thermal conduction-advection system with COMSOL Multiphysics. The system is shown in Fig. 19.2a, which has width L and height H. The left and right ends are set with a periodic boundary condition. Then, the wave vector can take on only discrete values, say, $\beta = 2\pi m/L$ with m being any positive integers. We take on $m = 5$, and initial temperature is set at $T = 40\cos(\beta x) + 323$ K (Fig. 19.2b, f).

We discuss two cases with $v//\beta$ (Fig. 19.2b–e) and $v \perp \beta$ (Fig. 19.2f–i). When $v//\beta$, Im (κ) appears due to $v \cdot \beta \neq 0$, as predicted by Eq. (19.7). Therefore, propagation occurs and the period of wave-like temperature profile is $t_0 = 2\pi/\text{Re}(\omega) = 2\pi/(v \cdot \beta) = 100$ s, as predicted by Eq. (19.10). The wave-like temperature profiles at $t = 0.5t_0 = 50$ s and $t = t_0 = 100$ s are shown in Fig. 19.2c, d, respectively. The temperature distributions along x axis in Fig. 19.2b–d are plotted in Fig. 19.2e. Clearly, the wave-like temperature profile has amplitude decrement because of the positive Re (κ). Meanwhile, the wave-like temperature profile propagates along x axis due to the positive Im (κ). After propagating for a period (100 s), the wave-like temperature profile approximately gains a phase difference of 2π, thus going back to the initial position (Fig. 19.2b, d).

When $v \perp \beta$, Im (κ) vanishes due to $v \cdot \beta = 0$. Therefore, propagation does not occur, and the period is $t_0 = 2\pi/(v \cdot \beta) = \infty$ s, as predicted by Eq. (19.10). The wave-like temperature profiles at $t = 50$ s and $t = 100$ s are presented in Fig. 19.2g, h, respectively. The temperature distributions along x axis in Fig. 19.2f–h are plotted in Fig. 19.2i. The wave-like temperature profile has also amplitude decrement due to the positive Re (κ). However, the wave-like temperature profile does not propagate because of the zero Im (κ). Therefore, the behaviors of thermal conduction and advection can be well described by using complex thermal conductivity. When wave vector and convective velocity are with an arbitrary angle α, the velocity component $v \cos\alpha$ contributes to propagation.

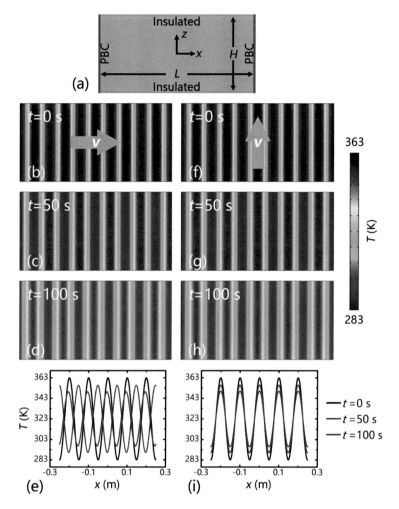

Fig. 19.2 **a** Schematic diagram. Temperature evolutions with **b–e** $v//\beta$ and **f–i** $v \perp \beta$. Parameters: $L = 0.5$ m, $H = 0.25$ m, $\rho C = 10^6$ J m^{-3} K^{-1}, $\sigma = 1$ W m^{-1} K^{-1}, and $v = 1$ mm/s. PBC in (**a**) means periodic boundary condition. Adapted from Ref. [20]

We further discuss the energy flow in Fig. 19.2b–d. Relative energy flow J' can be calculated with periodicity average,

$$J' = \frac{1}{\lambda} \int_0^\lambda (-\kappa \nabla T) \, dx = 0, \tag{19.12}$$

where $\lambda = 2\pi/\beta$ is wavelength. Here, we only take on the real part of J' because the imaginary part does not make sense. Although conductive flow is irrelevant to

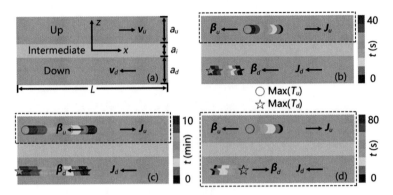

Fig. 19.3 Two-dimensional negative thermal transport. **a** Schematic diagram with $a_u = a_d = 2$ mm, $a_i = 1$ mm, $L = 0.5$ m, $\sigma_u = \sigma_d = 10$ W m^{-1} K^{-1}, $\sigma_i = 0.1$ W m^{-1} K^{-1}, and $\rho_u C_u = \rho_d C_d = \rho_i C_i = 10^6$ J m^{-3} K^{-1}. These parameters lead to $g/\beta = 4$ mm/s. **b** $v_u = -v_d = 5$ mm/s. **c** $v_u = 0.5$ mm/s and $v_d = -1.5$ mm/s. **d** $v_u = -v_d = 1$ mm/s. Circles and stars denote the trajectories of Max(T_u) and Max(T_d), respectively. Adapted from Ref. [20]

reference temperature, convective flow ($\rho C v T$) is closely associated with reference temperature [24–28]. Therefore, absolute energy flow J is

$$J = \rho C v T_0. \tag{19.13}$$

In what follows, we discuss absolute energy flow and neglect the expression of absolute for brevity. Clearly, J and v have the same direction. In other words, only thermal advection contributes to energy flow. As we can imagine from Fig. 19.2b–d, thermal advection results in the motion of wave-like temperature profile, so the direction of wave vector β follows that of convective velocity v, yielding positive thermal transport (Im (κ) > 0; see Fig. 19.1b). To some extent, positive thermal transport is the result of causality, so negative thermal transport (Im (κ) < 0; see Fig. 19.1b) is unique.

19.3   Negative Thermal Transport

It might be very difficult to reveal negative thermal transport in an isolated system (Fig. 19.2a), so we consider an open system (Fig. 19.3a) where an intermediate layer allows heat exchange between up and down layers. The complex thermal conductivities of up layer κ_u and down layer κ_d are denoted as

$$\kappa_u = \sigma_u + i\frac{\rho_u C_u v_u \cdot \beta_u}{\beta_u^2}, \tag{19.14a}$$

$$\kappa_d = \sigma_d + i\frac{\rho_d C_d v_d \cdot \beta_d}{\beta_d^2}, \tag{19.14b}$$

where the subscripts u and d denote the parameters in up and down layers, respectively. We set the wave-like temperature profiles in up layer T_u and down layer T_d as

$$T_u = A_u e^{i(\beta_u \cdot x - \omega t)} + T_0, \tag{19.15a}$$

$$T_d = A_d e^{i(\beta_d \cdot x - \omega t)} + T_0. \tag{19.15b}$$

The heat exchange between up and down layers is along z axis, which is not of our concern. Then, the energy flows along x axis in up layer $\boldsymbol{J}_u$ and down layer $\boldsymbol{J}_d$ can be calculated as

$$\boldsymbol{J}_u = \rho_u C_u \boldsymbol{v}_u T_0, \tag{19.16a}$$

$$\boldsymbol{J}_d = \rho_d C_d \boldsymbol{v}_d T_0. \tag{19.16b}$$

Clearly, the directions of energy flow in up and down layers are opposite due to $\boldsymbol{v}_u = -\boldsymbol{v}_d$.

The thermal conduction-advection processes in up and down layers can be described by the complex thermal conduction equation,

$$\rho_u C_u \partial T_u / \partial t + \nabla \cdot (-\kappa_u \nabla T_u) = s_u, \tag{19.17a}$$

$$\rho_d C_d \partial T_d / \partial t + \nabla \cdot (-\kappa_d \nabla T_d) = s_d, \tag{19.17b}$$

where s_u and s_d are two heat sources, reflecting the heat exchange between up and down layers [21, 22]. Since the three layers in Fig. 19.3a are thin enough ($L \gg a_{u,i,d}$), the temperature variance along z axis can be neglected, yielding $\partial^2 T / \partial z^2 = 0$. The energy flow from down layer to up layer j_u can be calculated as $j_u = -\sigma_i (T_u - T_d) / a_i$, where σ_i and a_i are the thermal conductivity and width of stationary intermediate layer, respectively. It is also reasonable to suppose that energy flow is uniformly distributed in up layer due to thin thickness, so the heat source in up layer s_u can be expressed as $s_u = j_u / a_u = -\sigma_i (T_u - T_d) / (a_i a_u)$, where a_u is the width of up layer. Similarly, the heat source in down layer s_d can be derived as $s_d = j_d / a_d = -\sigma_i (T_d - T_u) / (a_i a_d)$, where a_d is the width of down layer. With these analyses, Eq. (19.17) can be reduced to

$$\rho_u C_u \partial T_u / \partial t - \kappa_u \partial^2 T_u / \partial x^2 = h_u (T_d - T_u), \tag{19.18a}$$

$$\rho_d C_d \partial T_d / \partial t - \kappa_d \partial^2 T_d / \partial x^2 = h_d (T_u - T_d), \tag{19.18b}$$

where $h_u = \sigma_i / (a_i a_u)$ and $h_d = \sigma_i / (a_i a_d)$, reflecting the exchange rate of heat energy. We take on the same material parameters of up and down layers, say, $\sigma_u = \sigma_d (= \sigma)$, $\rho_u C_u = \rho_d C_d (= \rho C)$, $a_u = a_d (= a)$, and $h_u = h_d (= h)$. We also suppose $\boldsymbol{v}_u = -\boldsymbol{v}_d (= \boldsymbol{v})$ and $\boldsymbol{\beta}_u = \boldsymbol{\beta}_d (= \boldsymbol{\beta})$, thus yielding $\kappa_u = \overline{\kappa}_d (= \kappa)$ where $\overline{\kappa}_d$ is the conjugate of κ_d. The substitution of Eq. (19.15) into Eq. (19.18) yields an eigenequation $\hat{H}\boldsymbol{\psi} = \omega\boldsymbol{\psi}$, where the Hamiltonian $\hat{H}$ reads

$$\hat{H} = \begin{bmatrix} -i\left(g + \eta\beta^2\right) & ig \\ ig & -i\left(g + \overline{\eta}\beta^2\right) \end{bmatrix}, \tag{19.19}$$

where $\eta = \kappa/(\rho C)$ and $g = h/(\rho C)$. The eigenvalue of Eq. (19.19) is

$$\omega_{\pm} = -i\left[g + \text{Re}(\eta)\,\beta^2 \pm \sqrt{g^2 - \text{Im}^2(\eta)\,\beta^4}\right], \tag{19.20}$$

where $\text{Re}(\eta) = \sigma/(\rho C)$ and $\text{Im}(\eta)\,\beta^2 = v\beta$.

With Eq. (19.20), we can obtain three different cases of negative thermal transport. We discuss the first case with $g^2 - \text{Im}^2(\eta)\,\beta^4 < 0$, say, $v > g/\beta$. The eigenvector is

$$\psi_{+} = \left[1,\ e^{i\pi/2 - \delta}\right]^{\varsigma}, \tag{19.21a}$$

$$\psi_{-} = \left[1,\ e^{i\pi/2 + \delta}\right]^{\varsigma}, \tag{19.21b}$$

where $\delta = \cosh^{-1}\left[\text{Im}(\eta)\,\beta^2/g\right]$, and ς denotes transpose. The eigenvectors in Eq. (19.21) indicate that the wave-like temperature profiles in up and down layers move with a constant phase difference of $\pi/2$ but with different amplitudes. We take on $\beta = 2\pi m/L$ with $m = 1$ in what follows. The initial wave-like temperature profiles in up and down layers are set as the eigenvector described by Eq. (19.21b), say, $T_u = 40\cos(\beta x) + 323$ K and $T_d = e^{\delta} 40\cos(\beta x + \pi/2) + 323$ K. We track the motion of maximum temperature in up layer $\text{Max}(T_u)$ and down layer $\text{Max}(T_d)$ to observe the directions of wave vector. Since the amplitude of wave-like temperature profile in down layer (with $e^{\delta} > 1$) is larger than that in up layer, the directions of wave vector in up and down layers are both leftward. Therefore, negative thermal transport occurs in up layer, and the transport in down layer is still positive (Fig. 19.3b).

We discuss the second case with $g^2 - \text{Im}^2(\eta)\,\beta^4 > 0$, say, $0 < v < g/\beta$. The corresponding eigenvector is

$$\psi_{+} = \left[1,\ e^{i(\pi - \alpha)}\right]^{\varsigma}, \tag{19.22a}$$

$$\psi_{-} = \left[1,\ e^{i\alpha}\right]^{\varsigma}, \tag{19.22b}$$

where $\alpha = \sin^{-1}\left[\text{Im}(\eta)\,\beta^2/g\right]$. The eigenvectors in Eq. (19.22) indicate that the wave-like temperature profiles in up and down layers are motionless with a constant phase difference of $\pi - \alpha$ or α. To make the wave-like temperature profiles move, we give the system a reference velocity v_0, resulting in $v_u = v_u' + v_0$ and $v_d = v_d' + v_0$, where v_u' and v_d' are original convective velocities. This operation does not affect the essence of eigenvectors in Eq. (19.22), and only gives a reference velocity v_0 to wave-like temperature profiles. We set the initial wave-like temperature profiles in up and down layers to be the eigenvector described by Eq. (19.22b), say, $T_u = 40\cos(\beta x) + 323$ K and $T_d = 40\cos(\beta x + \alpha) + 323$ K. We also take on $v_0 = -0.5 v_u'$, so the wave-like temperature profiles in up and down layers still maintain a constant phase difference of α but with leftward motion. The trajectories of Max(T_u) and Max(T_d) are presented in Fig. 19.3c. Clearly, negative thermal transport occurs in up layer.

These two cases are related to eigenvectors, indicating that negative thermal transport occurs in one layer (say, up layer). However, thermal transport is still positive if we regard up and down layers as a whole. We further explore the third case, related to non-eigenvectors and their dynamics, to reveal negative thermal transport in up and down layers. For this purpose, we set the initial wave-like temperature profiles by adding the eigenvector described by Eq. (19.22b) with an extra phase γ, say, $\psi'_- = \left[1, \; e^{i(\alpha+\gamma)} \right]^\varsigma$, yielding $T_u = 40 \cos(\beta x) + 323$ K and $T_d = 40 \cos(\beta x + \alpha + \gamma) + 323$ K. In this way, even if we do not give a reference velocity to the system, the wave-like temperature profile still moves to reach the eigenvector. One principle of the evolution route is to make the temperature profile decay as slowly as possible. Therefore, the eigenvector ψ_+ with a phase difference of $\pi - \alpha$ described by Eq. (19.22a) becomes a key due to its large decay rate (say, the ω_+ of Eq. (19.20)). The evolution route should try to avoid ψ_+ to survive longer. When $\gamma \in (0, \pi - 2\alpha)$, negative thermal transport will not make the temperature profile go through ψ_+, but positive thermal transport will make the temperature profile go through ψ_+ twice. Therefore, the evolution route naturally chooses negative thermal transport in both up and down layers to decay more slowly (Fig. 19.3d). Nevertheless, the wave-like temperature profile remains motionless after reaching the eigenvector, so negative thermal transport is no longer present. In other words, negative thermal transport in both up and down layers is transient.

19.4 Experimental Suggestion

We also suggest experimental demonstration with a three-dimensional solid ring structure (Fig. 19.4a), which can naturally meet the requirement of periodic boundary conditions. Up ring (with width a_u) and down ring (with width a_d) rotate with opposite angular velocities (Ω_u and Ω_d), which are connected by a stationary intermediate layer (with width a_i). The inner and outer radii of the ring structure are r_1 and r_2, respectively. Like two dimensions, we track Max(T_u) and Max(T_d) on the interior edge of the solid ring structure. The parametric settings for Fig. 19.4b–d are basically the same as those for Fig. 19.3b–d, respectively. Therefore, the results are also similar. Negative thermal transport occurs in the up ring of Fig. 19.4b, c and occurs in both up and down rings of Fig. 19.4d. Three-dimensional and two-dimensional results agree well with theoretical analyses, confirming the feasibility of negative thermal transport in thermal conduction and advection.

Negative thermal transport may enlighten the inverse Doppler effect in thermal conduction and advection. Since energy flow is generated from the energy source, a detector with the opposite direction of energy flow is getting close to the energy source. Positive thermal transport makes wave vector (regarded as a thermal signal) follow the direction of energy flow. Therefore, the detector and wave vector directions

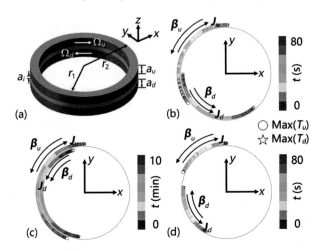

Fig. 19.4 Experimental suggestions with a three-dimensional solid ring structure. **a** Schematic diagram with $r_1 = 80$ mm, $r_2 = 82$ mm. Other parameters are kept the same as those for Fig. 19.3a. **b** $\Omega_u = -\Omega_d = 0.063$ rad/s. **c** $\Omega_u = 0.006$ rad/s and $\Omega_d = -0.019$ rad/s. **d** $\Omega_u = -\Omega_d = 0.013$ rad/s. Adapted from Ref. [20]

are opposite, yielding frequency increment (the Doppler effect). However, negative thermal transport leads to the same detector and wave vector directions. As a result, the frequency decreases even though the detector gets close to the energy source (the inverse Doppler effect). These results may also provide guidance to extend transformation thermotics [29–31] to complex regime [32] and regulate thermal imaging [33–39]. Other thermal systems, such as those with periodic structures [40–43], are also good candidates to explore negative thermal transport. Nevertheless, here we reveal negative thermal transport in an open system with energy exchange, so there is a difference from wave systems where no energy exchange is required to realize negative refraction. Whether negative thermal transport can exist in an isolated system remains studied.

19.5   Conclusion

In summary, we have established the thermal counterpart of a complex refractive index by coining a complex thermal conductivity. As a result, a negative imaginary thermal conductivity is just the thermal counterpart of a negative refractive index, featuring the opposite directions of energy flow and wave vector in thermal conduction and advection. Negative thermal transport seems to violate causality, but it can occur in an open system with heat exchange. We further reveal negative thermal transport in three cases and provide three-dimensional experimental suggestions, confirming its physical feasibility. These results provide a different perspective to cognize conduction and advection. They may enlighten outspread explorations of negative thermal transport, such as the inverse Doppler effect in thermal conduction and advection.

19.6 Exercise and Solution

Exercise

1. Explain the left half in Fig. 19.1b.

Solution

1. In the left half, the real part of a complex thermal conductivity is negative, so the temperature field amplitude increases. This effect does not occur naturally because the second law of thermodynamics is violated. However, it may happen with external energy sources.

References

1. Veselago, V.G.: The electrodynamics of substances with simultaneously negative values of ϵ and μ. Sov. Phys. Usp. **10**, 509 (1968)
2. Pendry, J.B., Holden, A.J., Stewart, W.J., Youngs, I.: Extremely low frequency plasmons in metallic mesostructures. Phys. Rev. Lett. **76**, 4773 (1996)
3. Pendry, J.B., Holden, A.J., Robbins, D.J., Stewart, W.J.: Magnetism from conductors and enhanced nonlinear phenomena. IEEE Trans. Microw. Theory Tech. **47**, 2075 (1999)
4. Smith, D.R., Padilla, W.J., Vier, D.C., Nemat-Nasser, S.C., Schultz, S.: Composite medium with simutaneously negative permeability and permittivity. Phys. Rev. Lett. **84**, 4184 (2000)
5. Smith, D.R., Kroll, N.: Negative refractive index in left-handed materials. Phys. Rev. Lett. **85**, 2933 (2000)
6. Shelby, R.A., Smith, D.R., Schultz, S.: Experimental verification of a negative index of refraction. Science **292**, 77 (2001)
7. Pendry, J.B.: Negative refraction makes a perfect lens. Phys. Rev. Lett. **85**, 3966 (2000)
8. Lagarkov, A.N., Kissel, V.N.: Near-perfect imaging in a focusing system based on a left-handed-material plate. Phys. Rev. Lett. **92**, 077401 (2004)
9. Grbic, A., Eleftheriades, G.V.: Overcoming the diffraction limit with a planar left-handed transmission-line lens. Phys. Rev. Lett. **92**, 117403 (2004)
10. Seddon, N., Bearpark, T.: Observation of the inverse Doppler effect. Science **302**, 1537 (2003)
11. Luo, C.Y., Ibanescu, M.H., Johnson, S.G., Joannopoulos, J.D.: Cerenkov radiation in photonic crystals. Science **299**, 368 (2003)
12. Ziolkowski, R.W.: Pulsed and CW Gaussian beam interactions with double negative metamaterial slabs. Opt. Express **11**, 662 (2003)
13. Shendeleva, M.L.: Thermal wave reflection and refraction at a plane interface: two-dimensional geometry. Phys. Rev. B **65**, 134209 (2002)
14. Vemuri, K.P., Bandaru, P.R.: Geometrical considerations in the control and manipulation of conductive heat flux in multilayered thermal metamaterials. Appl. Phys. Lett. **103**, 133111 (2013)

15. Vemuri, K.P., Bandaru, P.R.: Anomalous refraction of heat flux in thermal metamaterials. Appl. Phys. Lett. **104**, 083901 (2014)
16. Yang, T.Z., Vemuri, K.P., Bandaru, P.R.: Experimental evidence for the bending of heat flux in a thermal metamaterial. Appl. Phys. Lett. **105**, 083908 (2014)
17. Vemuri, K.P., Canbazoglu, F.M., Bandaru, P.R.: Guiding conductive heat flux through thermal metamaterials. Appl. Phys. Lett. **105**, 193904 (2014)
18. Kapadia, R.S., Bandaru, P.R.: Heat flux concentration through polymeric thermal lenses. Appl. Phys. Lett. **105**, 233903 (2014)
19. Hu, R., Xie, B., Hu, J.Y., Chen, Q., Luo, X.B.: Carpet thermal cloak realization based on the refraction law of heat flux. EPL **111**, 54003 (2015)
20. Xu, L.J., Huang, J.P.: Negative thermal transport in conduction and advection. Chin. Phys. Lett. **37**, 080502 (2020)
21. Li, Y., Peng, Y.G., Han, L., Miri, M.A., Li, W., Xiao, M., Zhu, X.F., Zhao, J.L., Alù, A., Fan, S.H., Qiu, C.W.: Anti-parity-time symmetry in diffusive systems. Science **364**, 170 (2019)
22. Cao, P.C., Li, Y., Peng, Y.G., Qiu, C.W., Zhu, X.F.: High-order exceptional points in diffusive systems: robust APT symmetry against perturbation and phase oscillation at APT symmetry breaking. ES Energy Environ. **7**, 48 (2020)
23. Torrent, D., Poncelet, O., Batsale, J.C.: Nonreciprocal thermal material by spatiotemporal modulation. Phys. Rev. Lett. **120**, 125501 (2018)
24. Guenneau, S., Petiteau, D., Zerrad, M., Amra, C., Puvirajesinghe, T.: Transformed Fourier and Fick equations for the control of heat and mass diffusion. AIP Adv. **5**, 053404 (2015)
25. Dai, G.L., Shang, J., Huang, J.P.: Theory of transformation thermal convection for creeping flow in porous media: cloaking, concentrating, and camouflage. Phys. Rev. E **97**, 022129 (2018)
26. Yang, F.B., Xu, L.J., Huang, J.P.: Thermal illusion of porous media with convection-diffusion process: transparency, concentrating, and cloaking. ES Energy Environ. **6**, 45 (2019)
27. Xu, L.J., Yang, S., Dai, G.L., Huang, J.P.: Transformation omnithermotics: simultaneous manipulation of three basic modes of heat transfer. ES Energy Environ. **7**, 65 (2020)
28. Xu, L.J., Huang, J.P.: Chameleonlike metashells in microfluidics: a passive approach to adaptive responses. Sci. China-Phys. Mech. Astron. **63**, 228711 (2020)
29. Fan, C.Z., Gao, Y., Huang, J.P.: Shaped graded materials with an apparent negative thermal conductivity. Appl. Phys. Lett. **92**, 251907 (2008)
30. Chen, T.Y., Weng, C.N., Chen, J.S.: Cloak for curvilinearly anisotropic media in conduction. Appl. Phys. Lett. **93**, 114103 (2008)
31. Xu, L.J., Dai, G.L., Huang, J.P.: Transformation multithermotics: controlling radiation and conduction simultaneously. Phys. Rev. Appl. **13**, 024063 (2020)
32. Xu, L.J., Huang, J.P.: Controlling thermal waves with transformation complex thermotics. Int. J. Heat Mass Transf. **159**, 120133 (2020)
33. Hu, R., Hu, J.Y., Wu, R.K., Xie, B., Yu, X.J., Luo, X.B.: Examination of the thermal cloaking effectiveness with layered engineering materials. Chin. Phys. Lett. **33**, 044401 (2016)
34. Hu, R., Zhou, S.L., Li, Y., Lei, D.-Y., Luo, X.B., Qiu, C.-W.: Illusion thermotics. Adv. Mater. **30**, 1707237 (2018)
35. Han, T.C., Yang, P., Li, Y., Lei, D.Y., Li, B.W., Hippalgaonkar, K., Qiu, C.-W.: Full-parameter omnidirectional thermal metadevices of anisotropic geometry. Adv. Mater. **30**, 1804019 (2018)
36. Hu, R., Huang, S.Y., Wang, M., Luo, X.B., Shiomi, J., Qiu, C.-W.: Encrypted thermal printing with regionalization transformation. Adv. Mater. **31**, 1807849 (2019)
37. Liu, Y.D., Song, J.L., Zhao, W.X., Ren, X.C., Cheng, Q., Luo, X.B., Fang, N.X.L., Hu, R.: Dynamic thermal camouflage via a liquid-crystal-based radiative metasurface. Nanophotonics **9**, 855 (2020)
38. Peng, Y.-G., Li, Y., Cao, P.-C., Zhu, X.-F., Qiu, C.-W.: 3D printed meta-helmet for wide-angle thermal camouflages. Adv. Funct. Mater. **30**, 2002061 (2020)
39. Hu, R., Xi, W., Liu, Y.D., Tang, K.C., Song, J.L., Luo, X.B., Wu, J.Q., Qiu, C.-W.: Thermal camouflaging metamaterials. Mater. Today **45**, 120 (2021)
40. Maldovan, M.: Narrow low-frequency spectrum and heat management by thermocrystals. Phys. Rev. Lett. **110**, 025902 (2013)

41. Xu, L.J., Yang, S., Huang, J.P.: Thermal transparency induced by periodic interparticle interaction. Phys. Rev. Appl. **11**, 034056 (2019)
42. Cai, Z., Huang, Y.Z., Liu, W.V.: Imaginary time crystal of thermal quantum matter. Chin. Phys. Lett. **37**, 050503 (2020)
43. Xu, L.J., Huang, J.P.: Thermal convection-diffusion crystal for prohibition and modulation of wave-like temperature profiles. Appl. Phys. Lett. **117**, 011905 (2020)

Chapter 20
Theory for Thermal Wave Nonreciprocity: Angular Momentum Bias

Abstract In this chapter, we demonstrate that an angular momentum bias generated by a volume force can also lead to modal splitting in convection-diffusion systems but with different features. We further reveal the thermal Zeeman effect by studying the temperature field propagation in an angular-momentum-biased ring with three ports (one for input and two for output). With an optimal volume force, temperature field propagation is allowed at one output port but isolated at the other, and the rectification coefficient can reach a maximum value of 1. The volume forces corresponding to the rectification coefficient peaks can also be quantitatively predicted by scalar (i.e., temperature) interference. Compared with existing mechanisms for thermal nonreciprocity, an angular momentum bias does not require temperature-dependent and phase-change materials, which has an advantage in wide-temperature-range applicability. These results may provide insights into thermal stabilization and thermal topology. The related mechanism is also universal for other convection-diffusion systems such as mass transport, chemical mixing, and colloid aggregation.

Keywords Thermal wave nonreciprocity · Angular momentum bias · Scalar interference

20.1 Opening Remarks

Nonreciprocity refers to asymmetric propagation in opposite directions, which has attracted broad interest in wave systems [1, 2]. A common approach to nonreciprocity is based on the modal splitting induced by an angular momentum bias. For example, magneto-optical media can realize electromagnetic nonreciprocity based on the electronic Zeeman effect. Inspired by the electronic Zeeman effect, the acoustic Zeeman effect was also proposed to obtain acoustic nonreciprocity with air circulation [3]. The origin of an angular momentum bias is various, which can be attributed to circular motions [3, 4], magnetic fields [5, 6], or spatiotemporal modulations [7–10]. However, the related mechanism is confronted with many challenges in convection-diffusion systems. On the one hand, it is unknown how to apply an angular momentum bias in convection-diffusion systems. On the other hand, convection-diffusion sys-

© The Author(s) 2023
L.-J. Xu and J.-P. Huang, *Transformation Thermotics and Extended Theories*,
https://doi.org/10.1007/978-981-19-5908-0_20

tems have many crucial differences from wave systems, which are discussed in detail when exhibiting our results.

Macroscopic thermal transport is a typical convection-diffusion system where breaking reciprocity is highly expected and widely explored [11]. Reciprocity generally refers to a physical quantity having the same properties in different directions. For thermal transport, the physical quantity can be heat flux, temperature amplitude, etc. Thermal nonreciprocity can be realized with temperature-dependent (i.e., nonlinear) or phase-change materials [12–14], but the strong temperature dependence restricts its wide-temperature-range applicability. Moreover, spatiotemporal modulations can also help achieve thermal nonreciprocity [15, 16], but thermal conductivities and mass densities require complicated and dynamic control. Therefore, it remains difficult to realize thermal nonreciprocity with linear, wide-temperature-range applicable, and easy-to-control materials.

Inspired by the electronic and acoustic Zeeman effects in wave systems [3–10], we introduce the thermal Zeeman effect with an angular momentum bias generated by a volume force (Fig. 20.1a, b). Here, a volume force is exerted on all fluid particles and is proportional to the mass of the fluid in that volume, such as the forces exerted on fluids in a gravitational field and ferrofluids in a magnetic field. We then study the temperature propagation in a three-port ring to achieve thermal nonreciprocity and isolation (Fig. 20.1c). Here, temperature propagation refers to the propagation of a periodic temperature profile [17–25], which can also be regarded as a temperature fluctuation. Scalar (i.e., temperature) interference is crucial to explain thermal nonreciprocity, which quantitatively predicts the rectification coefficient peaks in simulations. The present scheme is free from nonlinear and phase-change materials, thus applying to a wide temperature range. Moreover, complicated parameter control is also unnecessary, making it feasible. Following the idea that acoustic topology can be achieved by arranging three-port rings in a graphene-like array [26–31], we may also realize thermal topology with the proposed mechanism of thermal nonreciprocity (Fig. 20.1d).

20.2 Thermal Zeeman Effect

A thermal convection-diffusion process is dominated by $\rho C \partial T / \partial t + \nabla \cdot (-\kappa \nabla T + \rho C v T) = 0$, where ρ, C, κ, and v are the mass density, heat capacity, thermal conductivity, and convective velocity of a fluid, respectively [32]. T and t represent absolute temperature and time, respectively. Without loss of generality, we discuss a steady incompressible creeping flow [33–36] driven by a linear pressure field along the x axis. A convective velocity $v(y)$ has a quadratic distribution along the vertical direction [37]. We consider a small vertical height h and discuss an average convective velocity $v = -h^2 (\nabla P - f) / (12\mu)$, where μ is the dynamic viscosity of the fluid, P denotes pressure, and f is volume force [37]. In what follows, we also discuss the average values of velocities, temperatures, and heat fluxes.

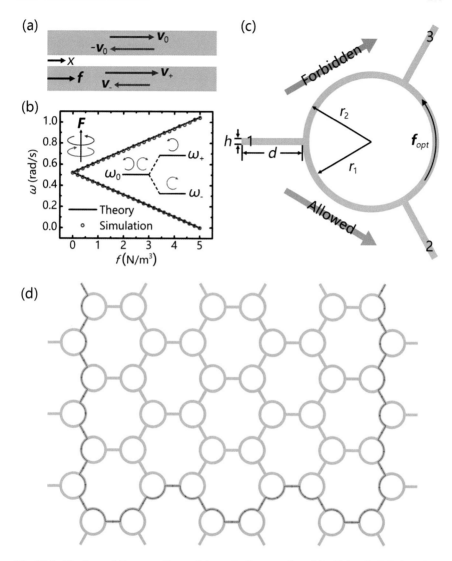

Fig. 20.1 The thermal Zeeman effect. **a** Schematic diagram of modal splitting. **b** Splitting of the real part of frequency as a function of volume force. **c** Angular-momentum-biased ring exhibiting thermal nonreciprocity and isolation. **d** Schematic diagram of thermal topology. Adapted from Ref. [38]

We then consider a periodic temperature profile $T = A\cos(\beta x - \omega t) + T_0$, where A, β, ω, and T_0 are temperature amplitude, wave number, circular frequency, and reference temperature, respectively. In the absence of a volume force f, a pressure field along $+x$ (or $-x$) generates an average convective velocity v_0 (or $-v_0$), see the upper inset of Fig. 20.1a. Therefore, circular frequencies are the same, i.e., $\omega_0 = \beta v_0 - i\beta^2 D$ with thermal diffusivity $D = \kappa/(\rho C)$. Re (ω_0) represents circular frequency and $-$Im (ω_0) denotes temporal decay rate. When there is a volume force f along $+x$, a pressure field along $+x$ (or $-x$) generates an average convective velocity v_+ (or v_-), see the lower inset of Fig. 20.1a. Circular frequencies are no longer the same but split into

$$\omega_\pm = \beta v_\pm - i\beta^2 D, \qquad (20.1)$$

with $v_\pm = v_0 \pm h^2 f/(12\mu)$. The difference between convection-diffusion systems and wave systems is reflected in the imaginary part of Eq. (20.1). Wave systems are generally Hermitian with energy conservation, so circular frequencies are real numbers without loss [3]. However, convection-diffusion systems are non-Hermitian with loss [19], so circular frequencies become complex.

For intuitive understanding, we can imagine periodic conditions on the left and right boundaries in Fig. 20.1a and regard the $+x$ direction as the anticlockwise azimuthal direction. An angular velocity $V = e_r \times v/r_0$ and an angular volume force $F = e_r \times f/r_0$ are introduced, where e_r is the radial unit vector and r_0 is an average radius. The ring allows only discrete wave numbers, i.e., $\beta = N/r_0$ where N is a positive integer [19]. The frequency splitting described by Eq. (20.1) can then be understood by the Zeeman effect, which results from an angular momentum bias generated by an angular volume force F, just like the energy splitting of atoms due to a magnetic bias or the frequency splitting of sounds due to an angular momentum bias [3]. We also confirm the frequency splitting with finite-element simulations based on the template of Heat Transfer in Fluids in COMSOL Multiphysics. Meshes are set as follows: the maximum element size is 5×10^{-4} m, the minimum element size is 10^{-6} m, the maximum element growth rate is 1.1, the curvature factor is 0.2, and the resolution of narrow regions is 1. The relative tolerance for a time-dependent solver is 10^{-4}. We use water parameters. We also set a pressure gradient of $|\nabla P| = 5$ N/m^3, a wave number of $\beta = 100\pi$ m^{-1}, and a height of $h = 2$ mm. The real part of Eq. (20.1) then becomes $\omega_\pm = \pi(5 \pm f)/30$. The simulation result agrees well with the theory (Fig. 20.1b).

20.3   Thermal Wave Nonreciprocity

We further consider a three-port ring to demonstrate thermal nonreciprocity with the thermal Zeeman effect, as shown in Fig. 20.1c. We set port 1 as an input port and ports 2 and 3 as output ports. We set a high pressure P_h at port 1 and two identical low pressures P_l at ports 2 and 3. We also set a periodic temperature source at port 1, i.e.,

$T_1 = A_1 \cos(-\omega t) + T_0$. Ports 2 and 3 are set with open conditions with no reflection. For a zero volume force, two symmetrical velocities are obtained in the ring, i.e., $v_{1 \to 2}$ along the counterclockwise direction and $v_{1 \to 3}$ along the clockwise direction. Therefore, temperature propagation at ports 2 and 3 are identical due to structural symmetry. However, when a volume force along the counterclockwise direction is applied, $v_{1 \to 2}$ increases but $v_{1 \to 3}$ decreases. Therefore, an angular momentum bias is achieved in the ring, and the temperature propagation from port 1 to port 3 is forbidden with an optimal volume force f_{opt}.

We then perform finite-element simulations with time steps of 0.5 s to observe thermal nonreciprocity. Two crucial parameters should be considered, i.e., the Peclet number and the Reynolds number. Since we use water for simulations, the Peclet number is Pe = 2800, demonstrating that convection is dominant. As a result, the convection-diffusion equation mainly exhibits hyperbolic features that can support the propagation of wave-like temperature profiles. The Reynolds number is Re = 4, that approximately corresponds to a creeping or laminar flow [33–36], so the effects of boundary layer behavior and singular perturbation can be ignored. In short terms, the expected phenomena require (I) a large Peclet number for convection≫diffusion and (II) a small Reynolds number without turbulent flow.

The properties of temperature propagation can be reflected in temperature amplitudes. A zero temperature amplitude indicates that temperature propagation is isolated. The temperature and velocity profiles without a volume force are shown in the first column of Fig. 20.2. Due to structural symmetry, the temperature amplitudes at ports 2 and 3 are identical. However, it is crucially different when the volume force reaches an optimal value $f_{opt} = 2$ N/m^3. The temperature amplitude at port 3 is dramatically reduced to zero, whereas that at port 2 still exists (see the second column in Fig. 20.2). In other words, we achieve the isolation of temperature propagation at port 3, and thermal nonreciprocity is maximized. We then continue to increase the volume force to 6 N/m^3. Although nonreciprocity still exists (see the third column in Fig. 20.2), the temperature amplitude at port 3 is no longer zero. The velocity profiles with different volume forces are shown in Fig. 20.2g–i. The velocities at three ports are irrelevant to the volume force, but those in the ring are affected linearly to realize an angular momentum bias.

After discussing temperature and velocity properties, we can explore heat flux properties further. We independently study conductive fluxes and convective fluxes for clarity. Temperature amplitudes decay spatially $T = Ae^{-\alpha x} \cos(\beta x - \omega t) + T_0$. Conductive fluxes are given by $J_{cond} = -\kappa \partial T/\partial x = \kappa Ae^{-\alpha x}[\alpha \cos(\beta x - \omega t) + \beta \sin(\beta x - \omega t)]$, which are proportional to T. Convective fluxes are determined by $J_{conv} = \rho C v T$, which are also proportional to T. Therefore, heat flux properties are similar to temperature properties due to $J \propto T$. Since conductive fluxes are related to spatial derivation, we discuss heat fluxes very close to ports 2 and 3 (with a 4-mm distance) to ensure accuracy. Finite-element simulations are presented in Fig. 20.3. When the volume force is zero, the conductive fluxes (or convective fluxes) at ports 2 and 3 are identical (see the first column of Fig. 20.3). When an optimal volume force $f_{opt} = 2$ N/m^3 is applied, the conductive flux vanishes at port 3 but still exists at port 2 (like an alternating conductive flux).

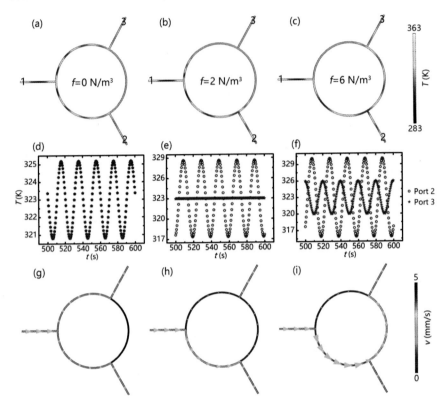

Fig. 20.2 Temperature and velocity profiles. **a–c** Temperature profiles at 600 s with volume forces of 0, 2, and 6 N/m³, respectively. **d–f** Average temperatures at ports 2 and 3 from 500 to 600 s. **g–i** Steady velocity profiles. Arrows denote convective velocities. The fluid is water, whose mass density, heat capacity, thermal conductivity, and dynamic viscosity are 1000 kg/m³, 4200 J kg⁻¹ K⁻¹, 0.6 W m⁻¹ K⁻¹, and 0.001 Pa s, respectively. The structure sizes are $r_1 = 49$ mm, $r_2 = 51$ mm, $h = r_2 - r_1 = 2$ mm, and $d = 49$ mm. Other parameters: $P_h = 1$ Pa, $P_l = 0$ Pa, and $T_1 = 40\cos(-\pi t/10) + 323$ K. Adapted from Ref. [38]

Therefore, the isolation of conductive fluxes is achieved. Although the convective flux at port 3 is nonzero, it does not vary temporally. The convective flux at port 2 still varies periodically (see the second column of Fig. 20.3). When the volume force is 6 N/m³, conductive and convective fluxes are also nonreciprocal (see the last column of Fig. 20.3).

20.4　Scalar Interference

We further discuss thermal nonreciprocity quantitatively, and six key positions Σ_1-Σ_6 are labeled in Fig. 20.4a. We define two transmission coefficients as $R_{1-2} = A_2/A_1$ and $R_{1-3} = A_3/A_1$, where A_1, A_2, and A_3 are the temperature amplitudes

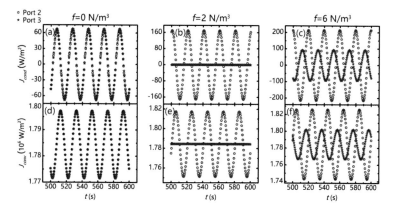

Fig. 20.3 Heat flux profiles. **a–c** Conductive fluxes and **d–f** convective fluxes with volume forces of 0, 2, and 6 N/m³, respectively. Conductive fluxes have negative values due to direction changes. Adapted from Ref. [38]

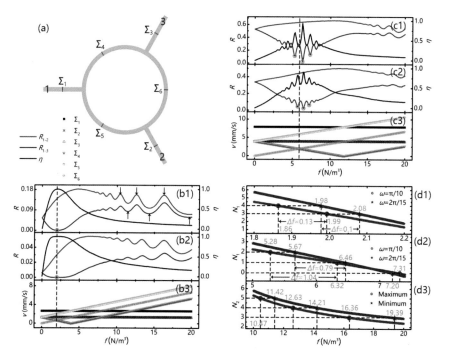

Fig. 20.4 Quantitative analyses of transmission coefficients and rectification coefficients. **a** Schematic diagram showing six key positions Σ_1-Σ_6. Transmission coefficient R, rectification coefficient η, and convective velocity v as a function of volume force with (b1)–(b3) $P_h = 1$ Pa and (c1)–(c3) $P_h = 3$ Pa. The circular frequencies of periodic temperature profiles are $\pi/10$ for (b1, c1) and $2\pi/15$ for (b2, c2). The $N_1 - f$ curves described by Eq. (20.2a) with (d1) $P_h = 1$ Pa and (d2) $P_h = 3$ Pa. (d3) The $N_2 - f$ curves described by Eqs. (20.5a) and (20.5b). Adapted from Ref. [38]

at ports 1, 2, and 3, respectively. We also define a rectification coefficient η as $(R_{1-2} - R_{1-3}) / (R_{1-2} + R_{1-3})$. R and η as a function of f are shown in Fig. 20.4b1 with volume force steps of $0.1\,\text{N/m}^3$. R_{1-2} first increases, then decreases, and finally varies quasiperiodically. R_{1-3} decreases initially, increases afterwards, and varies quasiperiodically at last. R_{1-2} and R_{1-3} lead to an initial increase and a final decrease in η, and $\eta_{max} = 1$ appears at $f_{opt} = 2\,\text{N/m}^3$, indicating the isolation of temperature propagation at port 3. Although R_{1-2} and R_{1-3} ultimately varies quasiperiodically, they are synchronous, so η still decreases. We also change the circular frequency to $2\pi/15$ rad/s, and the transmission results are shown in Fig. 20.4b2. $\eta_{max} = 1$ still appears at $f_{opt} = 2\,\text{N/m}^3$. We then explain two main phenomena quantitatively, i.e., the optimal volume force f_{opt} and the final quasiperiodic variations of R_{1-2} and R_{1-3}. For clarity, we also plot the average convective velocities at positions Σ_1-Σ_6 as a function of volume force in Fig. 20.4b3.

The optimal volume force f_{opt} can be quantitatively predicted by scalar (i.e., temperature) interference. Unlike the vector (say, electric or magnetic field) interference in wave systems, scalar interference cannot be explained by the principle of vector superposition. A key point to understanding scalar interference is the decay rate. Let us take a visual example. Constructive interference means that a high temperature meets another high temperature, but the mixed temperature is not doubled and decays as usual. Destructive interference means a high temperature meets a low temperature, and the mixed temperature decays immediately with a far larger decay rate.

We then use scalar interference to explain thermal nonreciprocity. The transmission at port 2 has only one route, i.e., Σ_1-Σ_5-Σ_2. However, the transmission at port 3 has two routes, i.e., Σ_1-Σ_5-Σ_6-Σ_3 and Σ_1-Σ_4-Σ_3. When two routes have a phase difference of $(2N_1 - 1)\pi$ with N_1 being an integer, destructive interference causes the transmission at port 3 to reach a local minimum value. To achieve a global minimum transmission at port 3, the temperature amplitudes of routes Σ_1-Σ_5-Σ_6-Σ_3 and Σ_1-Σ_4-Σ_3 should be comparable, which requires $v_{\Sigma_6} \gtrsim v_{\Sigma_4}$ ($\gtrsim$ means a little greater than). These requirements can be summarized as

$$\left[-\beta\left(v_{\Sigma_4}\right) + \beta\left(v_{\Sigma_5}\right) + \beta\left(v_{\Sigma_6}\right)\right]\pi\,(r_1 + r_2)\,/3 = (2N_1 - 1)\,\pi, \qquad (20.2a)$$

$$f_{opt} \gtrsim f_{v_{\Sigma_4} = v_{\Sigma_6}} = \frac{3\,(P_h - P_l)}{2\,(2\pi r_2 + 9d)}, \qquad (20.2b)$$

where Eq. (20.2a) ensures destructive interference and Eq. (20.2b) ensures comparable temperature amplitudes of routes Σ_1-Σ_5-Σ_6-Σ_3 and Σ_1-Σ_4-Σ_3. The additional requirement described by Eq. (20.2b) also reflects the difference between convection-diffusion systems and wave systems. Since wave systems are usually Hermitian without loss, it does not require to consider wave amplitudes. However, convection-diffusion systems are non-Hermitian with loss [19], so temperature amplitudes should be considered. The wave number β can be expressed as a function of convective velocity v,

$$\beta\,(v) = \frac{\sqrt{-2v^2 + 2\sqrt{v^4 + 16\omega^2 D^2}}}{4D}, \qquad (20.3)$$

where the convective velocities at positions Σ_4, Σ_5, and Σ_6 are

$$v_{\Sigma_4} = -\frac{h^2}{12\mu}\left[-\frac{3(P_h - P_l)}{2\pi r_2 + 9d} + f\right], \tag{20.4a}$$

$$v_{\Sigma_5} = -\frac{h^2}{12\mu}\left[-\frac{3(P_h - P_l)}{2\pi r_2 + 9d} - f\right], \tag{20.4b}$$

$$v_{\Sigma_6} = \frac{h^2}{12\mu}f. \tag{20.4c}$$

For the results in Fig. 20.4b1, b2, we plot the corresponding $N_1 - f$ curves described by Eq. (20.2a) in Fig. 20.4d1. The f corresponding to an integer N_1 is what we require. We can also derive $f_{v_{\Sigma_4}=v_{\Sigma_6}} = 1.97$ N/m^3 according to Eq. (20.2b). Therefore, theoretical predictions of the optimal volume force are $f_{opt} = 1.99$ N/m^3 ($N_1 = 3$) for Fig. 20.4b1 and $f_{opt} = 2.08$ N/m^3 ($N_1 = 3$) for Fig. 20.4b2, which agree well with $f = 2$ N/m^3 found in simulations. Moreover, only $N_1 = 3$ appears in simulations, and other values of N_1 vanish. This is because the volume force interval between two adjacent integers of N_1, i.e., $\Delta f \approx 0.1$ N/m^3 is too small to observe.

We then increase P_h to 3 Pa to observe the scalar interference at port 3, and R_{1-3} varies quasiperiodically near $f_{v_{\Sigma_4}=v_{\Sigma_6}}$ (Fig. 20.4c1, c2). We take the three valley R_{1-3} in Fig. 20.4c1, or 20.4c2 as an example. The corresponding volume forces are 5.3, 6.3, and 7.3 N/m^3 for Fig. 20.4c1, and 5.7, 6.4, and 7.2 N/m^3 for Fig. 20.4(c2). The theoretical predictions with Eq. (20.2a) are 5.28 ($N_1 = 2$), 6.32 ($N_1 = 1$), and 7.31 ($N_1 = 0$) N/m^3 for Fig. 20.4c1, and 5.67 ($N_1 = 2$), 6.46 ($N_1 = 1$), and 7.20 ($N_1 = 0$) N/m^3 for Fig. 20.4c2, which are clearly presented in Fig. 20.4d2. We can also derive $f_{v_{\Sigma_4}=v_{\Sigma_6}} = 5.91$ N/m^3 with Eq. (20.2b). Therefore, $f_{opt} = 6.32$ N/m^3 and $f_{opt} = 6.46$ N/m^3 correspond to the smallest transmissions in Fig. 20.4c1, c2, respectively. Meanwhile, $\eta_{max} = 0.94$ appears at $f = 6.3$ N/m^3 in Fig. 20.4c1, and $\eta_{max} = 0.90$ occurs at $f = 6.4$ N/m^3 in Fig. 20.4c2. Therefore, the optimal volume force f_{opt} derived with Eqs. (20.2a) and (20.2b) is in good agreement with simulations.

The final quasiperiodic variations of R_{1-2} and R_{1-3} can be attributed to the discrete modal of the ring [19]. We take the results in Fig. 20.4b1 as an example. The final variations begin at approximately $f = 8$ N/m^3, and the convective velocities at positions Σ_4, Σ_5, and Σ_6 are along the counterclockwise direction. Therefore, fluids flow counterclockwise in the ring with only a velocity difference. Since the ring can only support discrete wave numbers [19], R_{1-2} and R_{1-3} exhibit quasiperiodic variations with f. When f corresponds to an allowed (or forbidden) wave number of the ring, transmission reaches a local maximum (or minimum) value. Therefore, the volume force for a local maximum (or minimum) transmission should satisfy

$$\left[\beta\left(v_{\Sigma_4}\right) + \beta\left(v_{\Sigma_5}\right) + \beta\left(v_{\Sigma_6}\right)\right]\pi\left(r_1 + r_2\right)/3 = 2N_2\pi, \tag{20.5a}$$

$$\left[\beta\left(v_{\Sigma_4}\right) + \beta\left(v_{\Sigma_5}\right) + \beta\left(v_{\Sigma_6}\right)\right]\pi\left(r_1 + r_2\right)/3 = \left(2N_2 - 1\right)\pi, \tag{20.5b}$$

where N_2 is a positive integer.

We also compare theoretical predictions with finite-element simulations by taking the right three peaks of R_{1-2} and R_{1-3} in Fig. 20.4b1 as an example. Their corresponding volume forces are 10.4, 12.6, and 16.3 N/m^3, respectively. The theoretical predictions given by Eq. (20.5a) are 10.47 ($N_2 = 5$), 12.63 ($N_2 = 4$), and 16.36 ($N_2 = 3$) N/m^3, respectively (Fig. 20.4d3). We also take the right three valleys of R_{1-2} and R_{1-3} in Fig. 20.4b1 as another example. Their corresponding volume forces are 11.5, 14.3, and 19.6 N/m^3, respectively. The theoretical predictions given by Eq. (20.5b) are 11.42 ($N_2 = 5$), 14.21 ($N_2 = 4$), and 19.39 ($N_2 = 3$) N/m^3, respectively (see Fig. 20.4d3). Therefore, the simulations still match with the theoretical predictions with Eqs. (20.5a) and (20.5b).

We finally provide some experimental suggestions for completeness. A periodic temperature can be realized by alternately using a ceramic heater and a semiconductor cooler. Ferrofluids are a good candidate to realize a volume force, generally composed of ferromagnetic nanoparticles with a 10-nm diameter dispersed in carrier fluids [39]. Here, we may use aqueous ferrofluids containing Fe_3O_4 nanoparticles. Compared with the thermal conductivity and viscosity of water, those of aqueous ferrofluids are slightly enhanced [40] but still approximately applicable. Then we can apply an external magnetic field to guide ferromagnetic nanoparticles to move counterclockwise so that a volume force can be effectively realized. An infrared camera can detect the temperatures at ports 2 and 3. Therefore, it should be possible to observe thermal nonreciprocity experimentally.

20.5   Conclusion

We reveal thermal nonreciprocity based on the thermal Zeeman effect, referring to the modal splitting with an angular momentum bias generated by a volume force. The maximum rectification coefficient can reach 1, so the isolation of temperature propagation is achieved at one output port. Scalar interference can quantitatively explain these results, whose key lies in the decay rate. The proposed mechanism does not require nonlinear and phase-change materials, with a wide range of applicability. Thermal nonreciprocity may have not only potential applications to reduce thermal fluctuation and realize thermal stabilization but also open new directions in thermal metamaterials [41] such as topological thermotics, as schematically shown in Fig. 20.1d. Moreover, an angular momentum bias is also general for other convection-diffusion systems such as mass transport [42, 43], chemical mixing [44], and colloid aggregation [45, 46] where mass diffusivity and concentration correspond to thermal diffusivity and temperature in thermal transport, respectively.

20.6 Exercise and Solution

Exercise

1. Derive Eqs. (20.2)–(20.4).

Solution

1. We rewrite the periodic temperature profile $T = Ae^{-\alpha x} \cos(\beta x - \omega t) + T_0$ as $T = Ae^{i(\beta' x - \omega t)} + T_0$ with $\beta' = \beta + i\alpha$. By substituting the temperature solution into the convection-diffusion equation, we can derive

$$-i\rho C\omega + \kappa(\beta + i\alpha)^2 + i\rho Cv(\beta + i\alpha) = 0. \tag{20.6}$$

Since Eq. (20.6) is always valid, we can calculate the real part and imaginary part independently,

$$\kappa\beta^2 - \kappa\alpha^2 - \rho Cv\alpha = 0, \tag{20.7a}$$
$$\rho C\omega - 2\kappa\alpha\beta - \rho Cv\beta = 0. \tag{20.7b}$$

By solving Eqs. (20.7a) and (20.7b), we can obtain

$$\beta(v) = \frac{\sqrt{-2v^2 + 2\sqrt{v^4 + 16\omega^2 D^2}}}{4D}. \tag{20.8}$$

The convective velocities at positions Σ_4, Σ_5, and Σ_6 can be calculated as

$$v_{\Sigma_4} = -\frac{h^2}{12\mu}\left(-|\nabla P_{\Sigma_4}|_{f=0} + f\right), \tag{20.9a}$$

$$v_{\Sigma_5} = -\frac{h^2}{12\mu}\left(-|\nabla P_{\Sigma_5}|_{f=0} - f\right), \tag{20.9b}$$

$$v_{\Sigma_6} = -\frac{h^2}{12\mu}\left(-|\nabla P_{\Sigma_6}|_{f=0} - f\right), \tag{20.9c}$$

where we suppose $f < |\nabla P_{\Sigma_4}|_{f=0}$. If $f > |\nabla P_{\Sigma_4}|_{f=0}$, the minus sign on the right side of Eq. (20.9a) should become a plus sign. We then require to calculate $|\nabla P_{\Sigma_4}|_{f=0}, |\nabla P_{\Sigma_5}|_{f=0}$, and $|\nabla P_{\Sigma_6}|_{f=0}$. When $f = 0$, we can derive $|\nabla P_{\Sigma_4}|_{f=0} = |\nabla P_{\Sigma_5}|_{f=0}$ and $|\nabla P_{\Sigma_6}|_{f=0} = 0$ due to structural symmetry. We then define the pressure at the joint between port 1 and the ring as P_m and consider the route Σ_1–Σ_4–Σ_3. Convective velocities v_{Σ_1} ($f = 0$) and v_{Σ_4} ($f = 0$) can then be expressed as

$$v_{\Sigma_1}(f = 0) = \frac{h^2}{12\mu}\frac{P_h - P_m}{d}, \tag{20.10a}$$

$$v_{\Sigma_4}\,(f=0) = \frac{h^2}{12\mu}\frac{P_m - P_l}{d + 2\pi r_2/3}.\tag{20.10b}$$

Velocity conservation gives $v_{\Sigma_1}\,(f=0) = 2v_{\Sigma_4}\,(f=0)$, so we can express P_m as

$$P_m = \frac{(2\pi r_2 + 3d)\,P_h + 6d\,P_l}{2\pi r_2 + 9d}.\tag{20.11}$$

We can finally derive

$$|\boldsymbol{\nabla} P_{\Sigma_4}|_{f=0} = |\boldsymbol{\nabla} P_{\Sigma_5}|_{f=0} = \frac{P_m - P_l}{d + 2\pi r_2/3} = \frac{3\,(P_h - P_l)}{2\pi r_2 + 9d},\tag{20.12}$$

$$f_{v_{\Sigma_4} = v_{\Sigma_6}} = \frac{1}{2}|\boldsymbol{\nabla} P_{\Sigma_4}|_{f=0} = \frac{3\,(P_h - P_l)}{2\,(2\pi r_2 + 9d)}.\tag{20.13}$$

References

1. Caloz, C., Alù, A., Tretyakov, S., Sounas, D., Achouri, K., Deck-Léger, Z.-L.: Electromagnetic nonreciprocity. Phys. Rev. Appl. **10**, 047001 (2018)
2. Nassar, H., Yousefzadeh, B., Fleury, R., Ruzzene, M., Alù, A., Daraio, C., Norris, A.N., Huang, G.L., Haberman, M.R.: Nonreciprocity in acoustic and elastic materials. Nat. Rev. Mater. **5**, 667 (2020)
3. Fleury, R., Sounas, D.L., Sieck, C.F., Haberman, M.R., Alù, A.: Sound isolation and giant linear nonreciprocity in a compact acoustic circulator. Science **343**, 516 (2014)
4. Liu, X.X., Cai, X.B., Guo, Q.Q., Yang, J.: Robust nonreciprocal acoustic propagation in a compact acoustic circulator empowered by natural convection. New J. Phys. **21**, 053001 (2019)
5. He, C., Lu, M.-H., Heng, X., Feng, L., Chen, Y.-F.: Parity-time electromagnetic diodes in a two-dimensional nonreciprocal photonic crystal. Phys. Rev. B **83**, 075117 (2011)
6. Lian, J., Fu, J.-X., Gan, L., Li, Z.-Y.: Robust and disorder-immune magnetically tunable one-way waveguides in a gyromagnetic photonic crystal. Phys. Rev. B **85**, 125108 (2012)
7. Sounas, D.L., Caloz, C., Alù, A.: Giant non-reciprocity at the subwavelength scale using angular momentum-biased metamaterials. Nat. Commun. **4**, 2407 (2013)
8. Sounas, D.L., Alù, A.: Angular-momentum-biased nanorings to realize magnetic-free integrated optical isolation. ACS Photonics **1**, 198 (2014)
9. Fleury, R., Sounas, D.L., Alù, A.: Subwavelength ultrasonic circulator based on spatiotemporal modulation. Phys. Rev. B **91**, 174306 (2015)
10. Wang, Y.F., Yousefzadeh, B., Chen, H., Nassar, H., Huang, G.L., Daraio, C.: Observation of nonreciprocal wave propagation in a dynamic phononic lattice. Phys. Rev. B **121**, 194301 (2018)
11. Wong, M.Y., Tso, C.Y., Ho, T.C., Lee, H.H.: A review of state of the art thermal diodes and their potential applications. Int. J. Heat Mass Transf. **164**, 120607 (2021)
12. Li, Y., Shen, X.Y., Wu, Z.H., Huang, J.Y., Chen, Y.X., Ni, Y.S., Huang, J.P.: Temperature-dependent transformation thermotics: from switchable thermal cloaks to macroscopic thermal diodes. Phys. Rev. Lett. **115**, 195503 (2015)
13. Alexander, T.J.: High-heat-flux rectification due to a localized thermal diode. Phys. Rev. E **101**, 062122 (2020)
14. Li, Y., Li, J.X., Qi, M.H., Qiu, C.-W., Chen, H.S.: Diffusive nonreciprocity and thermal diode. Phys. Rev. B **103**, 014307 (2021)

15. Torrent, D., Poncelet, O., Batsale, J.-C.: Nonreciprocal thermal material by spatiotemporal modulation. Phys. Rev. Lett. **120**, 125501 (2018)
16. Xu, L.J., Huang, J.P., Ouyang, X.P.: Tunable thermal wave nonreciprocity by spatiotemporal modulation. Phys. Rev. E **103**, 032128 (2021)
17. Farhat, M., Guenneau, S., Chen, P.-Y., Alù, A., Salama, K.N.: Scattering cancellation-based cloaking for the Maxwell-Cattaneo heat waves. Phys. Rev. Appl. **11**, 044089 (2019)
18. Gandolfi, M., Giannetti, C., Banfi, F.: Temperonic crystal: a superlattice for temperature waves in graphene. Phys. Rev. Lett. **125**, 265901 (2020)
19. Li, Y., Peng, Y.-G., Han, L., Miri, M.-A., Li, W., Xiao, M., Zhu, X.-F., Zhao, J.L., Alù, A., Fan, S.H., Qiu, C.-W.: Anti-parity-time symmetry in diffusive systems. Science **364**, 170 (2019)
20. Cao, P.C., Li, Y., Peng, Y.G., Qiu, C.W., Zhu, X.F.: High-order exceptional points in diffusive systems: robust APT symmetry against perturbation and phase oscillation at APT symmetry breaking. ES Energy Environ. **7**, 48 (2020)
21. Xu, L.J., Wang, J., Dai, G.L., Yang, S., Yang, F.B., Wang, G., Huang, J.P.: Geometric phase, effective conductivity enhancement, and invisibility cloak in thermal convection-conduction. Int. J. Heat Mass Transf. **165**, 120659 (2021)
22. Xu, L.J., Huang, J.P.: Controlling thermal waves with transformation complex thermotics. Int. J. Heat Mass Transf. **159**, 120133 (2020)
23. Xu, L.J., Huang, J.P.: Thermal convection-diffusion crystal for prohibition and modulation of wave-like temperature profiles. Appl. Phys. Lett. **117**, 011905 (2020)
24. Xu, L.J., Huang, J.P.: Negative thermal transport in conduction and advection. Chin. Phys. Lett. **37**, 080502 (2020)
25. Xu, L.J., Huang, J.P.: Active thermal wave cloak. Chin. Phys. Lett. **37**, 120501 (2020)
26. Khanikaev, A.B., Fleury, R., Mousavi, S.H., Alù, A.: Topologically robust sound propagation in an angular-momentum-biased graphene-like resonator lattice. Nat. Commun. **6**, 8260 (2015)
27. Ni, X., He, C., Sun, X.-C., Liu, X.-P., Lu, M.-H., Feng, L., Chen, Y.-F.: Topologically protected one-way edge mode in networks of acoustic resonators with circulating air flow. New J. Phys. **17**, 053016 (2015)
28. Fleury, R., Khanikaev, A.B., Alù, A.: Floquet topological insulators for sound. Nat. Commun. **7**, 11744 (2016)
29. Ding, Y.J., Peng, Y.G., Zhu, Y.F., Fan, X.D., Yang, J., Liang, B., Zhu, X.F., Wan, X.G., Chen, J.C.: Experimental demonstration of acoustic Chern insulators. Phys. Rev. Lett. **122**, 014302 (2019)
30. Liu, X.X., Guo, Q.Q., Yang, J.: Miniaturization of Floquet topological insulators for airborne acoustics by thermal control. Appl. Phys. Lett. **114**, 054102 (2019)
31. Liu, X.X., Guo, Q.Q., Yang, J.: Tunable acoustic valley edge states in a flow-free resonator system. Appl. Phys. Lett. **115**, 074102 (2019)
32. Huang, J.P.: Theoretical Thermotics: Transformation Thermotics and Extended Theories for Thermal Metamaterials. Springer, Singapore (2020)
33. Dai, G.L., Shang, J., Huang, J.P.: Theory of transformation thermal convection for creeping flow in porous media: cloaking, concentrating, and camouflage. Phys. Rev. E **97**, 022129 (2018)
34. Park, J., Youn, J.R., Song, Y.S.: Hydrodynamic metamaterial cloak for drag-free flow. Phys. Rev. Lett. **123**, 074502 (2019)
35. Park, J., Youn, J.R., Song, Y.S.: Fluid-flow rotator based on hydrodynamic metamaterial. Phys. Rev. Appl. **12**, 061002 (2019)
36. Yeung, W.-S., Mai, V.-P., Yang, R.-J.: Cloaking: controlling thermal and hydrodynamic fields simultaneously. Phys. Rev. Appl. **13**, 064030 (2020)
37. Batchelor, G.K.: An Introduction to Fluid Dynamics. Cambridge University Press (2000)
38. Xu, L.J., Huang, J.P., Ouyang, X.P.: Nonreciprocity and isolation induced by an angular momentum bias in convection-diffusion systems. Appl. Phys. Lett. **118**, 221902 (2021)
39. Odenbach, S.: Magnetoviscous Effects in Ferrofluids. Springer, Berlin (2002)
40. Shima, P.D., Philip, J., Raj, B.: Magnetically controllable nanofluid with tunable thermal conductivity and viscosity. Appl. Phys. Lett. **95**, 133112 (2009)

41. Wang, J., Dai, G.L., Huang, J.P.: Thermal metamaterial: fundamental, application, and outlook. iScience **23**, 101637 (2020)
42. Restrepo-Florez, J.M., Maldovan, M.: Mass diffusion cloaking and focusing with metamaterials. Appl. Phys. Lett. **111**, 071903 (2017)
43. Xu, L.J., Dai, G.L., Wang, G., Huang, J.P.: Geometric phase and bilayer cloak in macroscopic particle-diffusion systems. Phys. Rev. E **102**, 032140 (2020)
44. Avanzini, F., Falasco, G., Esposito, M.: Chemical cloaking. Phys. Rev. E **102**, 032140 (2020)
45. Zaccone, A., Wu, H., Gentili, D., Morbidelli, M.: Theory of activated-rate processes under shear with application to shear-induced aggregation of colloids. Phys. Rev. E **80**, 051404 (2009)
46. Banetta, L., Zaccone, A.: Radial distribution function of Lennard-Jones fluids in shear flows from intermediate asymptotics. Phys. Rev. E **99**, 052606 (2019)

Chapter 21
Theory for Thermal Geometric Phases: Exceptional Point Encirclement

Abstract In this chapter, we experimentally demonstrate that the geometric phase can also emerge in a macroscopic thermal convection-conduction system. Following Li et al. [Science 364, 170–173 (2019)], we study two moving rings with equal-but-opposite velocities, joined together by a stationary intermediate layer. We first confirm an exceptional point of velocity that separates a stationary temperature profile and a moving one. We then investigate a cyclic path of time-varying velocity containing the exceptional point, and an extra phase difference of π appears (say, the geometric phase). These results broaden the scope of the geometric phase and provide insights into the thermal topology.

Keywords Thermal geometric phase · Exceptional point encirclement · Non-hermitian system

21.1 Opening Remarks

The Berry phase [1] was found in a quantum mechanical system, and now it has become a fundamental concept in various systems, including the classical one [2]. The Berry phase has significant impacts on electronic properties [3] and phononic properties [4, 5]. In addition to waves, diffusion is a widespread method for transferring energy or mass, such as heat conduction and the Brownian motion of classical particles. Geometric phase was also revealed in diffusion systems [6–11], which led to novel phenomena such as heat pumping [12] and geometric heat flux [13].

However, the related physics has not yet been established in a macroscopic thermal convection-conduction system, mainly resulting from the difficulty of defining phase. Inspired by pioneering studies, we introduce phase-related properties with thermal convection [14–21], and study two moving rings with equal-but-opposite velocities, joined together by a stationary intermediate layer. This macroscopic system is different from the microscopic one where phonon is the carrier of heat transfer [22–24], but it can also be effectively described by a non-Hermitian Hamiltonian [25–28]. Here, the Hamiltonian can be understood as a matrix, and non-Hermitian means that the conjugate transpose is not the matrix itself. As a result, an exceptional point of

L.-J. Xu and J.-P. Huang, *Transformation Thermotics and Extended Theories*,
https://doi.org/10.1007/978-981-19-5908-0_21

velocity appears, similar to optics and photonics [29, 30]. The exceptional point is related to anti-parity-time symmetry [15, 16], which is also widely explored in other systems [31–33]. The exceptional point further leads to the geometric phase for a cyclic path of time-varying velocity. If the cyclic path contains the exceptional point, a moving temperature profile can accumulate an extra phase difference of π (the geometric phase).

As revealed by a recent study [35] which provides an alternative method to explain the findings reported in Ref. [15], the competition between convection and conduction is the key to the exceptional point. Therefore, the two moving rings are the protagonists, and the intermediate layer plays a supporting role in allowing the competition.

21.2   Exceptional Point

As shown in Fig. 21.1a, we investigate two moving rings with equal-but-opposite velocities ($+u$ and $-u$), joined together by a stationary intermediate layer with inner radius r_1 and outer radius r_2. For the convenience of theoretical discussion, we unfold the three-dimensional model along $y - z$ plane, and expand the interior surface along x axis [with length $l = 2\pi r_1$, see Fig. 21.1b, where the left and right ends are applied with periodic boundary condition. This theoretical simplification does not affect the final conclusions, just for discussion convenience. We denote the temperature distributions in the upper ring, lower ring, and intermediate layer as T_1, T_2, and T_i, respectively. The macroscopic thermal convection-conduction process is dominated by

$$
\begin{cases}
\frac{\partial T_1}{\partial t} = D_1 \left(\frac{\partial^2 T_1}{\partial x^2} + \frac{\partial^2 T_1}{\partial z^2} \right) - u \frac{\partial T_1}{\partial x}, & w_i/2 \leq z \leq w_i/2 + w \\[2mm]
\frac{\partial T_i}{\partial t} = D_i \left(\frac{\partial^2 T_i}{\partial x^2} + \frac{\partial^2 T_i}{\partial z^2} \right), & -w_i/2 < z < w_i/2 \\[2mm]
\frac{\partial T_2}{\partial t} = D_2 \left(\frac{\partial^2 T_2}{\partial x^2} + \frac{\partial^2 T_2}{\partial z^2} \right) + u \frac{\partial T_2}{\partial x}, & -w_i/2 - w \leq z \leq -w_i/2
\end{cases}
\qquad (21.1)
$$

where $D_1 (= D + d)$, $D_2 (= D - d)$, and D_i are the diffusivities of the upper ring, lower ring, and intermediate layer, respectively. The two moving rings and the intermediate layer thicknesses are denoted as w and w_i, respectively. For generality, we extend Li et al.'s theory [15] by setting the two moving rings with different diffusivities. Meanwhile, we follow the Hamiltonian description of Li et al.'s theory, and one can also use a dimensionless description of Zhao et al.'s theory [35].

As a quasi one-dimensional model ($l \gg w$), it is reasonable to suppose that the temperature variance along z axis is negligible ($\partial^2 T / \partial z^2 = 0$). The intermediate layer allows energy exchange between the two moving rings, which can be regarded as two source terms. Therefore, the middle equation in Eq. (21.1) is replaced with two source terms, i.e., s_1 for the upper ring and s_2 for the lower ring. We can then

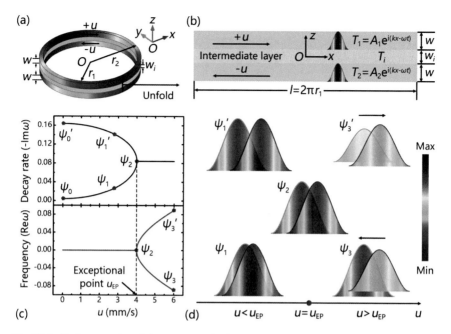

Fig. 21.1 Basic properties of the macroscopic thermal convection-conduction system. **a** Three-dimensional model. **b** Simplified two-dimensional model. **c** Decay rate (-Imω) and frequency (Reω) as a function of velocity u. Parameters: $w = 2.5$ mm, $w_i = 0.5$ mm, $r_1 = 50$ mm, $r_2 = 52$ mm, $D = 10^{-5}$ m^2/s, $\rho c = 10^6$ J m^{-3} K^{-1}, and $\kappa_i = 0.1$ W m^{-1} K^{-1}. These parameters lead to $u_{EP} = 4$ mm/s. The velocity of ψ_1 and ψ_1' is $2\sqrt{2}$ mm/s. **d** Five representative eigenstates. The phase difference of ψ_1 (or ψ_1') is $\pi/4$ (or $3\pi/4$), and that of ψ_2, ψ_3, and ψ_3' is $\pi/2$. The left and right temperature profiles of each eigenstate correspond to the lower and upper rings, respectively. Adapted from Ref. [34]

obtain

$$
\begin{cases}
\frac{\partial T_1}{\partial t} = D_1 \frac{\partial^2 T_1}{\partial x^2} - u \frac{\partial T_1}{\partial x} + \frac{s_1}{\rho c}, & w_i/2 \leq z \leq w_i/2 + w \\
\frac{\partial T_2}{\partial t} = D_2 \frac{\partial^2 T_2}{\partial x^2} + u \frac{\partial T_2}{\partial x} + \frac{s_2}{\rho c}, & -w_i/2 - w \leq z \leq -w_i/2
\end{cases}
\tag{21.2}
$$

We take the same density and heat capacity product of the upper and lower rings, i.e., ρc.

The boundary conditions are given by the continuities of temperature and heat flux,

$$
\begin{cases}
T_1 = T_i, & z = w_i/2 \\
T_2 = T_i, & z = -w_i/2 \\
j_1 = -\kappa_1 \frac{\partial T_1}{\partial z} = -\kappa_i \frac{\partial T_i}{\partial z}, & z = w_i/2 \\
j_2 = \kappa_2 \frac{\partial T_2}{\partial z} = \kappa_i \frac{\partial T_i}{\partial z}, & z = -w_i/2
\end{cases}
\tag{21.3}
$$

where j_1 and j_2 are the heat fluxes from the intermediate layer to the upper and lower rings, respectively. κ_1, κ_2, and κ_i are the thermal conductivities of the upper ring, lower ring, and intermediate layer, respectively. Since we have neglected the higher-order terms $\left(\partial^2 T / \partial z^2 = 0 \right)$, T_i is linear along z axis, thus yielding $\partial T_i / \partial z = (T_1 - T_2) / w_i$. The width of the two moving rings (w) is small enough, so we can assume that the two sources (s_1 and s_2) are uniformly distributed along the ring width, i.e., $s_1 = j_1/w = -\kappa_i(T_1 - T_2)/(ww_i)$ and $s_2 = j_2/w = -\kappa_i(T_2 - T_1)/(ww_i)$. Equation (21.2) can then be reduced to

$$\begin{cases} \frac{\partial T_1}{\partial t} = D_1 \frac{\partial^2 T_1}{\partial x^2} - u \frac{\partial T_1}{\partial x} + h\,(T_2 - T_1)\,, \ w_i/2 \le z \le w_i/2 + w \\ \\ \frac{\partial T_2}{\partial t} = D_2 \frac{\partial^2 T_2}{\partial x^2} + u \frac{\partial T_2}{\partial x} + h\,(T_1 - T_2)\,, \ -w_i/2 - w \le z \le -w_i/2 \end{cases}, \quad (21.4)$$

where $h = \kappa_i/(\rho c w w_i)$. Since h describes the heat exchange rate between the two moving rings, which is vertical to the velocity direction, it is independent of the velocity.

We use plane-wave solutions to introduce phase-related properties,

$$\begin{cases} T_1 = A_1 e^{i(kx-\omega t)} + T_0 \\ T_2 = A_2 e^{i(kx-\omega t)} + T_0 \end{cases}, \quad (21.5)$$

where A_1 (or A_2) is the temperature amplitude in the upper (or lower) ring, k is wave number, ω is complex frequency, and T_0 is reference temperature which is set to zero for brevity. Only the real parts of Eq. (21.5) make sense. By substituting Eq. (21.5) into Eq. (21.4), we can obtain

$$\begin{cases} \omega A_1 = -ik^2 D_1 A_1 + ku A_1 + ih\,(A_2 - A_1)\,, \ w_i/2 \le z \le w_i/2 + w \\ \omega A_2 = -ik^2 D_2 A_2 - ku A_2 + ih\,(A_1 - A_2)\,, \ -w_i/2 - w \le z \le -w_i/2 \end{cases}. \quad (21.6)$$

Equation (21.6) can also be expressed as

$$\hat{H}|\psi\rangle = \omega|\psi\rangle, \quad (21.7)$$

where $|\psi\rangle = [A_1,\ A_2]^\tau$ is eigenstate, and τ denotes transpose. The Hamiltonian $\hat{H}$ reads

$$\hat{H} = \begin{bmatrix} -i\left(k^2 D_1 + h\right) + ku & ih \\ ih & -i\left(k^2 D_2 + h\right) - ku \end{bmatrix}. \quad (21.8)$$

Equation (21.8) is a general expression. Here, we discuss the case of $d = 0$ (i.e., $D_1 = D_2 = D$), and Eq. (21.8) becomes

$$\hat{H} = \begin{bmatrix} -i\left(k^2 D + h\right) + ku & ih \\ ih & -i\left(k^2 D + h\right) - ku \end{bmatrix}, \quad (21.9)$$

where $D = \kappa_1/(\rho c) = \kappa_2/(\rho c)$.

The eigenvalues of the Hamiltonian (Eq. (21.9)) take on the form

$$\omega_{\pm} = -i\left[(k^2 D + h) \pm \sqrt{h^2 - k^2 u^2}\right], \tag{21.10}$$

which are complex numbers. The system exhibits two different properties as u varies. The point $u_{EP} = h/k$ determines the transition of two different properties, thus serving as an exceptional point. As required by the periodic boundary condition, wave numbers can only take on discrete values, i.e., $k = 2\pi n/l = nr_1^{-1}$ with n being positive integers. We discuss the fundamental modes with $n = 1$ because their decay rates are the lowest.

In the region $u < u_{EP}$, the complex frequencies ($\omega_{\pm}$) exhibit two different branches with purely imaginary values (Fig. 21.1c), indicating that the waves described by Eq. (21.5) only decay but do not propagate. The difference between ω_+ and ω_- is the decay rate: the decay rate of ω_- is smaller than that of ω_+. Therefore, ω_+ is also observable, but it decays much faster than ω_-. The corresponding eigenstates are

$$|\psi_+\rangle = \left[1, e^{i(\pi - \alpha)}\right]^\tau, \quad |\psi_-\rangle = \left[1, e^{i\alpha}\right]^\tau, \tag{21.11}$$

where $\alpha = \sin^{-1}(ku/h)$. Therefore, the temperature profiles of the two moving rings maintain a constant phase difference ($\pi - \alpha$ for ω_+ and α for ω_-) and decay motionlessly (see ψ_1 and ψ_1' in Fig. 21.1d).

When the velocity reaches the exceptional point, the difference between ω_+ and ω_- disappears (Fig. 21.1c). The two eigenstates have the same phase difference of $\pi/2$ and decay motionlessly (see ψ_2 in Fig. 21.1d).

When $u > u_{EP}$, the complex frequencies ($\omega_{\pm}$) take on real components (Fig. 21.1c), indicating that the waves described by Eq. (21.5) not only decay but also propagate. The corresponding eigenstates become

$$|\psi_+\rangle = \left[e^{-\delta}, e^{i\pi/2 - 2\delta}\right]^\tau, \quad |\psi_-\rangle = \left[e^{-\delta}, e^{i\pi/2}\right]^\tau, \tag{21.12}$$

where $\delta = \cosh^{-1}(ku/h)$. Therefore, the two eigenstates maintain the same phase difference of $\pi/2$ but decay with motion. The moving direction follows the ring with a larger temperature amplitude (see ψ_3 and ψ_3' in Fig. 21.1d).

We use COMSOL Multiphysics to perform finite-element simulations based on a three-dimensional model (Fig. 21.2). We define T_1 and T_2 as the temperature distributions along the upper and lower interior edges of the two moving rings. We track the evolutions of temperature profile by following maximum-temperature points, i.e., $\text{Max}(T_1)$ and $\text{Max}(T_2)$. The initial states are set as the three eigenstates ψ_0, ψ_0', and ψ_2 (Fig. 21.2a–c).

If we set the velocity to $2\sqrt{2}$ mm/s ($< u_{EP}$), the initial state moves to a certain position and remains stationary thereafter (Fig. 21.2d–f). All three final states (with $\alpha \approx \pi/4$) are the eigenstate corresponding to eigenvalue ω_- (i.e., ψ_1, see Fig. 21.2g–i). This phenomenon occurs because of the non-orthogonality of the eigenstates at different branches [25]. Meanwhile, the decay rate at the upper branch is much higher than that at the lower branch, so the eigenstate at the lower branch becomes the observable one associated with eigenvalue ω_-.

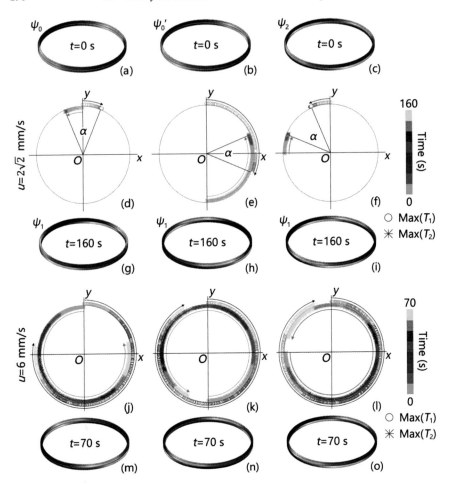

Fig. 21.2 Evolutions of the temperature profiles. Two types of color maps denote temperature (**a–c**, **g–i** and **m–o**) and time (**d–f** and **j–l**), respectively. The temperature scale is the same as that in Fig. 21.1. The initial states are presented in **a–c** with the form in the Cartesian coordinates as $T_1 = Ay/\sqrt{x^2 + y^2} + B$ and $T_2 = Ay/\sqrt{x^2 + y^2} + B$ for ψ_0, $T_1 = Ay/\sqrt{x^2 + y^2} + B$ and $T_2 = -Ay/\sqrt{x^2 + y^2} + B$ for ψ'_0, $T_1 = Ay/\sqrt{x^2 + y^2} + B$ and $T_2 = -Ax/\sqrt{x^2 + y^2} + B$ for ψ_2, and $T_i = B$, where $A = 100$ and $B = 400$. The trajectories of Max(T_1) and Max(T_2) along interior edges are shown in **d–i** with $u = 2\sqrt{2}$ mm/s and (**j**)-(**o**) with $u = 6$ mm/s. The parameters are the same as those for Fig. 21.1. The meshing parameters are as follows. The maximum and minimum element sizes are 5×10^{-4} and 10^{-5} m, respectively. The maximum element growth rate is 1.3, the curvature factor is 0.2, and the resolution of narrow regions is 1. Adapted from Ref. [34]

Moreover, we also care about how these initial states evolve to final states (i.e., evolution routes). One principle is that the evolution routes should try to keep away from the eigenstate corresponding to eigenvalue ω_+ (i.e., $|\psi_+\rangle$, with a much higher decay rate) to survive longer. For this purpose, the moving direction of the maximum-temperature point in Fig. 21.2f is even against the background velocities of respective moving rings.

If we set the velocity to 6 mm/s ($> u_{EP}$), the trajectories of Max(T_1) and Max(T_2) are always moving because the eigenvalue has a real component (Fig. 21.2j–l). The corresponding states at 70 s are presented in Fig. 21.2m–o. Here, the duration for $u = 6$ mm/s (70 s) is shorter than that for $u = 2\sqrt{2}$ mm/s (160 s) because the decay rate for $u = 6$ mm/s is higher than that for $u = 2\sqrt{2}$ mm/s.

The results presented in Fig. 21.2d–i can help us draw some conclusions. When the velocity is smaller than the exceptional point, the final eigenstate prefers to stay at the lower branch corresponding to eigenvalue ω_-. The evolution route should keep away from the eigenstate corresponding to eigenvalue ω_+. The evolution routes should ensure that the temperature profiles decay as slowly as possible before reaching the final states.

21.3 Thermal Geometric Phase

After understanding the exceptional point and evolution routes, we can reveal the geometric phase in our system. The Hamiltonian (Eq. (21.9)) is a function of multiple parameters. We resort to a time-varying velocity $u(t)$ which is experimentally controllable, instead of other parameters such as the thickness of the intermediate layer. Although the Hamiltonian $\hat{H}$ is not Hermitian, we can check that

$$\hat{H}^\dagger |\overline{\psi}_\pm\rangle = \overline{\omega}_\pm |\overline{\psi}_\pm\rangle, \tag{21.13}$$

where $\hat{H}^\dagger$ is the Hermitian transpose of $\hat{H}$. $|\overline{\psi}_\pm\rangle$ and $\overline{\omega}_\pm$ are the complex conjugates of $|\psi_\pm\rangle$ and $\omega_\pm$, respectively. The eigenstates satisfy

$$\left\langle \overline{\psi}_\pm | \psi_\mp \right\rangle = 0. \tag{21.14}$$

$\left\langle \overline{\psi}_\pm | \psi_\mp \right\rangle$ denotes the complex inner product of the vectors $|\overline{\psi}_\pm\rangle$ and $|\psi_\mp\rangle$. As discussed in Fig. 21.2, the final states always go back to the eigenstate associated with eigenvalue ω_- (i.e., the initial state) after experiencing a cyclic evolution, thus ensuring an adiabatic process. Therefore, we can write down the complex geometric phase under an adiabatic approximation as

$$\varphi_\pm = i \int \frac{\left\langle \overline{\psi}_\pm(u) | d\psi_\pm(u) \right\rangle}{\left\langle \overline{\psi}_\pm(u) | \psi_\pm(u) \right\rangle}, \tag{21.15}$$

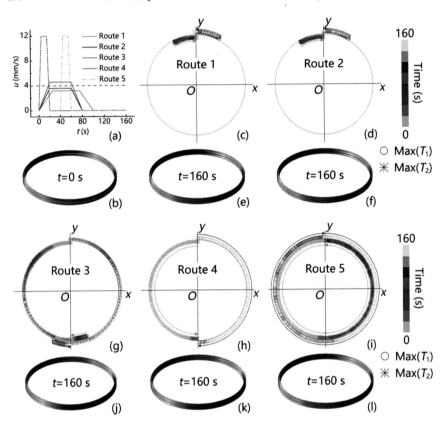

Fig. 21.3 Finite-element simulations of the geometric phase. The parameters are the same as those for Fig. 21.1. **a** Five evolution routes. **b** Initial state. The trajectories of $Max(T_1)$ and $Max(T_2)$ corresponding to the five evolution routes are presented in **c**, **d**, **g**, **h**, and **i**. The final states are presented in **e**, **f**, **j**, **k**, and **l**, respectively. Adapted from Ref. [34]

which is in accordance with the results of non-Hermitian quantum systems [36]. We find that $\langle \overline{\psi}_\pm(u)|\psi_\pm(u)\rangle = 0$ at the exceptional point because the two eigenstates coalesce. Therefore, the exceptional point is a pole in the complex integral. We can rewrite Eq. (21.15) in a closed loop around the exceptional point as [37]

$$\varphi_\pm = \frac{i}{2}\oint d\ln\langle \overline{\psi}_\pm(u)|\psi_\pm(u)\rangle. \qquad (21.16)$$

According to the residue theorem, we know that $\varphi_\pm = \pi$ or $-\pi$, and the sign depends on the direction of the closed loop. If the evolution route does not contain the exceptional point, the geometric phase in a cyclic evolution equals zero.

We also perform finite-element simulations to observe the geometric phase. For this purpose, we apply a cyclic path of time-varying velocity, which is governed by the Hamiltonian $\hat{H}[u(t)]$. Note that a fiduciary marker on the surface of the ring would not necessarily return to its original location after the cyclic evolution of velocity. We explore five different paths of velocity, as shown in Fig. 21.3a. The initial velocity is $u = 0$ mm/s, and the initial state is set to the eigenstate associated with eigenvalue ω_- (i.e., ψ_0, see Fig. 21.3b).

In Route 1 or 2, the eigenvalue is purely imaginary because the velocity is smaller than the exceptional point, indicating no extra phase difference accumulated. As a result, the two evolution routes bring the final state back to the initial state exactly (see Fig. 21.3c, d or Fig. 21.3e, f).

However, the situation is different when the cyclic evolution of velocity contains the exceptional point (Fig. 21.3g, h). As the velocity increases and exceeds the exceptional point, the eigenvalue obtains a real component, indicating that an extra phase difference accumulates. This property means that the initial state moves smoothly from one branch to another. When the state goes through the eigenstate corresponding to another eigenvalue ω_+ (i.e., ψ_0'), the state cannot return, as discussed in Fig. 21.2. The phase difference then continuously increases to reach a different place with eigenvalue ω_- (i.e., ψ_0). Therefore, the temperature profile stops at a different position with a phase difference of π compared with the initial position (Fig. 21.3j, k). This phenomenon is evidence of the geometric phase.

Finally, we repeat Route 4 twice (indicated as Route 5). Since the state goes through the eigenstate corresponding to eigenvalue ω_+ (i.e., ψ_0') twice, Route 5 brings the state back to the initial state without any global phase change (Fig. 21.3i, l).

Although the macroscopic thermal convection-conduction system can also be effectively described by a non-Hermitian Hamiltonian (Eq. (21.9)), its properties are distinct from wave systems. In the case of non-Hermitian wave systems, the cyclic evolution around the exceptional point may interchange the instantaneous eigenvectors [30]. However, the present eigenstates at the upper branch (especially with small velocities) are metastable because of their large decay rates. Therefore, they do not dominate the end of evolutions. In contrast, these metastable eigenstates are more similar to a "wall", which should be avoided when the velocity is smaller than the exceptional point, thus determining the evolution route and final position. When the velocity is larger than the exceptional point, it provides an opportunity to cross the eigenstate corresponding to the eigenvalue ω_+, once to accumulate an extra phase difference of π. Therefore, the geometric property is reflected on $|\psi_+\rangle$ to some extent.

We also conduct experiments to validate the theoretical analyses and finite-element simulations. The different views of the experimental setup are shown in Fig. 21.4a. The aluminum frames are wrapped with black tape to avoid the environmental reflection of thermal radiation and ensure the accuracy of temperature profiles. We also use small woodblocks (with thermal conductivity of 0.1 W m^{-1} K^{-1}) to separate the rings and aluminum frames (with thermal conductivity of 218 W m^{-1} K^{-1}) to reduce thermal dissipation. With these preparations, the environmental dissipation is small enough to be neglected. The two rings are driven by two motors with different rotation directions, as shown by blue arrows. The two motors are controlled by one circuit, thus ensuring the equal-but-opposite rotation speed. The slide rheostat

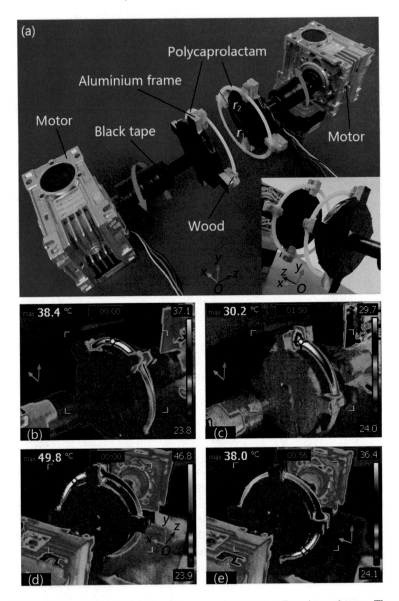

Fig. 21.4 Experimental demonstration of the geometric phase. **a** Experimental setup. The sizes of the two moving rings are $r_1 = 60$, $r_2 = 65$, and $w = 2$ mm, respectively. The rings are made of polycaprolactam with $D = 7.3 \times 10^{-7}$ m^2/s and $\rho c = 1.8 \times 10^6$ J m^{-3} K^{-1}. The intermediate layer is 1-mm-thick grease with $\kappa_i = 0.1$ W m^{-1} K^{-1}. These parameters cause the exceptional value of rotation speed (Ω_{EP}) to be approximately 0.25 rmp. **b** and **c** Initial and final states with rotation speed not crossing the exceptional point. **d** and **e** Initial and final states with the rotation speed crossing the exceptional point. Adapted from Ref. [34]

Fig. 21.5 Detailed
information of Fig. 21.4. In **a**
and **c**, dashed lines represent
the exceptional value of
rotation speed
($\Omega_{EP} = 0.25$ rmp), and solid
lines exhibit the cyclic
evolutions of rotation speed
(Ω). The measured and
simulated results are
presented in **b** and **d**. The
phase difference is $\varphi = 0$ for
the cyclic evolution shown in
a, whereas it is $\varphi = \pi$ for the
cyclic evolution shown in **c**.
Adapted from Ref. [34]

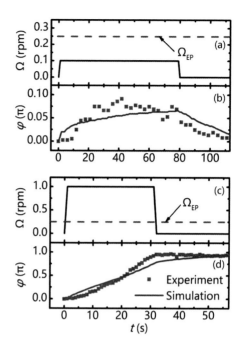

can manually adjust the rotation speed. We use the Flir E60 infrared camera to detect
temperature profiles and find maximum temperatures in a particular region.

We use a copper plate with one end heated to generate the initial temperature
profiles of the two moving rings (see Fig. 21.4b, d). We quickly remove the copper
plate and push the two rings together to observe the evolution. We further tune the
rotation speed (Ω) from zero to a small value ($\sim$0.1 rpm) or a large one ($\sim$1 rpm) and
then set it back to zero. If the rotation speed does not cross the exceptional point, the
final state (Fig. 21.4c) is precisely the same as the initial state (Fig. 21.4b). However,
if the rotation speed crosses the exceptional point, the final state (Fig. 21.4e) exhibits
a phase difference of π compared with the initial position (Fig. 21.4d).

We also analyze the experimental results and compare them with finite-element
simulations. For this purpose, we track the maximum temperature of the front ring.
The initial position is set as a rotation angle of 0, and φ is the rotation angle of the
maximum temperature of the front ring, which denotes the geometric phase of our
interest. When applying a cyclic evolution avoiding the exceptional point (Fig. 21.5a),
the final state goes back to the initial state exactly, resulting in $\varphi = 0$ (Fig. 21.5b).
When applying a cyclic evolution containing the exceptional point (Fig. 21.5c), the
final state has a geometric phase compared with the initial state, resulting in $\varphi = \pi$
(Fig. 21.5d).

21.4   Conclusion

Besides various wave systems, macroscopic thermal convection-conduction systems (essentially non-Hermitian systems) can also exhibit the geometric phase by encircling an exceptional point. More relevant properties such as topological invariance (or winding number) in thermotics can be further explored with a similar method applied in non-Hermitian systems [38–42]. These results may also provide insights into heat regulation with exceptional points.

21.5   Exercise and Solution

Exercise

1. Calculate the eigenvalues and eigenvectors of Eq. (21.8).

Solution

1. The eigenvalues of Eq. (21.8) can be expressed as

$$\omega_{\pm} = -i\left[\left(k^2 D + h\right) \pm \sqrt{h^2 + k^4 d^2 - k^2 u^2 + 2k^3 dui}\right], \tag{21.17}$$

and the corresponding eigenstates are

$$|\psi_{\pm}\rangle = \left[1, \frac{-h}{k^2 d + kui \pm \sqrt{h^2 + k^4 d^2 - k^2 u^2 + 2k^3 dui}}\right]^{\tau}. \tag{21.18}$$

References

1. Berry, M.V.: Quantal phase factors accompanying adiabatic changes. Proc. R. Soc. A **392**, 45 (1984)
2. Hannay, J.H.: Angle variable holonomy in adiabatic excursion of an integrable Hamiltonian. J. Phys. A-Math. Gen. **18**, 221 (1985)
3. Xiao, D., Chang, M.C., Niu, Q.: Berry phase effects on electronic properties. Rev. Mod. Phys. **82**, 1959 (2010)
4. Strohm, C., Rikken, G.L.J.A., Wyder, P.: Phenomenological evidence for the phonon Hall effect. Phys. Rev. Lett. **95**, 155901 (2005)
5. Zhang, L.F., Ren, J., Wang, J.-S., Li, B.W.: Topological nature of the phonon Hall effect. Phys. Rev. Lett. **105**, 225901 (2010)
6. Mead, C.A., Truhlar, D.G.: On the determination of Born-Oppenheimer nuclear motion wave functions including complications due to conical intersections and identical nuclei. J. Chem. Phys. **70**, 2284 (1979)
7. Ning, C.Z., Haken, H.: Geometrical phase and amplitude accumulations in dissipative systems with cyclic attractors. Phys. Rev. Lett. **68**, 2109 (1992)

8. Ao, P.: Potential in stochastic differential equations: Novel construction. J. Phys. A-Math. Gen. **37**, 25 (2004)

9. Olson, J.C., Ao, P.: Nonequilibrium approach to Bloch-Peierls-Berry dynamics. Phys. Rev. B **75**, 035114 (2007)

10. Misaki, K., Miyashita, S., Nagaosa, N.: Diffusive real-time dynamics of a particle with Berry curvature. Phys. Rev. B **97**, 075122 (2018)

11. Xu, L.J., Dai, G.L., Wang, G., Huang, J.P.: Geometric phase and bilayer cloak in macroscopic particle-diffusion systems. Phys. Rev. E **102**, 032140 (2020)

12. Ren, J., Hanggi, P., Li, B.W.: Berry-phase-induced heat pumping and its impact on the fluctuation theorem. Phys. Rev. Lett. **104**, 170601 (2010)

13. Ren, J., Liu, S., Li, B.W.: Geometric heat flux for classical thermal transport in interacting open systems. Phys. Rev. Lett. **108**, 210603 (2012)

14. Li, Y., Zhu, K.-J., Peng, Y.-G., Li, W., Yang, T.Z., Xu, H.-X., Chen, H., Zhu, X.-F., Fan, S.H., Qiu, C.-W.: Thermal meta-device in analogue of zero-index photonics. Nat. Mater. **18**, 48 (2019)

15. Li, Y., Peng, Y.-G., Han, L., Miri, M.-A., Li, W., Xiao, M., Zhu, X.-F., Zhao, J.L., Alù, A., Fan, S.H., Qiu, C.-W.: Anti-parity-time symmetry in diffusive systems. Science **364**, 170 (2019)

16. Cao, P.C., Li, Y., Peng, Y.G., Qiu, C.W., Zhu, X.F.: High-order exceptional points in diffusive systems: robust APT symmetry against perturbation and phase oscillation at APT symmetry breaking. ES Energy Environ. **7**, 48 (2020)

17. Xu, L.J., Huang, J.P., Ouyang, X.P.: Nonreciprocity and isolation induced by an angular momentum bias in convection-diffusion systems. Appl. Phys. Lett. **118**, 221902 (2021)

18. Xu, L.J., Huang, J.P.: Controlling thermal waves with transformation complex thermotics. Int. J. Heat Mass Transf. **159**, 120133 (2020)

19. Xu, L.J., Yang, S., Huang, J.P.: Controlling thermal waves of conduction and convection. EPL **133**, 20006 (2021)

20. Xu, L.J., Huang, J.P.: Negative thermal transport in conduction and advection. Chin. Phys. Lett. **37**, 080502 (2020)

21. Xu, L.J., Huang, J.P.: Active thermal wave cloak. Chin. Phys. Lett. **37**, 120501 (2020)

22. Li, N.B., Ren, J., Wang, L., Zhang, G., Hanggi, P., Li, B.W.: Colloquium: Phononics: manipulating heat flow with electronic analogs and beyond. Rev. Mod. Phys. **84**, 1045 (2012)

23. Hu, R., Luo, X.B.: Two-dimensional phonon engineering triggers microscale thermal functionalities. Natl. Sci. Rev. **6**, 1071 (2019)

24. Hu, R., Iwamoto, S., Feng, L., Ju, S.H., Hu, S.Q., Ohnishi, M., Nagai, N., Hirakawa, K., Shiomi, J.: Machine-learning-optimized aperiodic superlattice minimizes coherent phonon heat conduction. Phys. Rev. X **10**, 021050 (2020)

25. Bender, C.M.: Making sense of non-Hermitian Hamiltonians. Rep. Prog. Phys. **70**, 947 (2007)

26. Gao, T., Estrecho, E., Bliokh, K.Y., Liew, T.C.H., Fraser, M.D., Brodbeck, S., Kamp, M., Schneider, C., Hofling, S., Yamamoto, Y., Nori, F., Kivshar, Y.S., Truscott, A.G., Dall, R.G., Ostrovskaya, E.A.: Observation of non-Hermitian degeneracies in a chaotic exciton-polariton billiard. Nature **526**, 554 (2015)

27. Leykam, D., Bliokh, K.Y., Huang, C.L., Chong, Y.D., Nori, F.: Edge modes, degeneracies, and topological numbers in non-Hermitian systems. Phys. Rev. Lett. **118**, 040401 (2017)

28. Gong, Z.P., Ashida, Y., Kawabata, K., Takasan, K., Higashikawa, S., Ueda, M.: Topological phases of non-Hermitian systems. Phys. Rev. X **8**, 031079 (2018)

29. Xu, H., Mason, D., Jiang, L.Y., Harris, J.G.E.: Topological energy transfer in an optomechanical system with exceptional points. Nature **537**, 80 (2016)

30. Miri, M.-A., Alù, a.: Exceptional points in optics and photonics. Science **363**, eaar7709 (2019)

31. Zhu, X.F., Ramezani, H., Shi, C.Z., Zhu, J., Zhang, X.: PT-symmetric acoustics. Phys. Rev. X **4**, 031042 (2014)

32. Assawaworrarit, S., Yu, X.F., Fan, S.H.: Robust wireless power transfer using a nonlinear parity-time-symmetric circuit. Nature **546**, 387 (2017)

33. Liu, T., Zhu, X.F., Chen, F., Liang, S.J., Zhu, J.: Unidirectional wave vector manipulation in two-dimensional space with an all passive acoustic parity-time-symmetric metamaterials crystal. Phys. Rev. Lett. **120**, 124502 (2018)

34. Xu, L.J., Wang, J., Dai, G.L., Yang, S., Yang, F., Wang, G., Huang, J.P.: Geometric phase, effective conductivity enhancement, and invisibility cloak in thermal convection-conduction. Int. J. Heat Mass Transf. **165**, 120659 (2021)
35. Zhao, L., Zhang, L.N., Bhatia, B., Wang, E.N.: Understanding anti-parity-time symmetric systems with a conventional heat transfer framework — Comment on "Anti-parity-time symmetry in diffusive systems" (2019). arXiv: 1906.08431v1
36. Garrison, J.C., Wright, E.M.: Complex geometrical phases for dissipative systems. Phys. Lett. A **128**, 171 (1988)
37. Mailybaev, A.A., Kirillov, O.N., Seyranian, A.P.: Geometric phase around exceptional points. Phys. Rev. A **72**, 014104 (2005)
38. Parto, M., Wittek, S., Hodaei, H., Harari, G., Bandres, M.A., Ren, J.H., Rechtsman, M.C., Segev, M., Christodoulides, D.N., Khajavikhan, M.: Edge-mode lasing in 1D topological active arrays. Phys. Rev. Lett. **120**, 113901 (2018)
39. Luo, X.-W., Zhang, C.W.: Higher-order topological corner states induced by gain and loss. Phys. Rev. Lett. **123**, 073601 (2019)
40. Kawabata, K., Shiozaki, K., Ueda, M., Sato, M.: Symmetry and topology in non-Hermitian physics. Phys. Rev. X **9**, 041015 (2019)
41. Okuma, N., Kawabata, K., Shiozaki, K., Sato, M.: Topological origin of non-Hermitian skin effects. Phys. Rev. Lett. **124**, 086801 (2020)
42. Borgnia, D.S., Kruchkov, A.J., Slager, R.-J.: Non-Hermitian boundary modes and topology. Phys. Rev. Lett. **124**, 056802 (2020)

Chapter 22
Theory for Thermal Edge States: Graphene-Like Convective Lattice

Abstract In this chapter, we reveal that edge states are not necessarily limited to wave systems but can also exist in convection-diffusion systems that are essentially different from wave systems. For this purpose, we study heat transfer in a graphene-like (or honeycomb) lattice to demonstrate thermal edge states with robustness against defects and disorders. Convection is compared to electron cyclotron, which breaks space-reversal symmetry and determines the direction of thermal edge propagation. Diffusion leads to interference-like behavior between opposite convections, preventing bulk temperature propagation. We also display thermal unidirectional interface states between two lattices with opposite convection. These results extend the physics of edge states beyond wave systems.

Keywords Thermal edge state · Thermal interface state · Conduction and convection

22.1 Opening Remarks

Topological insulators were initially revealed in quantum mechanics systems [1, 2], which are insulated in bulk but conductive on the surface. Since the foundation of quantum physics is Schrödinger wave mechanics, there is a similarity between quantum waves and classical waves in equation forms. Therefore, the concept of topological insulators has also been extended to classical wave systems [3], including but not limited to electromagnetics [4–11] and acoustics [12–21]. The related research was commonly conducted in nonreciprocal systems with the broken time-reversal symmetry induced by an external magnetic bias for electromagnetics [4–6] or an external momentum bias for acoustics [12–16]. Regardless of the quantum or classical description, topological insulators can support edge states on the surface, with broad applications for isolators and sensors.

Although edge states have been intensively studied in wave systems, they have received almost no attention in diffusion systems. Unlike wave systems with time-reversal symmetry, diffusion systems feature space-reversal symmetry, indicating that diffusion is identical in two opposite directions. Inspired by topological wave

© The Author(s) 2023
L.-J. Xu and J.-P. Huang, *Transformation Thermotics and Extended Theories*,
https://doi.org/10.1007/978-981-19-5908-0_22

insulators [3] with the broken time-reversal symmetry, it is natural to consider the broken space-reversal symmetry of diffusion systems. Fortunately, several methods are available to break space-reversal symmetry, such as applying asymmetric structures and nonlinear materials [22–25] and considering spatiotemporal modulations [26–29]. Besides symmetry differences, diffusion systems also lack the concept of phase because diffusion generally occurs from high to low potentials, such as from high to low temperatures for heat transfer and from high to low concentrations for mass transfer. To solve the problem, we can introduce a periodic temperature [30–36] for heat transfer or a periodic concentration [37] for mass transfer, which has been experimentally validated [30, 31]. We can discuss thermal edge states with these preliminary analyses by considering heat transfer with conduction and convection for breaking space-reversal symmetry.

22.2 Theoretical Foundation

Two basic structures with counterclockwise and clockwise convection are presented in Fig. 22.1a, b, respectively. Convection is an analog of electron cyclotron, which determines the direction of thermal edge propagation, thus called thermal spin. We regard counterclockwise convection as spin-up and clockwise convection as spin-down for brevity. The vertex regions in Fig. 22.1 are solid pumps with high thermal conductivities to drive fluids with a convective velocity of v. Besides fluids, convection can also be effectively realized with spatiotemporal modulations of thermal conductivity and density [26–29], which has been experimentally verified to break space-reversal symmetry [28]. Therefore, what we discuss is a simple and practical system of heat transfer whose governing equation is [39]

$$\rho C \frac{\partial T}{\partial t} + \nabla \cdot (-\kappa \nabla T + \rho C v T) = 0, \tag{22.1}$$

where ρ, C, κ, and v denote density, heat capacity, thermal conductivity, and convective velocity, respectively. T and t represent temperature and time, respectively. $\rho C v$ is the convective term that breaks space-reversal symmetry.

We then need to introduce the concept of phase. For this purpose, we consider a periodic temperature source whose temperature is

$$T = A e^{-i\omega t} + B, \tag{22.2}$$

where A, ω, and B are the temperature amplitude, circular frequency, and reference temperature of the temperature source. The real part of Eq. (22.2) denotes the actual temperature. The temperature source can generate a temperature profile with spatiotemporal periodicity,

$$T = A e^{i(\alpha \cdot r - \omega t)} + B, \tag{22.3}$$

with wave vector α and position vector r.

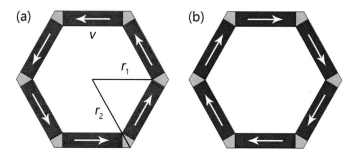

Fig. 22.1 Two basic structures with inner radius r_1 and outer radius r_2. Vertex regions are solid pumps to drive fluids with **a** counterclockwise velocity and **b** clockwise velocity of v. Adapted from Ref. [38]

To understand the broken space-reversal symmetry induced by convection, we discuss a one-dimensional case along the x axis. Since conduction has dissipation, the wave vector should be a complex number, i.e., $\alpha = \beta + i\gamma$ with wave number β and decay rate γ. The substitution of Eq. (22.3) into Eq. (22.1) yields

$$\beta = \frac{\sqrt{2}\varepsilon}{4}, \tag{22.4a}$$

$$\gamma = \frac{-8v\omega + 2\sqrt{2}v^2\varepsilon + \sqrt{2}D^2\varepsilon^3}{16\omega D}, \tag{22.4b}$$

with definitions of $\varepsilon = \sqrt{-v^2/D^2 + \sqrt{v^4/D^4 + 16\omega^2/D^2}}$ and $D = \kappa/(\rho C)$. When $v = 0$, it is identical along two opposite directions. If $v \neq 0$, a change from v to $-v$ yields the same ε and β but different γ, indicating different decay rates along two opposite directions. Therefore, nonreciprocal temperature propagation can be achieved with convection, which offers an opportunity to realize one-way temperature propagation. Generally speaking, two parameters mainly affect temperature propagation: thermal diffusivity determines the decay rate; and convective velocity determines the temperature propagation speed. The chosen parameters are based on water that has a relatively small thermal diffusivity, so dissipation is not that intense and the expected phenomena can still be observed.

22.3 Finite-Element Simulation

We then design a graphene-like (or honeycomb) lattice composed of spin-up units, as presented in Fig. 22.2a. We first discuss the bulk property and put a temperature source in the center (Fig. 22.2a). Each side contains opposite convection in bulk, so temperature propagation decays far more quickly, as described by Eq. (22.4b). Therefore, the bulk cannot support temperature propagation and becomes insulated

(Fig. 22.2b). We then discuss the surface property and put a temperature source at the bottom left corner (Fig. 22.2c). Each side contains only unidirectional convection on the surface, so the decay rate is far lower than the bulk. Since the graphene-like lattice is composed of spin-up units, the surface can support only counterclockwise temperature propagation (Fig. 22.2d), which is direct evidence of thermal edge states. Unlike the edge states in wave systems, those in convection-diffusion systems have diffusion-induced dissipation. A simple physical image to understand thermal edge states is that the surface decay rate is far lower than the bulk decay rate, so temperature propagation is allowed only on the surface. To confirm that thermal edge states have directionality, we further construct a graphene-like lattice with spin-down units (Fig. 22.2e). The simulation shows that thermal edge propagation still exists but with a clockwise direction (Fig. 22.2f). The results in Fig. 22.2 are in accordance with electron edge states whose propagation directions are determined by electron spins. Therefore, it is reasonable to compare convection to electron spin despite different mechanisms. In other words, the directions of thermal edge states are locked by thermal spins (i.e., convective directions).

Since edge states are unidirectional, defects and disorders cannot cause backscattering. Analogously, thermal edge states should also be robust against defects and disorders. We perform extended simulations based on the graphene-like lattice to confirm this robustness. We first change one unit from spin-up to spin-down (Fig. 22.3a). The result indicates that the thermal edge state still exists (Fig. 22.3b), but it has a slightly higher decay rate than Fig. 22.2d. We then stop one unit from rotating (Fig. 22.3c), and the result is presented in Fig. 22.3d, demonstrating that the thermal edge state remains unchanged. We finally remove six units from the graphene-like lattice, as displayed in Fig. 22.3e. Temperature propagation is still allowed only on the surface (Fig. 22.3f). Therefore, the results in Fig. 22.3 prove that thermal edge states are robust against defects and disorders. Moreover, since thermal edge states are robust, the graphene-like lattice is not mandatory, and other lattices are also applicable, such as a square lattice.

We further discuss thermal interface states. In quantum mechanics and classical wave systems, the interface between two materials with different topological phases can support topological interface states. Therefore, similar properties should also apply to convection-diffusion systems. To reveal thermal interface states, we combine two graphene-like lattices composed of spin-up and spin-down units (Fig. 22.4a, c). Since two lattices have different spin directions, unique sides exist at their interface with the same convective directions. Therefore, the decay rate at the interface is the smallest, which can support temperature propagation (Fig. 22.4b). We also prove the unidirectionality of temperature propagation by putting the temperature source at the output of Fig. 22.4b, and temperature propagation is forbidden (Fig. 22.4d). Therefore, thermal interface states exist between two lattices with different spin directions. The results in Fig. 22.4 also agree well with the understanding of electron interface states that the interface of two materials with different topological phases is conductive. The simulations in Figs. 22.2, 22.3 and 22.4 prove that the edge states in convection-diffusion systems have properties similar to wave systems.

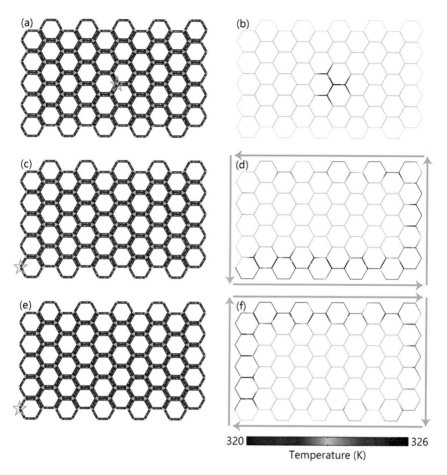

Fig. 22.2 Thermal edge states. Left and right columns display the structures and simulations at 500 s, respectively. The stars in **a**, **c**, and **e** denote the positions of periodic temperature sources whose temperatures are $T = 40\cos(-\pi t/5) + 323$ K. The arrows in **d** and **f** show the direction of temperature propagation. The fluids are water with a thermal conductivity of 0.6 W m^{-1} K^{-1}, a heat capacity of 4200 J kg^{-1} K^{-1}, and a density of 1000 kg/m^3. The solid pumps are copper with a thermal conductivity of 400 W m^{-1} K^{-1}, a heat capacity of 390 J kg^{-1} K^{-1}, and a density of 8900 kg/m^3. $r_1 = 2 - 2\sqrt{3}/30$ mm and $r_2 = 2$ mm. Adapted from Ref. [38]

We finally discuss the transition of thermal edge states. For this purpose, we change two parameters of the graphene-like lattice, i.e., the thermal conductivity of the fluid and the circular frequency of the temperature source (Fig. 22.5a). We first change the thermal conductivity of the fluid from 0.6 W m^{-1} K^{-1} (water) to 0.001 W m^{-1} K^{-1} and 400 W m^{-1} K^{-1}, and the results are shown in Fig. 22.5b, c, respectively. Both cases become conductive in bulk, and thermal edge states no longer exist. The decay rate explains this phenomenon. When the thermal conductivity of the fluid is small (0.001 W m^{-1} K^{-1}), the heat exchange between opposite convection

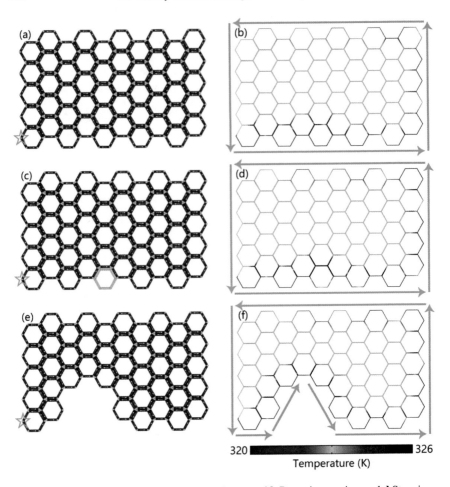

320 ████████████████████████ 326
Temperature (K)

Fig. 22.3 Robustness against defects and disorders. **a** and **b** Reversing a unit. **c** and **d** Stopping a unit. **e** and **f** Removing six units. The other parameters are the same as those in Fig. 22.2. Adapted from Ref. [38]

is insufficient. Thus, the decay rate in bulk is similar to that on the surface, and temperature propagation is allowed both in bulk and on the surface. When the thermal conductivity of the fluid is large (400 W m^{-1} K^{-1}), the convective term becomes relatively weak and can be ignored. In this way, the broken space-reversal symmetry induced by convection is not obvious, so nonreciprocal propagation almost does not exist. Therefore, the graphene-like lattice supports edge states and bulk states. We further discuss the frequency of the temperature source. The periodicity in Fig. 22.5d–f is 5, 50, and 100 s, respectively. As predicted by Eq. (22.4b), a smaller ω (i.e., a larger periodicity) yields a lower decay rate. Therefore, as ω decreases, the thermal edge state has a larger penetration depth. When the periodicity reaches 100 s, the bulk is almost conductive (Fig. 22.5f). It can also be imagined that when the periodicity tends

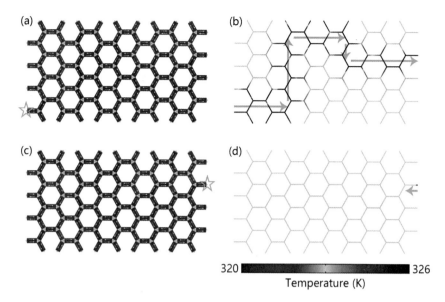

Fig. 22.4 Thermal interface states. **a** and **b** Temperature source at the bottom left corner. **c** and **d** Temperature source at the top right corner. The other parameters are the same as those in Fig. 22.2. Adapted from Ref. [38]

to infinity ($\omega \to 0$), the graphene-like lattice also supports bulk states, demonstrating the necessity to consider the role of phase (or periodicity).

22.4 Discussion

Thermal edge states are closely related to three factors. (I) Convection strength, which should be neither too weak nor too strong. If convection is too weak, the broken space-reversal symmetry is not apparent. If convection is too strong, conduction-induced heat exchange between opposite convection is insufficient. (II) Temperature frequency. The opposite convection can prevent temperature propagation due to the interference-like behavior of two temperature waves. If a near-zero temperature frequency is applied (tending to steady states without phase features), the interference-like behavior is not obvious, and bulk states can be supported. (III) System size. Since temperature amplitude features decay along the propagation direction, a large size causes a large decay. Therefore, we should carefully design the convection strength, temperature frequency, and system size. Meanwhile, thermal edge states are based on practical materials like water and copper, which can be experimentally realized in principle.

The edge states in convection-diffusion systems are also compared with those in wave systems, and these two edge states show similar properties. Therefore, the

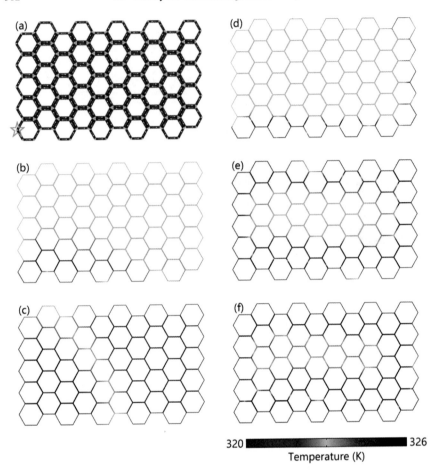

Fig. 22.5 Transition of thermal edge states. **a** Schematic diagram. Temperature profile with the thermal conductivity of the fluid being **b** 0.001 W m^{-1} K^{-1} and **c** 400 W m^{-1} K^{-1}. Temperature profile with the frequency of the temperature source being **d** $2\pi/5$ rad/s, **e** $\pi/25$ rad/s, and **f** $\pi/50$ rad/s. Adapted from Ref. [38]

fundamental origin of thermal edge states might also be topology. However, it is not simple to calculate band structures or Chern numbers in convection-diffusion systems because there is no obvious correspondence between the diffusion equation and the Schrödinger equation. Recent interest in non-Hermitian topology [40–44] may provide some insights. A common approach to a non-Hermitian Hamiltonian is introducing gain and loss to the Hermitian Hamiltonian. In contrast, diffusion itself features loss, so our system of heat transfer is essentially non-Hermitian [30], which can also be confirmed by the spatial decay of temperature propagation in Figs. 22.2, 22.3, 22.4 and 22.5. For simplicity, further explorations on non-Hermitian thermal topology might focus on one-dimensional systems at first.

22.5   Conclusion

We reveal that robust one-way edge states can also exist in convection-diffusion systems. Convection breaks space-reversal symmetry and contributes to one-way temperature propagation. Convection can also be compared to electron cyclotron, which determines the direction of thermal edge propagation. We further confirm the robustness of thermal edge states against defects and disorders. Moreover, we identify thermal interface states between two lattices with different spin directions. These findings may also guide exploring topological properties with diffusive dynamics and open a new topological diffusion research field, especially topological thermotics.

22.6   Exercise and Solution

Exercise

1. Derive Eq. (22.4).

Solution

1. The substitution of Eq. (22.3) into Eq. (22.1) yields

$$-i\omega\rho C + \alpha^2\kappa + i\alpha\rho Cv = 0. \tag{22.5}$$

The wave number α is a complex number, which can be written as $\alpha = \beta + i\gamma$. Therefore, Eq. (22.5) can be rewritten as

$$-i\omega\rho C + (\beta + i\gamma)^2\kappa + i(\beta + i\gamma)\rho Cv = 0. \tag{22.6}$$

Then, Eq. (22.6) can be divided into two parts

$$\left(\beta^2 - \gamma^2\right)\kappa - \gamma\rho Cv = 0, \tag{22.7a}$$
$$-\omega\rho C + 2\beta\gamma\kappa + \beta\rho Cv = 0. \tag{22.7b}$$

Solving these two equations, we can derive Eq. (22.4).

References

1. Hasan, M.Z., Kane, C.L.: Colloquium: topological insulators. Rev. Mod. Phys. **82**, 3045 (2010)
2. Qi, X.-L., Zhang, S.-C.: Topological insulators and superconductors. Rev. Mod. Phys. **83**, 1057 (2011)
3. Zangeneh-Nejad, F., Alù, A., Fleury, R.: Topological wave insulators: a review. C. R. Phys. **21**, 467 (2020)

4. Haldane, F.D.M., Raghu, S.: Possible realization of directional optical waveguides in photonic crystals with broken time-reversal symmetry. Phys. Rev. Lett. **100**, 013904 (2008)

5. Wang, Z., Chong, Y.D., Joannopoulos, J.D.: and Marin Soljačić, Observation of unidirectional backscattering-immune topological electromagnetic states. Nature **461**, 772 (2009)

6. Poo, Y., Wu, R.-X., Lin, Z.F., Yang, Y., Chan, C.T.: Experimental realization of self-guiding unidirectional electromagnetic edge states. Phys. Rev. Lett. **106**, 093903 (2011)

7. Lu, L., Joannopoulos, J.D., Soljačić, M.: Topological photonics. Nat. Photonics **8**, 821 (2014)

8. Christiansen, R.E., Wang, F.W., Sigmund, O.: Topological insulators by topology optimization. Phys. Rev. Lett. **122**, 234502 (2019)

9. Ozawa, T., Price, H.M., Amo, A., Goldman, N., Hafezi, M., Lu, L., Rechtsman, M.C., Schuster, D., Simon, J., Zilberberg, O., Carusotto, I.: Topological photonics. Rev. Mod. Phys. **91**, 015006 (2019)

10. Yang, Y.H., Gao, Z., Xue, H.R., Zhang, L., He, M.J., Yang, Z.J., Singh, R., Chong, Y.D., Zhang, B.L., Chen, H.S.: Realization of a three-dimensional photonic topological insulator. Nature **565**, 622 (2019)

11. Hu, G.W., Ou, Q.D., Si, G.Y., Wu, Y.J., Wu, J., Dai, Z.G., Krasnok, A., Mazor, Y., Zhang, Q., Bao, Q.L., Qiu, C.-W., Alù, A.: Topological polaritons and photonic magic angles in twisted α-MoO$_3$ bilayers. Nature **582**, 209 (2020)

12. Khanikaev, A.B., Fleury, R., Mousavi, S.H., Alù, A.: Topologically robust sound propagation in an angular-momentum-biased graphene-like resonator lattice. Nat. Commun. **6**, 8260 (2015)

13. Yang, Z.J., Gao, F., Shi, X.H., Lin, X., Gao, Z., Chong, Y.D., Zhang, B.L.: Topological acoustics. Phys. Rev. Lett. **114**, 114301 (2015)

14. Ni, X., He, C., Sun, X.-C., Liu, X.-P., Lu, M.-H., Feng, L., Chen, Y.-F.: Topologically protected one-way edge mode in networks of acoustic resonators with circulating air flow. New J. Phys. **17**, 053016 (2015)

15. Fleury, R., Khanikaev, A.B., Alù, A.: Floquet topological insulators for sound. Nat. Commun. **7**, 11744 (2016)

16. Ding, Y.J., Peng, Y.G., Zhu, Y.F., Fan, X.D., Yang, J., Liang, B., Zhu, X.F., Wan, X.G., Chen, J.C.: Experimental demonstration of acoustic Chern insulators. Phys. Rev. Lett. **122**, 014302 (2019)

17. Lu, J.Y., Qiu, C.Y., Ye, L.P., Fan, X.Y., Ke, M.Z., Zhang, F., Liu, Z.Y.: Observation of topological valley transport of sound in sonic crystals. Nat. Phys. **13**, 369 (2017)

18. Wen, X.H., Qiu, C.Y., Qi, Y.J., Ye, L.P., Ke, M.Z., Zhang, F., Liu, Z.Y.: Acoustic Landau quantization and quantum Hall-like edge states. Nat. Phys. **15**, 352 (2019)

19. Souslov, A., Dasbiswas, K., Fruchart, M., Vaikuntanathan, S., Vitelli, V.: Topological waves in fluids with odd viscosity. Phys. Rev. Lett. **122**, 128001 (2019)

20. Fan, H.Y., Xia, B.Z., Tong, L., Zheng, S.J., Yu, D.J.: Elastic higher-order topological insulator with topologically protected corner states. Phys. Rev. Lett. **122**, 204301 (2019)

21. Qi, Y.J., Qiu, C.Y., Xiao, M., He, H.L., Ke, M.Z., Liu, Z.Y.: Acoustic realization of quadrupole topological insulators. Phys. Rev. Lett. **124**, 206601 (2020)

22. Li, B.W., Wang, L., Casati, G.: Thermal diode: Rectification of heat flux. Phys. Rev. Lett. **93**, 184301 (2004)

23. Li, Y., Shen, X.Y., Wu, Z.H., Huang, J.Y., Chen, Y.X., Ni, Y.S., Huang, J.P.: Temperature-dependent transformation thermotics: from switchable thermal cloaks to macroscopic thermal diodes. Phys. Rev. Lett. **115**, 195503 (2015)

24. Huang, S.Y., Zhang, J.W., Wang, M., Lan, W., Hu, R., Luo, X.B.: Macroscale thermal diode-like black box with high transient rectification ratio. ES Energy Environ. **6**, 51 (2019)

25. Su, C., Xu, L.J., Huang, J.P.: Nonlinear thermal conductivities of core-shell metamaterials: rigorous theory and intelligent application. EPL **130**, 34001 (2020)

26. Edwards, B., Engheta, N.: Asymmetrical diffusion through time-varying material parameters. In: Conference on Lasers and Electro-Optics JTu5A.34. Optical Society of America (2017)

27. Torrent, D., Poncelet, O., Batsale, J.-C.: Nonreciprocal thermal material by spatiotemporal modulation. Phys. Rev. Lett. **120**, 125501 (2018)

28. Camacho, M., Edwards, B., Engheta, N.: Achieving asymmetry and trapping in diffusion with spatiotemporal metamaterials. Nat. Commun. **11**, 3733 (2020)
29. Xu, L.J., Huang, J.P., Ouyang, X.P.: Tunable thermal wave nonreciprocity by spatiotemporal modulation. Phys. Rev. E **103**, 032128 (2021)
30. Li, Y., Peng, Y.-G., Han, L., Miri, M.-A., Li, W., Xiao, M., Zhu, X.-F., Zhao, J.L., Alù, A., Fan, S.H., Qiu, C.-W.: Anti-parity-time symmetry in diffusive systems. Science **364**, 170 (2019)
31. Xu, L.J., Wang, J., Dai, G.L., Yang, S., Yang, F.B., Wang, G., Huang, J.P.: Geometric phase, effective conductivity enhancement, and invisibility cloak in thermal convection-conduction. Int. J. Heat Mass Transf. **165**, 120659 (2021)
32. Cao, P.C., Li, Y., Peng, Y.G., Qiu, C.W., Zhu, X.F.: High-order exceptional points in diffusive systems: robust APT symmetry against perturbation and phase oscillation at APT symmetry breaking. ES Energy Environ. **7**, 48 (2020)
33. Xu, L.J., Huang, J.P.: Thermal convection-diffusion crystal for prohibition and modulation of wave-like temperature profiles. Appl. Phys. Lett. **117**, 011905 (2020)
34. Xu, L.J., Huang, J.P.: Controlling thermal waves with transformation complex thermotics. Int. J. Heat Mass Transf. **159**, 120133 (2020)
35. Xu, L.J., Huang, J.P.: Negative thermal transport in conduction and advection. Chin. Phys. Lett. **37**, 080502 (2020)
36. Xu, L.J., Huang, J.P.: Active thermal wave cloak. Chin. Phys. Lett. **37**, 120501 (2020)
37. Xu, L.J., Dai, G.L., Wang, G., Huang, J.P.: Geometric phase and bilayer cloak in macroscopic particle-diffusion systems. Phys. Rev. E **102**, 032140 (2020)
38. Xu, L.J., Huang, J.P.: Robust one-way edge state in convection-diffusion systems. EPL **134**, 60001 (2021)
39. Huang, J.P.: Theoretical Thermotics: Transformation Thermotics and Extended Theories for Thermal Metamaterials. Springer, Singapore (2020)
40. Parto, M., Wittek, S., Hodaei, H., Harari, G., Bandres, M.A., Ren, J.H., Rechtsman, M.C., Segev, M., Christodoulides, D.N., Khajavikhan, M.: Edge-mode lasing in 1D topological active arrays. Phys. Rev. Lett. **120**, 113901 (2018)
41. Luo, X.-W., Zhang, C.W.: Higher-order topological corner states induced by gain and loss. Phys. Rev. Lett. **123**, 073601 (2019)
42. Kawabata, K., Shiozaki, K., Ueda, M., Sato, M.: Symmetry and topology in non-Hermitian physics. Phys. Rev. X **9**, 041015 (2019)
43. Okuma, N., Kawabata, K., Shiozaki, K., Sato, M.: Topological origin of non-Hermitian skin effects. Phys. Rev. Lett. **124**, 086801 (2020)
44. Borgnia, D.S., Kruchkov, A.J., Slager, R.-J.: Non-Hermitian boundary modes and topology. Phys. Rev. Lett. **124**, 056802 (2020)

Chapter 23
Summary and Outlook

Abstract In this chapter, we summarize this book and look to the future. In particular, we raise several key scientific questions for future directions of theoretical thermotics and potential applications in heat regulation.

Keywords Theoretical thermotics · Engineering applications · Future development

23.1 Summary

In this book, we present twenty theories of theoretical thermotics, divided into two parts, i.e., inside and outside metamaterials. The major difference is the characteristic length. There is an explicit characteristic length in heat transfer for those fourteen theories inside metamaterials, (much) larger than the structural unit size. The other six theories are beyond the scope of characteristic lengths (outside metamaterials). Therefore, theoretical thermotics can guide the design of both metamaterial-based and metamaterial-free phenomena and functions. Theoretical thermotics is not limited to theories, and we also present simulations and experiments for mutual confirmation. Practical applications, such as invisibility, camouflage, nonreciprocity, and bistability, are also demonstrated. These results may provide insights into novel and advanced thermal regulation.

23.2 Outlook

Although theoretical thermotics has made significant progress during the last decade, many key scientific problems remain explored. For example, nonreciprocal heat transfer is a recent focus. On the one hand, spatiotemporal modulation becomes an intriguing mechanism for achieving diffusive nonreciprocity [1–4] due to the advectionlike effect. On the other hand, an isolated thermal system with mass conservation prohibits the advectionlike effect [5]. Therefore, it becomes particularly

L.-J. Xu and J.-P. Huang, *Transformation Thermotics and Extended Theories*,
https://doi.org/10.1007/978-981-19-5908-0_23

elusive whether spatiotemporal modulation can yield nonreciprocity in heat transfer. The answer may lie in transient heat transfer due to the novel mechanism of Willis coupling [6]. Therefore, it is promising to reveal more asymmetric diffusion mechanisms in transient heat transfer, especially based on wavelike temperature fields. Moreover, topological heat transfer is another research focus. Many pioneering works related to thermal geometric phases [7], thermal Su-Schrieffer-Heeger models [8–11], thermal edge states [12], thermal skin effects [13, 14], and thermal topological transitions [15–17] have been proposed. However, compared with topological wave propagation [18, 19], the related research in heat transfer is just getting started, and much profound physics remains studied, such as high-order thermal topology.

Theoretical thermotics mainly includes fundamental theories, but we should develop more practical applications. In particular, heat regulation is a critical issue in daily life and industrial production. Hence, theoretical thermotics also needs to focus on practical problems and provide guidance for heat regulation. For example, with the miniaturization of chips, heat dissipation becomes increasingly significant for device protection. Moreover, cooling with energy savings is also a crucial problem, and passive radiative cooling has become a powerful tool [20–22]. Therefore, theoretical thermotics should also play a role in solving these urgent requirements.

Last but not least, though theoretical thermotics aims to solve thermal problems, its influence should exceed thermotics. Since heat transfer is a branch of diffusion systems, the research paradigms of theoretical thermotics can also be extended to other diffusive systems, such as particle and plasma diffusions, thereby enriching the means of diffusion regulation. Furthermore, could theoretical thermotics inspire the research in wave systems? This question is very challenging but also very rewarding. In fact, a considerable part of the existing content of theoretical thermotics is inspired by the related research in wave systems, such as from transformation optics [23, 24] to transformation thermotics [25, 26] and from photonic crystals to thermal crystals [27]. It is worth pondering how to make theoretical thermotics more enlightening and impact non-thermal fields. For example, the pioneering attempt to control multiphysical fields originates from theoretical thermotics (thermal plus DC fields [28]), which has been extended to wave control, such as electromagnetic, acoustic, plus water waves [29] and magnetic plus acoustic fields [30, 31]. More research could be expected to extend the paradigms of theoretical thermotics to other non-thermal fields.

Undoubtedly, the future of theoretical thermotics is promising, whether in terms of fundamental research, practical applications, or potential impacts.

References

1. Torrent, D., Poncelet, O., Batsale, J.C.: Nonreciprocal thermal material by spatiotemporal modulation. Phys. Rev. Lett. **120**, 125501 (2018)
2. Camacho, M., Edwards, B., Engheta, N.: Achieving asymmetry and trapping in diffusion with spatiotemporal metamaterials. Nat. Commun. **11**, 3733 (2020)
3. Xu, L.J., Huang, J.P., Ouyang, X.P.: Tunable thermal wave nonreciprocity by spatiotemporal modulation. Phys. Rev. E **103**, 032128 (2021)

4. Ordonez-Miranda, J., Guo, Y.Y., Alvarado-Gil, J.J., Volz, S., Nomura, M.: Thermal-wave diode. Phys. Rev. Appl. **16**, L041002 (2021)
5. Li, J.X., Li, Y., Cao, P.-C., Qi, M.H., Zheng, X., Peng, Y.-G., Li, B.W., Zhu, X.-F., Alù, A., Chen, H.S., Qiu, C.-W.: Reciprocity of thermal diffusion in time-modulated systems. Nat. Commun. **13**, 167 (2022)
6. Xu, L.J., Xu, G.Q., Huang, J.P., Qiu, C.-W.: Diffusive Fizeau drag in spatiotemporal thermal metamaterials. Phys. Rev. Lett. **128**, 145901 (2022)
7. Xu, L.J., Wang, J., Dai, G.L., Yang, S., Yang, F., Wang, G., Huang, J.P.: Geometric phase, effective conductivity enhancement, and invisibility cloak in thermal convection-conduction. Int. J. Heat Mass Transf. **165**, 120659 (2021)
8. Yoshida, T., Hatsugai, Y.: Bulk-edge correspondence of classical diffusion phenomena. Sci. Rep. **11**, 888 (2021)
9. Makino, S., Fukui, T., Yoshida, T., Hatsugai, Y.: Edge states of a diffusion equation in one dimension: rapid heat conduction to the heat bath. Phys. Rev. E **105**, 024137 (2022)
10. Qi, M.H., Wang, D., Cao, P.-C., Zhu, X.-F., Qiu, C.-W., Chen, H.S., Li, Y.: Localized heat diffusion in topological thermal materials (2021). arXiv: 2107.05231
11. Hu, H., Han, S., Yang, Y.H., Liu, D.J., Xue, H.R., Liu, G.-G., Cheng, Z.Y., Wang, Q.J., Zhang, S., Zhang, B.L., Luo, Y.: Observation of topological edge states in thermal diffusion (2021). arXiv: 2107.05811v1
12. Xu, L.J., Huang, J.P.: Robust one-way edge state in convection-diffusion systems. EPL **134**, 60001 (2021)
13. Cao, P.-C., Li, Y., Peng, Y.-G., Qi, M.H., Huang, W.-X., Li, P.-Q., Zhu, X.-F.: Diffusive skin effect and topological heat funneling. Commun. Phys. **4**, 230–237 (2021)
14. Cao, P.-C., Peng, Y.-G., Li, Y., Zhu, X.-F.: Phase-locking diffusive skin effect. Chin. Phys. Lett. **39**, 057801 (2022)
15. Xu, G.Q., Li, Y., Li, W., Fan, S.H., Qiu, C.-W.: Configurable phase transitions in a topological thermal material. Phys. Rev. Lett. **127**, 105901 (2021)
16. Xu, G.Q., Yang, Y.H., Zhou, X., Chen, H.S., Alù, A., Qiu, C.-W.: Diffusive topological transport in spatiotemporal thermal lattices. Nat. Phys. **18**, 450 (2022)
17. Xu, G.Q., Li, W., Zhou, X., Li, H.G., Li, Y., Fan, S.H., Zhang, S., Christodoulides, D.N., Qiu, C.-W.: Observation of Weyl exceptional rings in thermal diffusion. Proc. Natl. Acad. Sci. U. S. A. **119**, e2110018119 (2022)
18. Ozawa, T., Price, H.M., Amo, A., Goldman, N., Hafezi, M., Lu, L., Rechtsman, M.C., Schuster, D., Simon, J., Zilberberg, O., Carusotto, I.: Topological photonics. Rev. Mod. Phys. **91**, 015006 (2019)
19. Ma, G.C., Xiao, M., Chan, C.T.: Topological phases in acoustic and mechanical systems. Nat. Rev. Phys. **1**, 281 (2019)
20. Yin, X.B., Yang, R.G., Tan, G., Fan, S.H.: Terrestrial radiative cooling: using the cold universe as a renewable and sustainable energy source. Science **370**, 786 (2020)
21. Raman, A.P., Anoma, M.A., Zhu, L.X., Rephaeli, E., Fan, S.H.: Passive radiative cooling below ambient air temperature under direct sunlight. Nature **515**, 540 (2014)
22. Zhai, Y., Ma, Y.G., David, S.N., Zhao, D.L., Lou, R.N., Tan, G., Yang, R.G., Yin, X.B.: Scalable-manufactured randomized glass-polymer hybrid metamaterial for daytime radiative cooling. Science **355**, 1062 (2017)
23. Leonhardt, U.: Optical conformal mapping. Science **312**, 1777 (2006)
24. Pendry, J.B., Schurig, D., Smith, D.R.: Controlling electromagnetic fields. Science **312**, 1780 (2006)
25. Fan, C.Z., Gao, Y., Huang, J.P.: Shaped graded materials with an apparent negative thermal conductivity. Appl. Phys. Lett. **92**, 251907 (2008)
26. Chen, T.Y., Weng, C.-N., Chen, J.-S.: Cloak for curvilinearly anisotropic media in conduction. Appl. Phys. Lett. **93**, 114103 (2008)
27. Maldovan, M.: Narrow low-frequency spectrum and heat management by thermocrystals. Phys. Rev. Lett. **110**, 025902 (2013)

28. Li, J.Y., Gao, Y., Huang, J.P.: A bifunctional cloak using transformation media. J. Appl. Phys. **108**, 074504 (2010)
29. Yang, Y.H., Wang, H.P., Yu, F.X., Xu, Z.W., Chen, H.S.: A metasurface carpet cloak for electromagnetic, acoustic and water waves. Sci. Rep. **6**, 20219 (2016)
30. Zhou, Y., Chen, J., Liu, L., Fan, Z., Ma, Y.G.: Magnetic Cacoustic biphysical invisible coats for underwater objects. NPG Asia Mater. **12**, 27 (2020)
31. Zhan, J.J., Mei, Y.J., Li, K., Zhou, Y., Chen, J., Ma, Y.G.: Conformal metamaterial coats for underwater magnetic-acoustic bi-invisibility. Appl. Phys. Lett. **120**, 094104 (2022)

Appendix A
Particle Diffusion: Exceptional Points, Geometric Phases, and Bilayer Cloaks

Opening Remarks

The Berry phase [1] was revealed in quantum wave systems, resulting from an adiabatic evolution along a closed path in parameter space. A similar phase was also explored in molecular physics with the Born-Oppenheimer approximation and perturbation theory [2]. Beyond these systems, the Berry phase also appears in diffusion systems with quantum corrections [3, 4]. Meanwhile, varieties of novel phenomena were found to be crucially associated with the Berry phase, such as topological materials [5, 6, 7, 8, 9, 10, 11, 12, 13], heat pumping [14], and geometric heat flux [15].

As a more general concept than the Berry phase, the geometric phase has also attracted broad research interest, which also appears in diffusion systems related to photons [16]. However, such a concept does not exist in macroscopic particle-diffusion systems because of the difficulty in defining the frequency and phase in such systems.

Here, we study a macroscopic particle-diffusion system to promote the exploration of the geometric phase. For this purpose, we introduce frequency and phase properties by applying two moving channels [17, 18, 19, 20] with equal-but-opposite velocities, connected by a stationary intermediate layer with permeability for particles (see the top inset of Fig. A.1). Although this structure is similar to that adopted for heat-diffusion systems in Ref. [20], a crucial difference is between energy transport (heat diffusion) and mass transport (particle diffusion). A non-Hermitian Hamiltonian can effectively describe the macroscopic particle-diffusion process. Concretely speaking, the introduced frequency and phase properties act as a Hermitian term, and the inherent diffusion properties serve as a non-Hermitian term [21, 22, 23, 24].

As a result, an exceptional point of velocity exists in the macroscopic particle-diffusion system just like that in wave systems of optics and photonics [25]. If the velocity exceeds the exceptional point, a stationary concentration profile will turn into a moving one due to the broken anti-parity-time symmetry [20, 26, 27, 28, 29, 30] (see the left inset of Fig. A.1). Therefore, a cyclic velocity path containing the exceptional point can give birth to an extra phase difference of π (say, the geometric

© The Editor(s) (if applicable) and The Author(s) 2023
L.-J. Xu and J.-P. Huang, *Transformation Thermotics and Extended Theories*,
https://doi.org/10.1007/978-981-19-5908-0

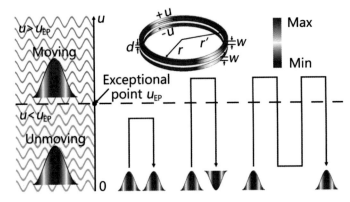

Fig. A.1 Constructing a non-Hermitian Hamiltonian and an adiabatic cyclic path in a macroscopic particle-diffusion system. The schematic model is shown on top. The proposed structure involves two moving channels separated by a permeable layer. Dependent on the velocity u, the concentration profile can be unmoving or moving. After a cyclic path (brown lines) in the velocity space, the particle concentration profile (shown in color from red to blue with reference to the color bar in the upper right corner) may return to the inial state with or without a π-phase shift. Adapted from Ref. [31]

phase). A schematic diagram can be seen in the bottom inset of Fig. A.1. Let us start from the basic properties of the macroscopic particle-diffusion system, the exceptional point of velocity.

Exceptional Point

We study a two-dimensional system as shown in Fig. A.2a. The perimeter and width of the two moving channels are l and w, respectively. The particles in the upper and lower channels move with equal-but-opposite velocities ($+u$ and $-u$). A stationary permeable layer separates the two moving channels, whose thickness (d) and diffusivity (D_m) determine the exchange rate of particles between the upper and lower channels. Considering the concentration distributions in the upper channel, lower channel, and intermediate layer (denoted as C_1, C_2, and C_m, respectively), the macroscopic particle-diffusion process is dominated by

$$
\begin{cases}
\frac{\partial C_1}{\partial t} = D\left(\frac{\partial^2 C_1}{\partial x^2} + \frac{\partial^2 C_1}{\partial z^2}\right) - u\frac{\partial C_1}{\partial x}, & d/2 \leq z \leq w + d/2 \\[2mm]
\frac{\partial C_m}{\partial t} = D_m\left(\frac{\partial^2 C_m}{\partial x^2} + \frac{\partial^2 C_m}{\partial z^2}\right), & -d/2 < z < d/2 \\[2mm]
\frac{\partial C_2}{\partial t} = D\left(\frac{\partial^2 C_2}{\partial x^2} + \frac{\partial^2 C_2}{\partial z^2}\right) + u\frac{\partial C_2}{\partial x}, & -w - d/2 \leq z \leq -d/2
\end{cases}
\qquad (A.1)
$$

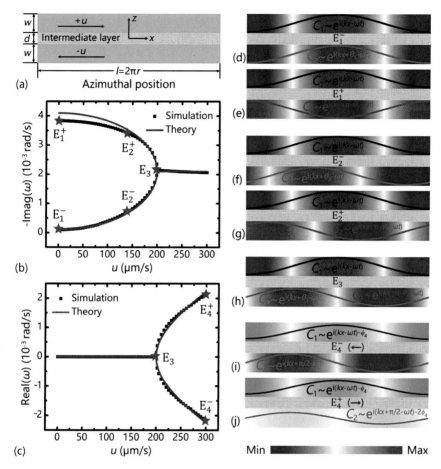

Fig. A.2 Emergence of the exceptional point. **a** The schematic diagram of the diffusion system with $r = 10$, $w = 0.5$, and $d = 0.1$ cm. **b** and **c** Imaginary and real parts of the eigenvalues ω of $H(u)$ as a function of velocity u. Red lines are analytical results of Eq. (A.4), and black squares are results from finite-element simulations. **d–j** Eigenstates at different positions indicated by blue stars in (**b**) and (**c**). $D = 10^{-6}$ m^2/s and $D_m = 10^{-8}$ m^2/s. Adapted from Ref. [31]

where D is the diffusivity of the particles in the two moving channels. Equation (A.1) indicates the mass conservation of the macroscopic particle-diffusion process. The intermediate layer allows the particles to exchange between the two moving channels, so the middle equation in Eq. (A.1) can be treated as two particle sources. Therefore, Eq. (A.1) can be rewritten as

$$
\begin{cases}
\frac{\partial C_1}{\partial t} = D \frac{\partial^2 C_1}{\partial x^2} - u \frac{\partial C_1}{\partial x} + \frac{D_m}{wd} (C_2 - C_1), & d/2 \leq z \leq w + d/2 \\
\frac{\partial C_2}{\partial t} = D \frac{\partial^2 C_2}{\partial x^2} + u \frac{\partial C_2}{\partial x} + \frac{D_m}{wd} (C_1 - C_2), & -w - d/2 \leq z \leq -d/2
\end{cases}
\tag{A.2}
$$

where we have taken $\partial^2 C_1/\partial z^2 = \partial^2 C_2/\partial z^2 = 0$ because the width (w) is supposed to be thin enough. By taking a periodic boundary condition on the left and right sides of the structure, we can insert the ansatz $A e^{i(kx-\omega t)} + B$ where A is the amplitude, B is a reference value, k is the wave number, and ω is the frequency of the particle concentration. Since particle concentrations cannot be complex, only the real part of the ansatz makes sense. By substituting the ansatz into Eq. (A.2), we can reach an eigenequation of the macroscopic particle-diffusion process $H\psi = \omega\psi$, where $\psi = [A_1, A_2]^T$ is the eigenstate, thus yielding $C_1 \sim A_1 e^{i(kx-\omega t)}$ and $C_2 \sim A_2 e^{i(kx-\omega t)}$. The superscript T denotes transpose, and we ignore the reference value B for brevity. The effective Hamiltonian H reads

$$H = \begin{bmatrix} -i\left(k^2 D + h\right) + ku & ih \\ ih & -i\left(k^2 D + h\right) - ku \end{bmatrix}, \tag{A.3}$$

where $h = D_m/(w\,d)$ reflects the exchange rate of the particles between the upper and lower channels. The eigenvalues of H are

$$\omega = -i\left[\left(k^2 D + h\right) \pm \sqrt{h^2 - k^2 u^2}\right]. \tag{A.4}$$

For clarity, we plot the real and imaginary parts of eigenvalues in the u-space, as illustrated in Fig. A.2b, c. In the region $u < u_{EP} (= h/k)$, H occupies two branches of purely imaginary eigenvalues, and the system is in the anti-parity-time symmetric region [20]. Especially at the exceptional point $u = u_{EP}$, the system marks the merging of the two eigenvalues. When $u > u_{EP}$, the real part of eigenvalues appears due to the broken anti-parity-time symmetry. Therefore, this point $u_{EP} (= h/k)$ serves as an exceptional point of velocity.

For a vanishing exchange rate ($h = 0$), the eigenvalues always possess nonzero real parts at $u \neq 0$, so there is only one moving profile. That is, the two concentration profiles in the upper and lower channels propagate independently. Differently, when the two channels are coupled together ($h \neq 0$), the system exhibits two different profiles as u varies. Therefore, the particle exchange between the two moving channels is the key factor which can be regarded as the interference in the macroscopic particle-diffusion system. When $u < h/k$, the eigenvalues are purely imaginary, indicating that concentration profiles always decay and do not propagate. The corresponding eigenstates are

$$\psi_+ = \left[1, \ e^{i(\pi-\theta)}\right]^T, \ \psi_- = \left[1, \ e^{i\theta}\right]^T, \tag{A.5}$$

where $\theta = \sin^{-1}(ku/h)$. The concentration profiles in the two moving channels maintain a constant phase difference ($\pi - \theta$ for ω_+ and θ for ω_-) and decay motionlessly. When $u > h/k$, the real parts of eigenvalues appear, indicating that concentration profiles not only decay but also propagate. The corresponding eigenstates are

$$\psi_+ = \left[e^{-\phi}, \ e^{i\pi/2-2\phi}\right]^T, \ \psi_- = \left[e^{-\phi}, \ e^{i\pi/2}\right]^T, \tag{A.6}$$

where $\phi = \cosh^{-1}(ku/h)$. The concentration profiles in the two moving channels also maintain a constant phase difference ($\pi/2$ for both ω_+ and ω_-), but decay with motion.

To confirm the theory, we use COMSOL Multiphysics to perform finite-element simulations. Due to the periodic boundary condition, the two moving channels with length $l = 2\pi r$ allow only discrete wave numbers $k = nl/2\pi = nr^{-1}$ with n being any positive integers. We focus on the fundamental modes ($n = 1$) which have the lowest decay rates. The simulation eigenvalues are obtained by setting the initial states to be exactly the corresponding eigenstates, say, $C_1 = A_1 \cos(kx) + B_1$ and $C_2 = A_2 \cos(kx + \theta) + B_2$ corresponding to ω_- [or $C_2 = A_2 \cos(kx + \pi - \theta) + B_2$ corresponding to ω_+] with $A_1 = A_2 = 200$, $B_1 = B_2 = 300$, and $\theta = \sin^{-1}(ku/h)$. We fit the amplitude decaying from 250 to 60000 s with the function of $Ae^{-\lambda t} + B$, where $\lambda = -\text{Imag}(\omega)$ is decay rate. When the velocity exceeds the exceptional point, we also track the motion before 2000 s and calculate the frequency [$\text{Real}(\omega)$]. The simulation results are plotted by discrete dots in Fig. A.2b, c which agree with the solid lines predicted by Eq. (A.4). Since the eigenstates at the upper branch of $-\text{Imag}(\omega)$ have larger decay rates, they are metastable, resulting in the smaller simulation eigenvalues of $-\text{Imag}(\omega)$ than the theoretical ones.

We also plot the eigenstates indicated by the stars in Fig. A.2b, c (see Fig. A.2d–j). When plotting the eigenstates, we adjust the length l and width w to keep an appropriate ratio for the clarity of presentation. When $u = 0$, the phase differences are $\theta_1 = 0$ for the eigenstate E_1^- and $\pi - \theta_1 = \pi$ for the eigenstate E_1^+. As u increases to the exceptional point, $\theta = \sin^{-1}(ku/h)$ also increases to $\pi/2$. Thus, the two profiles of C_1 and C_2 coincide with each other with a phase difference of $\pi/2$. The eigenstates in Fig. A.2d–h decay motionlessly. When $u > u_{\text{EP}}$, the concentration profiles keep a phase difference of $\pi/2$ and propagate together. The propagation follows the direction of the velocity of the channel with a larger concentration amplitude, say, backward in Fig. A.2i and forward in Fig. A.2j. The finite-element simulations agree with the eigenstates predicted by Eqs. (A.5) and (A.6).

We also care about the dynamics of particle concentrations. Here we use annular channels to perform finite-element simulations which can naturally satisfy the periodic boundary condition adopted in Fig. A.2. We also define C_1 and C_2 as the concentration distributions along the upper and lower interior edges of the channels. The initial states are set to be the five eigenstates indicated by the stars in Fig. A.2b with the forms in the Cartesian coordinates as $C_1 = A_1 y/\sqrt{x^2 + y^2} + B_1$ and $C_2 = A_2(y/\sqrt{x^2 + y^2} \cos\theta - x/\sqrt{x^2 + y^2} \sin\theta) + B_2$ for ω_- (or $C_2 = A_2(-y/\sqrt{x^2 + y^2} \cos\theta - x/\sqrt{x^2 + y^2} \sin\theta) + B_2$ for ω_+) with $A_1 = A_2 = 200$, $B_1 = B_2 = 300$, and $\theta = \sin^{-1}(ku/h)$. Then, we set the velocities to be $100 (< u_{\text{EP}})$ and $300 (> u_{\text{EP}})$ μm/s and study the evolutions. The directions of the velocities are clockwise for the upper ring and anticlockwise for the lower ring. The theoretical phase differences with $u = 100$ μm/s are $\pi/6$ for ω_- and $5\pi/6$ for ω_+. We track the evolutions of C_1 and C_2 by following their maximum points. The initial and final states are shown in the left column of Fig. A.3. The trajectories of Max(C_1)

and $\text{Max}(C_2)$ with two different velocities are plotted in the right two columns of Fig. A.3.

Since the initial states are not the eigenstates corresponding to $u = 100 \ \mu\text{m/s}$, these noneigenstates start moving to eigenstates. Finally, all five initial states move to the same final state with $\theta \approx \pi/6$, which is the eigenstate corresponding to the eigenvalue ω_-. This occurs because of the nonorthogonality of the two eigenstates at different branches (for example, the eigenstates E_1^+ and E_2^- are not orthogonal) [21]. Meanwhile, the decay rate of the upper branch is much larger than that of the lower branch, so the eigenstate at the lower branch becomes the final observable one associated with the eigenvalue ω_-.

Besides the final eigenstate, we also care about the evolution route. For example, the moving directions of the maximum points in Fig. A.3h, k, n are all against the velocities of respective channels. This phenomenon occurs because the evolution route should avoid going through the eigenstate corresponding to the eigenvalue ω_+ (with a far larger decay rate) to survive longer. Therefore, when the velocity is smaller than the exceptional point, the final state is always the eigenstate at the lower branch corresponding to ω_-, and the evolution route should try to avoid going through the eigenstate corresponding to the eigenvalue ω_+. A principle is to ensure concentration profiles survive as long as possible. When the velocity is larger than the exceptional point, the concentration profiles are always moving because the real parts of eigenvalues ω appear (see the right column in Fig. A.3).

Geometric Phase

We can check that the non-Hermitian Hamiltonian $\boldsymbol{H}$ satisfies $\boldsymbol{H}^\dagger \overline{\boldsymbol{\psi}}_\pm = \overline{\omega}_\pm \overline{\boldsymbol{\psi}}_\pm$, where $\boldsymbol{H}^\dagger$ is the Hermitian transpose of $\boldsymbol{H}$. $\overline{\boldsymbol{\psi}}_\pm$ and $\overline{\omega}_\pm$ are the complex conjugate of $\boldsymbol{\psi}_\pm$ and $\omega_\pm$, respectively. The eigenstates satisfy $\left\langle \overline{\boldsymbol{\psi}}_\pm, \ \boldsymbol{\psi}_\mp \right\rangle = 0$, where $\left\langle \overline{\boldsymbol{\psi}}_\pm, \ \boldsymbol{\psi}_\mp \right\rangle$ denotes the complex inner product of vectors $\overline{\boldsymbol{\psi}}_\pm$ and $\boldsymbol{\psi}_\mp$. Considering a time-varying velocity u, we can write down the complex geometric phase under adiabatic approximation as

$$\gamma_\pm = i \int \frac{\left\langle \overline{\boldsymbol{\psi}}_\pm(u), \ d\boldsymbol{\psi}_\pm(u) \right\rangle}{\left\langle \overline{\boldsymbol{\psi}}_\pm(u), \ \boldsymbol{\psi}_\pm(u) \right\rangle}, \tag{A.7}$$

which agrees with the results of non-Hermitian quantum systems [32]. Then, the exceptional point yields $\left\langle \overline{\boldsymbol{\psi}}_\pm(u_{\text{EP}}), \ \boldsymbol{\psi}_\pm(u_{\text{EP}}) \right\rangle = 0$ due to the coalescence of the two eigenstates. As a result, the exceptional point serves as a pole in the complex integral. We can rewrite Eq. (A.7) in a closed loop around the exceptional point as [33]

$$\gamma_\pm = \frac{i}{2} \oint d \ln \left\langle \overline{\boldsymbol{\psi}}_\pm(u), \ \boldsymbol{\psi}_\pm(u) \right\rangle. \tag{A.8}$$

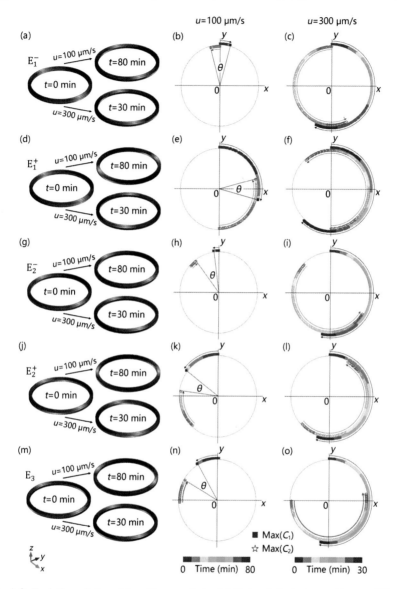

Fig. A.3 Evolution of eigenstates. The snapshots of initial and final states with two different velocities are presented in the left column, where the red (or blue) color represents the large (or small) concentration. The trajectories of Max(C_1) and Max(C_2) along the interior edges of the channels with $u = 100$ and $u = 300$ μm/s are shown in the middle and right columns, respectively. The width of the two channels is 0.5 cm, the thickness of the intermediate layer is 0.1 cm, and the inner and outer radii are 10 and 11 cm. $D = 10^{-6}$ m^2/s and $D_m = 10^{-8}$ m^2/s. Adapted from Ref. [31]

The geometric phase takes $\gamma_{\pm} = \pi$ or $-\pi$ according to the residue theorem, and the direction of the closed loop determines the sign. If the evolution route does not contain the exceptional point, the integral or geometric phase in a closed loop is naturally equal to zero.

We also perform finite-element simulations to visualize the geometric phase. We firstly consider a cyclic path of velocity without including the exceptional point. The initial velocity is $u = 100 \; \mu\text{m/s}$, and the initial state is set at the eigenstate associated with ω_- (say, a phase difference of $\pi/6$). Then, we evolve the velocity according to the curve shown in Fig. A.4a. In this process, the eigenvalue is always purely imaginary, indicating that no extra phase difference is accumulated. As a result, this path brings the final state back to the initial concentration profile exactly (see Fig. A.4b, c).

However, it is different when the path of velocity includes the exceptional point [see Fig. A.4d–f, g–i. As the velocity increases and exceeds the exceptional point, the real parts of eigenvalues appear, indicating that an extra phase difference starts to accumulate. Meanwhile, the initial state moves smoothly from one branch to the other. When the accumulated phase difference makes the profile go through the eigenstate corresponding to another eigenvalue ω_+ (with a phase difference of $5\pi/6$), the profile can no longer go back to the initial position, as discussed in Fig. A.3. Then, the phase difference continuously increases to reach a different position associated with eigenvalue ω_- and phase difference $\pi/6$. Fortunately, the particle concentration profile is flipped after one loop, and a phase of π is accumulated (see Fig. A.4e, f or Fig. A.4h, i). This case is just the indicator of the geometric phase.

Finally, when the cyclic evolution crosses the exceptional point twice (see Fig. A.4j, say, crossing the eigenstate corresponding to the eigenvalue ω_+ twice), it brings back to the initial state without any global phase change (see Fig. A.4k, l), as one can imagine from the geometric phase.

Bilayer Cloak

We also use this structure for practical applications to design a particle-diffusion cloak [34, 35, 36, 37]. Cloaking is one of the most attractive functions to protect objects from being detected. Particle-diffusion cloaking also has potential applications in various physical systems, such as chemical and biological systems, where mass transport is one of the most basic mechanisms. The feasibility of this idea results from the unique property of the present structure. On the one hand, the intermediate layer allows the particles to exchange between the two moving channels. On the other hand, the two moving channels also drive the particles in the intermediate layer, thus resulting in a larger effective diffusivity. Therefore, the diffusivity of the intermediate layer can be significantly enhanced due to the two moving channels. The detailed design is as follows.

We combine our structure with a square plate with the same thickness (d) and diffusivity (D_m) as the intermediate layer, whose different views are presented in

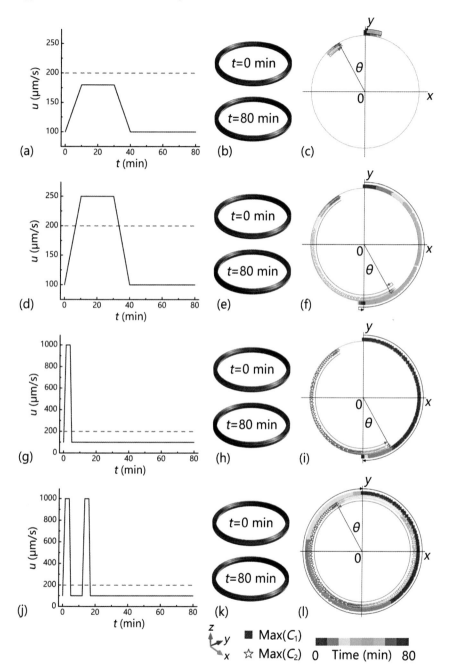

Fig. A.4 Emergence of the geometric phase. The left column describes the paths of the time-varying velocity. The middle column shows the initial and final states. The right column illustrates the trajectories of Max(C_1) and Max(C_2) along the interior edges of the channels. The parameters are the same as those for Fig. A.3. Adapted from Ref. [31]

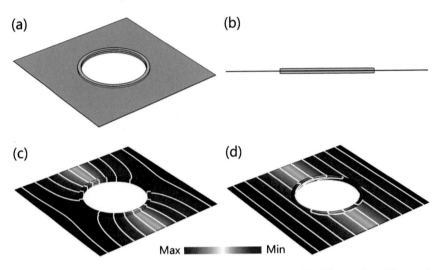

Fig. A.5 Particle-diffusion cloak. **a** and **b** Schematic diagrams with different views. The whole size is 44×44 cm^2, and the other sizes are the same as those adopted in Fig. A.3. $D = 10^{-8}$ m^2/s and $D_m = 10^{-6}$ m^2/s. **c** Simulation result with $u = 0$ μm/s. **d** Simulation result with $u = 37$ μm/s. Adapted from Ref. [31]

Fig. A.5a, b. The left and right sides are fixed at high and low concentrations. The other two sides are associated with no-particle conditions. When the velocity of the upper and lower channels is zero, the intermediate layer has no diffusivity enhancement. In this case, the concentration profile in the background is contracted, and cloaking cannot appear (see Fig. A.5c). Given the design of bilayer cloaks in thermotics [38, 39, 40, 41, 42], if we enhance the diffusivity of the intermediate layer up to $D'_m = D_m \left(1 + r^2/r'^2\right) / \left(1 - r^2/r'^2\right)$ (where r and r' are respectively the inner and outer radii of the moving channels, and D'_m is the enhanced diffusivity), the cloaking effect can be achieved. For this purpose, we set the velocity of two moving channels as 37 μm/s, and the enhanced diffusivity can satisfy the requirement of a bilayer cloak. As a result, a particle-diffusion cloak is realized (see Fig. A.5d). Now, one can place any object inside the cloak without distorting the concentration profile in the background. Such a scheme can avoid anisotropic, singular, and inhomogeneous parameters derived from transformation theory [34, 35, 36, 37]. Meanwhile, cloak-on and cloak-off can be controlled easily by adjusting the velocity.

Incidentally, all the parameters adopted in the finite-element simulations above are delicately chosen to match practical conditions. For example, the diffusivity of the two moving channels is set at the magnitude of pure gas diffusion, and that of the intermediate layer is at the magnitude of gas diffusion in porous media. A larger porosity means a stronger exchange rate since gas can penetrate the two moving channels more efficiently. Additionally, the gas in the channels can be driven to rotate by connecting rotary motors to the channels.

Conclusion

In summary, macroscopic particle-diffusion systems can exhibit exceptional points and geometric phases besides existing systems. Certainly, these results may also be extrapolated to other macroscopic diffusion systems like electrostatics and magnetostatics. We have also designed a particle-diffusion cloak with the present structure, extending the geometric phase to cloaking. These properties may pave a new way for studying topologically protected phenomena by designing the particle-diffusion counterparts of quantum Hall effects or topological insulators/superconductors. Many relevant open questions can be immediately prompted, such as those related to the ion-exchange behavior between membranes or the manipulation of particle diffusion.

References

1. Berry, M.V.: Quantal phase factors accompanying adiabatic changes. P. R. Soc. A Math. Phys. **392**, 45 (1984)
2. Mead, C.A., Truhlar, D.G.: On the determination of Born-Oppenheimer nuclear motion wave functions including complications due to conical intersections and identical nuclei. J. Chem. Phys. **70**, 2284 (1979)
3. Xiao, D., Chang, M.C., Niu, Q.: Berry phase effects on electronic properties. Rev. Mod. Phys. **82**, 1959 (2010)
4. Misaki, K., Miyashita, S., Nagaosa, N.: Diffusive real-time dynamics of a particle with Berry curvature. Phys. Rev. B **97**, 075122 (2018)
5. Fu, L., Kane, C.L., Mele, E.J.: Topological insulators in three dimensions. Phys. Rev. Lett. **98**, 106803 (2007)
6. Fu, L., Kane, C.L.: Superconducting proximity effect and Majorana Fermions at the surface of a topological insulator. Phys. Rev. Lett. **100**, 096407 (2008)
7. Hasan, M.Z., Kane, C.L.: Colloquium: topological insulators. Rev. Mod. Phys. **82**, 3045 (2010)
8. Qi, X.L., Zhang, S.C.: Topological insulators and superconductors. Rev. Mod. Phys. **83**, 1057 (2011)
9. Peng, Y.G., Qin, C.Z., Zhao, D.G., Shen, Y.X., Xu, X.Y., Bao, M., Jia, H., Zhu, X.F.: Experimental demonstration of anomalous Floquet topological insulator for sound. Nat. Commun. **7**, 13368 (2016)
10. Lu, J.Y., Qiu, C.Y., Ye, L.P., Fan, X.Y., Ke, M.Z., Zhang, F., Liu, Z.Y.: Observation of topological valley transport of sound in sonic crystals. Nat. Phys. **13**, 369 (2017)
11. He, H.L., Qiu, C.Y., Ye, L.P., Cai, X.X., Fan, X.Y., Ke, M.Z., Zhang, F., Liu, Z.Y.: Topological negative refraction of surface acoustic waves in a Weyl phononic crystal. Nature **560**, 61 (2018)
12. Xue, H.R., Yang, Y.H., Gao, F., Chong, Y.D., Zhang, B.L.: Acoustic higher-order topological insulator on a kagome lattice. Nat. Mater. **18**, 108 (2019)
13. Fan, X.Y., Qiu, C.Y., Shen, Y.Y., He, H.L., Xiao, M., Ke, M.Z., Liu, Z.Y.: Probing Weyl physics with one-dimensional sonic crystals. Phys. Rev. Lett. **122**, 136802 (2019)
14. Ren, J., Hanggi, P., Li, B.W.: Berry-phase-induced heat pumping and its impact on the fluctuation theorem. Phys. Rev. Lett. **104**, 170601 (2010)

15. Ren, J., Liu, S., Li, B.W.: Geometric heat flux for classical thermal transport in interacting open systems. Phys. Rev. Lett. **108**, 210603 (2012)
16. Ning, C.Z., Haken, H.: Geometrical phase and amplitude accumulations in dissipative systems with cyclic attractors. Phys. Rev. Lett. **68**, 2109 (1992)
17. Fleury, R., Sounas, D.L., Sieck, C.F., Haberman, M.R., Alu, A.: Sound isolation and giant linear nonreciprocity in a compact acoustic circulator. Science **343**, 516 (2014)
18. Torrent, D., Poncelet, O., Batsale, J.C.: Nonreciprocal thermal material by spatiotemporal modulation. Phys. Rev. Lett. **120**, 125501 (2018)
19. Li, Y., Zhu, K.J., Peng, Y.G., Li, W., Yang, T.Z., Xu, H.X., Chen, H., Zhu, X.F., Fan, S.H., Qiu, C.W.: Thermal meta-device in analogue of zero-index photonics. Nat. Mater. **18**, 48 (2019)
20. Li, Y., Peng, Y.G., Han, L., Miri, M.A., Li, W., Xiao, M., Zhu, X.F., Zhao, J.L., Alu, A., Fan, S.H., Qiu, C.W.: Anti-parity-time symmetry in diffusive systems. Science **364**, 170 (2019)
21. Bender, C.M.: Ghost busting: Making sense of non-Hermitian Hamiltonians. Rep. Prog. Phys. **70**, 947 (2007)
22. Wu, J.H., Artoni, M., La Rocca, G.C.: Non-hermitian degeneracies and unidirectional reflectionless atomic lattices. Phys. Rev. Lett. **113**, 123004 (2014)
23. Gao, T., Estrecho, E., Bliokh, K.Y., Liew, T.C.H., Fraser, M.D., Brodbeck, S., Kamp, M., Schneider, C., Hofling, S., Yamamoto, Y., Nori, F., Kivshar, Y.S., Truscott, A.G., Dall, R.G., Ostrovskaya, E.A.: Observation of non-Hermitian degeneracies in a chaotic exciton-polariton billiard. Nature **526**, 554 (2015)
24. Assawaworrarit, S., Yu, X.F., Fan, S.H.: Robust wireless power transfer using a nonlinear parity-time-symmetric circuit. Nature **546**, 387 (2017)
25. Miri, M.A., Alu, A.: Exceptional points in optics and photonics. Science **363**, eaar7709 (2019)
26. Wu, J.H., Artoni, M., La Rocca, G.C.: Parity-time-antisymmetric atomic lattices without gain. Phys. Rev. A **91**, 033811 (2015)
27. Peng, P., Cao, W.X., Shen, C., Qu, W.Z., Wen, J.M., Jiang, L., Xiao, Y.H.: Anti-parity-time symmetry with flying atoms. Nat. Phys. **12**, 1139 (2016)
28. Yang, F., Liu, Y.C., You, L.: Anti-PT symmetry in dissipatively coupled optical systems. Phys. Rev. A **96**, 053845 (2017)
29. Choi, Y., Hahn, C., Yoon, J.W., Song, S.H.: Observation of an anti-PT-symmetric exceptional point and energy-difference conserving dynamics in electrical circuit resonators. Nat. Commun. **9**, 2182 (2018)
30. Konotop, V.V., Zezyulin, D.A.: Odd-time reversal PT symmetry induced by an anti-PT-symmetric medium. Phys. Rev. Lett. **120**, 123902 (2018)
31. Xu, L.J., Dai, G.L., Wang, G., Huang, J.P.: Geometric phase and bilayer cloak in macroscopic particle-diffusion systems. Phys. Rev. E **102**, 032140 (2020)
32. Garrison, J.C., Wright, E.M.: Complex geometrical phases for dissipative systems. Phys. Lett. A **128**, 171 (1988)
33. Mailybaev, A.A., Kirillov, O.N., Seyranian, A.P.: Geometric phase around exceptional points. Phys. Rev. A **72**, 014104 (2005)
34. Guenneau, S., Puvirajesinghe, T.M.: Fick's second law transformed: one path to cloaking in mass diffusion. J. R. Soc. Interface **10**, 20130106 (2013)
35. Guenneau, S., Petiteau, D., Zerrad, M., Amra, C., Puvirajesinghe, T.: Transformed Fourier and Fick equations for the control of heat and mass diffusion. AIP Adv. **5**, 053404 (2015)
36. Restrepo-Florez, J.M., Maldovan, M.: Mass separation by metamaterials. Sci. Rep. **6**, 21971 (2016)
37. Restrepo-Florez, J.M., Maldovan, M.: Mass diffusion cloaking and focusing with metamaterials. Appl. Phys. Lett. **111**, 071903 (2017)
38. Xu, H.Y., Shi, X.H., Gao, F., Sun, H.D., Zhang, B.L.: Ultrathin three-dimensional thermal cloak. Phys. Rev. Lett. **112**, 054301 (2014)

39. Han, T.C., Bai, X., Gao, D.L., Thong, J.T.L., Li, B.W., Qiu, C.-W.: Experimental demonstration of a bilayer thermal cloak. Phys. Rev. Lett. **112**, 054302 (2014)
40. Ma, Y.G., Liu, Y.C., Raza, M., Wang, Y.D., He, S.L.: Experimental demonstration of a multiphysics cloak: Manipulating heat flux and electric current simultaneously. Phys. Rev. Lett. **113**, 205501 (2014)
41. Han, T.C., Yang, P., Li, Y., Lei, D.Y., Li, B.W., Hippalgaonkar, K., Qiu, C.-W.: Full-parameter omnidirectional thermal metadevices of anisotropic geometry. Adv. Mater. **30**, 1804019 (2018)
42. Xu, L.J., Huang, J.P.: Metamaterials for manipulating thermal radiation: transparency, cloak, and expander. Phys. Rev. Appl. **12**, 044048 (2019)

Appendix B
Plasma Diffusion: Transformation Scheme

Opening Remarks

Plasma, the fourth state of matter, is a gaseous mixture of unbound ions, electrons, and reactive radicals that becomes highly electrically conductive [1]. Although it is not common on the earth's surface, plasma can be obtained artificially by charging gases with direct/alternating current or radio/microwave sources. Due to the unique composition of the plasma, plasma technology plays a vital role in many fields spanning micro/nanoelectronics, chemistry, bio-medicine, aerospace, and material science [2, 3, 4, 5].

Despite large quantities of theoretical and experimental studies, manipulating plasma transport still faces critical challenges. Conventional control of charged particles depends on external magnetic fields. This simple method may limit the accuracy of manipulation. Since the past decade, transformation theory, an approach to replace space transformation with material transformation, has attracted wide attention in wave and diffusion systems as a reliable and powerful method of controlling matter [6, 7, 8] or energy flow [9, 10, 11, 12, 13, 14, 15, 16, 17]. However, it has not yet been introduced to plasma transport which can be regarded as a unique diffusion process. A possible reason might lie in the particularly complex motion process in plasmas.

We utilize a toy model (diffusion-migration model) to describe plasma transport and design three conceptual devices, i.e., cloak, concentrator, and rotator, to control transient plasma flow based on the transformation theory. Here, a "cloak" can provide a zero-density gradient inside the device; a "concentrator" gives a larger density gradient inside the device; a "rotator" can deflect the transport direction of the plasma inside the device. Most importantly, the devices do not disturb the density profiles of plasmas in the background. Our results might broaden the horizon of manipulating transient plasma transport and might be helpful to inspire further improvements in plasma physics.

L.-J. Xu and J.-P. Huang, *Transformation Thermotics and Extended Theories*,
https://doi.org/10.1007/978-981-19-5908-0

Theoretical Foundation

Compared to the conventional diffusion system, the realistic plasma transport is much more complicated. Because the interaction between charged particles and intrinsic local electromagnetic fields affects the transport process a lot. In addition, the ionization reaction in plasma can also have a significant effect on the momentum and the energy transfer of particles. In general, the transport of charged particles in plasma is dominated by [18]

$$\partial_t n - \nabla \cdot (D\nabla n) \pm \nabla \cdot \left(\mu \vec{E} n \right) + \nabla \cdot (\vec{v} n) = S, \tag{B.1}$$

where n, D, μ, $\vec{E}$, $\vec{v}$, and S are the density, diffusivity, mobility, electric field, advective velocity, and external source, respectively. In particular, the sign of the third term (i.e., migration term) is positive for positive particles and negative for negative particles. For brevity, we only considered electric fields with ignoring the advective process and the gaseous reaction [19]. Hence, the plasma transport can be simplified as a diffusion-migration process. Then according to the Einstein relation, Eq. (B.1) could be written as

$$\partial_t n - \nabla \cdot (D\nabla n) \pm \nabla \cdot \left[\left(\frac{D\vec{E}}{T} \right) n \right] = S, \tag{B.2}$$

in which T (in the unit of V) is assumed to be a constant plasma temperature. Then according to the transformation theory, when the controlling equation is form-invariant under a coordinate transformation, the plasma flow can be manipulated by transforming the corresponding parameters. Next, we will demonstrate that Eq. (B.2) at steady state strictly keeps form invariance after transforming coordinates.

For the steady state, the equation is

$$- \nabla \cdot (D\nabla n) \pm \nabla \cdot \left[\left(\frac{D\vec{E}}{T} \right) n \right] = S, \tag{B.3}$$

where we have replaced mobility with diffusivity. Then to obtain intuitive transformed results, we write down the component form of the diffusion-migration equation in a curvilinear space with the corresponding coordinate x_i [20],

$$- \partial_i \left(\sqrt{g} D^{ij} \partial_j n \right) \pm \partial_i \left[\left(\frac{\sqrt{g} D^{ij} \vec{E}_j}{T} \right) n \right] = \sqrt{g} S, \tag{B.4}$$

where g is the determinant of $g_i \cdot g_j$, with g_i and g_j being covariant bases of the curvilinear space. Then we write Eq. (B.4) in the physical space with coordinate x_i',

$$- \partial_{i'} \frac{\partial x_i'}{\partial x_i} \left[\sqrt{g} D^{ij} \frac{\partial x_j'}{\partial x_j} \partial_{j'} n \mp \left(\frac{\sqrt{g} D^{ij} \vec{E}_j}{T} \right) n \right] = \sqrt{g} S, \tag{B.5}$$

in which $\partial x_i'/\partial x_i$ and $\partial x_j'/\partial x_j$ are the components of the Jacobian matrix J whose determinant det J is equal to $1/\sqrt{g}$. Hence, we may reduce Eq. (B.5) to [22]

$$- \nabla' \cdot \left(D' \nabla' n \right) \pm \nabla' \cdot \left[\left(\frac{D' \vec{E}'}{T} \right) n \right] = S', \tag{B.6}$$

with $D' = JDJ^\tau / \det J$ and $\vec{E}' = J^{-\tau} \vec{E}$, and $S' = S/\det J$. Here, ∇' refers to the differential in the new coordinates x_i'. J is the Jacobian matrix with components $J_{ij} = \partial x_i'/\partial x_j$, J^τ is the transpose of J, while $J^{-\tau}$ is the inverse transpose of J. det J equals the determinant of J. Thus, the steady diffusion-migration equation strictly keeps form-invariance for arbitrary coordinate transformations.

However, the case is distinctly different in transient state. Eq. (B.2) at transient state could be reduced to

$$\frac{1}{\det J} \partial_t n - \nabla' \cdot \left(D' \nabla' n \right) \pm \nabla' \cdot \left[\left(\frac{D' \vec{E}'}{T} \right) n \right] = S'. \tag{B.7}$$

Compared with Eq. (B.2), the metric induced by the coordinate transformation in front of $\partial_t n$ in Eq. (B.7) can not be absorbed by materials or field parameters. Therefore, the transient plasma transport is not strictly form-invariant under a coordinate transformation except for det $J = 1$. Nevertheless, by taking an approximation, we can remove the unwanted metric and rewrite Eq. (B.7) as

$$\partial_t n - \nabla' \cdot \left(D'' \nabla' n \right) \pm \nabla' \cdot \left[\left(\frac{D'' \vec{E}''}{T} \right) n \right] = S'', \tag{B.8}$$

whose transformation rules are $D'' = JDJ^\tau$, $\vec{E}'' = J^{-\tau} \vec{E}$, and $S'' = S$. In this way, the transformed equation could keep form-invariant. It should be noted that Eq. (B.8) is generally an approximation of Eq. (B.7) because det J is position-dependent. To clearly understand this approximation, we rewrite its original transformed form Eq. (B.7) as

$$\frac{1}{\det J} \partial_t n - \nabla' \cdot \left(\frac{JDJ^\tau}{\det J} \nabla' n \right) \pm \nabla' \cdot \left[\left(\frac{JDJ^\tau}{\det J} \frac{J^{-\tau} \vec{E}}{T} \right) n \right] = \frac{S}{\det J}, \tag{B.9}$$

Then we multiply det J to both sides of Eq. (B.9) and decompose it,

$$\partial_t n - \nabla' \cdot \left(JDJ^\tau \nabla' n \right) \pm \nabla' \cdot \left[\left(\frac{JDJ^\tau J^{-\tau} \vec{E}}{T} \right) n \right] \pm \Delta = S, \tag{B.10}$$

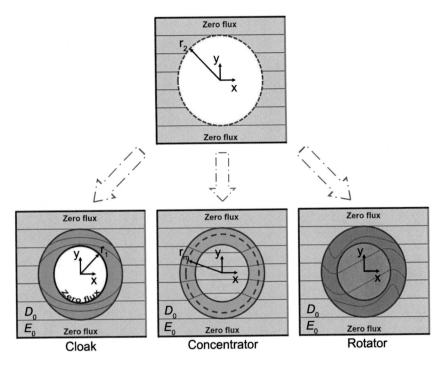

Fig. B.1 Schematic diagrams of conceptual devices. The solid blue line represents the plasma flow. We set the side length of the background matrix as $l = 0.12$ m. Other parameters: $D_0 = 9.2 \times 10^{-7}$ m s^{-1}, $\vec{E}_0 = [1.04 \times 10^4, 0]$ V m^{-1}, $r_1 = 0.020$ m, $r_2 = 0.030$ m, $r_m = 0.025$ m, $\theta_0 = \pi/3$, and $T_0 = 2.0$ V. Adapted from Ref. [21]

where $\Delta = \det J \nabla' (1/\det J) \left[J D \vec{E}/T \mp J D J^\tau \nabla' n \right]$. Comparing Eq. (B.10) with Eq. (B.8), the error caused by the approximation exactly depends on Δ. the error Δ is closely related to $\det J$ and cannot be eliminated if $\det J \neq 1$. Thus, the values of diffusivity and electric field intensity become crucial to the effect of the theory. As a result, small values of diffusivity and electric field intensity help prevent the devices from seriously disturbing the background plasma. Thus, the quantities of D and $\vec{E}$ need to be small enough to avoid a large error. The following simulation results show that this approximation is feasible and reasonable.

To confirm the theory, we propose three conceptional model devices to realize cloaking, concentrating, and rotating transient plasma transport without (obviously) disturbing the plasma distribution in the background. See Fig. B.1. When the plasma is input from the left-hand side, it remains unchanged on the right-hand side, as if there were no device in the middle. Concretely, the cloak can hide objects in the central region. The concentrator can increase the density gradient in the core region, while the rotator can flexibly rotate the propagation direction of plasma flow [23]. Next, we introduce the cloak first.

To realize the plasma cloak, the coordinate transformation from a virtual space r_i to the physical space r_i' is set as [23]

$$\begin{cases} r' = \dfrac{r_2 - r_1}{r_2} r + r_1, \\ \theta' = \theta. \end{cases} \tag{B.11}$$

See Fig. B.1. Here, r_1 and r_2 are the radii of the inner and outer boundaries of the cloak, respectively. This coordinate transformation can be physically explained as stretching the center dot into a circle with a radius of r_1 in the virtual space. Then we derive transformed parameters to achieve the cloaking of plasma flow according to the transformation rules.

Similarly, the coordinate transformations for realizing plasma concentrator and rotator can be written mathematically as [23]

$$\begin{cases} r' = \dfrac{r_1}{r_m} r, & r < r_m \\ r' = \dfrac{r_1 - r_m}{r_2 - r_m} r_2 + \dfrac{r_2 - r_1}{r_2 - r_m} r, & r_m < r < r_2 \\ \theta' = \theta. \end{cases} \tag{B.12}$$

$$\begin{cases} r' = r, \\ \theta' = \theta + \theta_0, & r < r_1 \\ \theta' = \theta + \theta_0 \dfrac{r - r_2}{r_1 - r_2}, & r_1 < r < r_2 \end{cases} \tag{B.13}$$

where r_m ($r_1 < r_m < r_2$) and θ_0 are the constant radius and angle, respectively. Equation (B.12) describes a physical picture that compresses a circle with a radius of r_m to a smaller circle with a radius of r_1 in the virtual space. In contrast, Eq. (B.13) gives a picture that in the virtual space, a series of circles with different radii are twisted by different angles determined by the values of the corresponding radii. Therefore, we obtain transformed parameters to converge or rotate plasma flow. Moreover, it should be noted that Eq. (B.8) for rotator is an accurate form instead of an approximation because of det $J = 1$ in this case.

Results and Discussion

Now we are in a position to employ COMSOL Multiphysics to perform finite element simulations. As reflected in Fig. B.1, a periodic plasma source, set as $n = n_1 \cos \omega_0 t + n_0$, is applied to the left boundary (the red one) of the background matrix. Here $n_1 = 5.0 \times 10^{15}$ m^{-3}, $\omega_0 = 2\pi/10$ s^{-1}, and $n_0 = 1.0 \times 10^{17}$ m^{-3}. We set the opposite (right) side as an outflow boundary (the blue one). The upper and

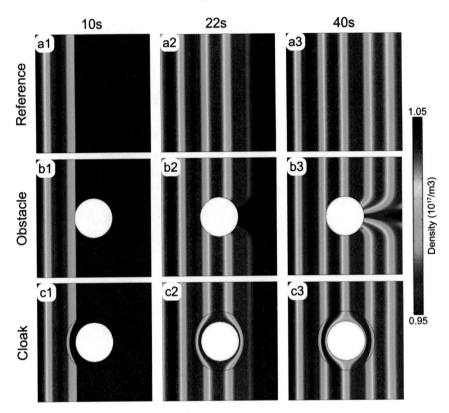

Fig. B.2 Simulation results of the cloak at transient states. **a1–a3** Density profiles for pure background at 10 s, 22 s, and 40 s, respectively. **b1–b3** Density profiles for background with an obstacle at 10 s, 22 s, and 40 s, respectively. **c1–c3** Density profiles for background with the cloak at 10 s, 22 s, and 40 s, respectively. Adapted from Ref. [21]

lower sides (boundaries) are set with zero-flux conditions. Besides, the zero-flux condition is additionally applied to the inner circle boundary of the cloak. The whole background matrix possesses a constant diffusivity D_0 and a uniform electric field $\vec{E}_0$. Then all the parameters can be designed according to the above transformation rules, and the simulation results of cloaking, concentrating, and rotating are shown in Figs. B.2, B.3 and B.4, respectively.

Figure B.2 illustrates the transient simulation of plasma transport under three conditions, namely, transporting in a pure background medium (set as the reference), in a background medium with a bare obstacle, and in a background medium with an obstacle covered by the cloak. The columns from left to right are screenshots of distributions of the plasma density at 10 s, 22 s, and 40 s, respectively. Due to the boundary condition of harmonically oscillating density, the plasma streams forward in a wave-like form. Moreover, the amplitude attenuation of the plasma flow reflected from the figures is caused by the diffusion, whose decay rate is codetermined

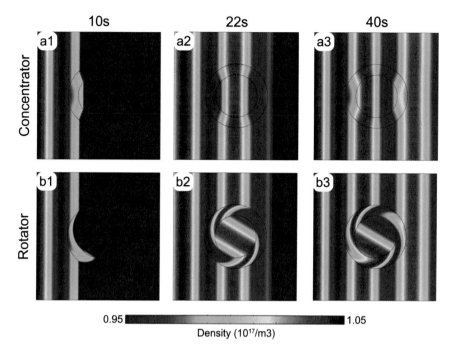

Fig. B.3 Simulation results of concentrator and rotator at transient states. **a1–a3** Density profiles for the concentrator at 10 s, 22 s, and 40 s, respectively. **b1–b3** Density profiles for the rotator at 10 s, 22 s, and 40 s, respectively. Adapted from Ref. [21]

by the oscillation frequency, diffusivity, and electric field. As a result, we carefully choose suitable values to make the results more intuitive. The cloak designed with the transformation theory helps to cancel the scattering induced by the obstacle. Therefore, the density profiles of the background plasma keep nearly undisturbed, which shows the validity of the theory.

The transient simulation results for the concentrator and rotator are shown in Fig. B.3. The first row of snapshots shows the converging effect of the plasma density gradient. In addition, as a determinant of the converging effect, a bigger ratio (r_m/r_1) would bring a higher converging effect. And the maximum ratio is r_2/r_1. For the rotator, the rotation of plasma flow appears in Fig. B.3b1–b3. Linearly deflecting concentric circles in the virtual space can account for the gradual deflection of the density profiles. The target rotation angle in the core region is determined by ω_0 in Eq. (B.13). Particularly, $\det J = 1$ for rotators helps eliminate the disturbance to background plasma density.

To further explore the performance of the devices, we extract the density values along a horizontal line (denoted by the yellow dashed lines in Fig. B.4) from the results at 40 s and compare the density distribution of functional devices with that of reference. See Fig. B.4b1–b3. Two regions should be remarked. One is the core region of the device, and the other is the background. All the red dashed lines in

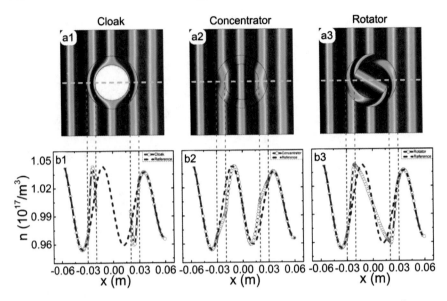

Fig. B.4 a1–a3 Color mapping of density profiles at 40 s with a cloak, concentrator, and rotator, respectively. **b1–b3** Comparisons between density profiles in the pure background (reference) and those with a cloak, concentrator, and rotator, respectively. The grey dashed lines denote the position of the devices. The data are extracted along the yellow dashed line ($y = 0$) in (**a1**)–(**a3**). Adapted from Ref. [21]

Fig. B.4b1–b3 denote the data of the reference, while the blue dotted lines represent the data of the cloak, concentrator, and rotator, respectively. In Fig. B.4b1, it is clear that the data are well overlapped in the background, and the plasma is excluded well from the core region. Moreover, the relative difference in the plasma density in the background region was less than 0.15%. In Fig. B.4b2, the dotted line is denser than the dashed line in the core region without being seriously dislocated in the background. And the relative difference was less than 0.13%. In Fig. B.4b3, the relative difference was less than 0.01%, which is far smaller than the value of the cloak or concentrator. As mentioned above, the accurate transformation form of Eq. (B.8) may account for this nearly zero difference. Overall, the simulation can confirm the feasibility and reliability of the theory.

Next, we can foresee some potential applications of the devices designed according to the transformation theory. For example, the cloak, whose core region is isolated, can be used to protect healthy tissue in plasma-curing infected wounds. In the field of catalyst preparation, converged plasma flow, which usually has a denser density of the active particle clusters, is beneficial to the interaction between plasma and catalyst. Hence, the concentrator can be used to improve catalytic efficiency. Or, in the aerospace industry, the concentrator possesses possibly an ability to improve the performance of plasma-assisted engines. Besides potential functions discussed above, separating or guiding plasma [24] could also be achieved by constructing appropriate coordinate transformation, which might be applicable to control plasma etching or plasma depositing. Moreover, the transformation theory might also help design plasma metamaterial, which is proposed to adjust electromagnetic waves [25, 26]. Consequently, the proposed methodology based on the transformation theory does make sense. Furthermore, despite the difficulties of achieving the transformed diffusivities and electric fields, it is still possible to realize the same effect by employing other methods. There are many kinds of research about customizing particle diffusivities. For example, using the scattering cancellation method, a bilayer diffusive cloak can be fabricated by two homogenous materials [27]. The complex diffusivity may also be realized according to the effective medium theory [28] or the machine learning method [29]. As for manipulating electric fields, studies of the electrostatic cloak and magnetic cloak could offer helpful inspiration [30, 31].

Many new mechanisms need to be studied despite unavoidable challenges. Under more general conditions, the influence of magnetic field and gas-phase reaction in plasma should be considered accordingly. It is full of difficulties to manipulate diffusivities and electric fields flexibly since the complicated interactions between charged particles and electromagnetic fields are too hard to control at will. Therefore, it is essential to introduce additional theories or methods, like particle-in-cell/Monte Carlo collision model [32] or nonequilibrium Green's function approach [33]. Additionally, the temperature of plasmas is usually time-varied or space-varied at transient states, thus leading to different transformation rules. In some cases, the advection might also happen in the plasma transport. Considering the advection term will make the regulation of plasmas more diverse. Moreover, the spatiotemporal modulation, a recent hot spot in heat diffusion [34], may bring fruitful properties to plasma physics. In short, improving the transformation theory for plasmas deserves more studies, attention, and effort.

Conclusion

We have employed a toy model, i.e., the diffusion-migration model, to describe plasma transport. We have shown the feasibility of the transformation theory. As a result, we have found that the transformed diffusion-migration equation is strictly form-invariant at steady states but not at transient states. Nevertheless, we have demonstrated that the transformed transient equation can be approximately form-invariant by setting small diffusivities. Then we designed three conceptual model devices, which function as a plasma cloak, concentrator, or rotator for transient plasma transport. Our results may broaden the approach to manipulating plasma flow and have potential applications in various fields, like medicine, the aerospace industry, etc.

References

1. Lieberman, M.A., Lichtenberg, A.J.: Principles of Plasma Discharges and Materials Processing. Wiley Interscience, New Jersey (2005)
2. Li, M., Wang, Z., Xu, R., Zhang, X., Chen, Z., Wang, Q.: Advances in plasma-assisted ignition and combustion for combustors of aerospace engines. Aerosp. Sci. Technol. **117**, 106952 (2021)
3. Liang, H., Ming, F., Alshareef, H.N.: Applications of plasma in energy conversion and storage materials. Adv. Energy Mater. **8**, 1801804 (2018)
4. Samal, S.: Thermal plasma technology: the prospective future in material processing. J. Clean Prod. **142**, 3131 (2017)
5. Tamura, H., Tetsuka, T., Kuwahara, D., Shinohara, S.: Study on uniform plasma generation mechanism of electron cyclotron resonance etching reactor. IEEE T. Plasma Sci. **48**, 3606 (2020)
6. Pendry, J.B., Schurig, D., Smith, D.R.: Controlling electromagnetic fields. Science **312**, 1780 (2006)
7. Leonhardt, U.: Optical conformal mapping. Science **312**, 1777 (2006)
8. Guenneau, S., Puvirajesinghe, T.M.: Fick's second law transformed: one path to cloaking in mass diffusion. J. R. Soc. Interface **10**, 20130106 (2013)
9. Fan, C.Z., Gao, Y., Huang, J.P.: Shaped graded materials with an apparent negative thermal conductivity. Appl. Phys. Lett. **92**, 251907 (2008)
10. Chen, T., Weng, C., Chen, J.-S.: Cloak for curvilinearly anisotropic media in conduction. Appl. Phys. Lett. **93**, 114103 (2008)
11. Xu, L.-J., Huang, J.-P.: Active thermal wave cloak. Chin. Phys. Lett. **37**, 120501 (2020)
12. Xu, L., Huang, J.: Negative thermal transport in conduction and advection. Chin. Phys. Lett. **37**, 080502 (2020)
13. Huang, J.: Thermal metamaterials make it possible to control the flow of heat at will. ES Energy Environ. **7**, 1 (2020)
14. Xu, L., Yang, S., Dai, G., Huang, J.: Transformation omnithermotics: simultaneous manipulation of three basic modes of heat transfer. ES Energy Environ. **7**, 65 (2020)
15. Hu, R., Zhou, S., Li, Y., Lei, D.Y., Luo, X., Qiu, C.W.: Illusion thermotics. Adv. Mater. **30**, 1707237 (2018)
16. Hu, R., Huang, S., Wang, M., Luo, X., Shiomi, J., Qiu, C.W.: Encrypted thermal printing with regionalization transformation. Adv. Mater. **31**, 1807849 (2019)
17. Zhang, J., Huang, S., Hu, R.: Adaptive radiative thermal camouflage via synchronous heat conduction. Chin. Phys. Lett. **38**, 010502 (2021)
18. Chen, F.F: Introduction to Plasma Physics and Controlled Fusion. Springer, Switzerland (1974)
19. Cui, S., Wu, Z., Lin, H., Xiao, S., Zheng, B., Liu, L., An, X., Fu, R.K.Y., Tian, X., Tan, W., Chu, P.K.: Hollow cathode effect modified time-dependent global model and high-power impulse magnetron sputtering discharge and transport in cylindrical cathode. J. Appl. Phys. **125**, 063302 (2019)
20. Dai, G.-L.: Designing nonlinear thermal devices and metamaterials under the Fourier law: a route to nonlinear thermotics. Front. Phys. **16**, 53301 (2021)
21. Zhang, Z.R., Huang, J.P.: Transformation plasma physics. Chin. Phys. Lett. **39**, 075201 (2022)
22. Zhang, Z., Xu, L., Huang, J.: Controlling chemical waves by transforming transient mass transfer. Adv. Theory Simul. **5**, 2100375 (2021)
23. Huang, J.P.: Theoretical Thermotics: Transformation Thermotics and Extended Theories for Thermal Metamaterials. Springer, Singapore (2020)

24. Lu, X., Ostrikov, K.: Guided ionization waves: the physics of repeatability. Appl. Phys. Rev. **5**, 031102 (2018)
25. Rodríguez, J.A., Cappelli, M.A.: Inverse design of plasma metamaterial devices with realistic elements (2022). https://arxiv.org/arXiv:2203.02572v1
26. Inami, C., Kabe, Y., Noyori, Y., Iwai, A., Bambina, A., Miyagi, S., Sakai, O.: Experimental observation of multi-functional plasma-metamaterial composite for manipulation of electromagnetic-wave propagation. J. Appl. Phys. **130**, 043301 (2021)
27. Zhou, X., Xu, G.Q., Zhang, H.Y.: Binary masses manipulation with composite bilayer metamaterial. Compos. Struct. **267**, 113866 (2021)
28. Restrepo-Flórez, J.M., Maldovan, M.: Mass separation by metamaterials. Sci. Rep. **6**, 21971 (2016)
29. Hu, R., Iwamoto, S., Feng, L., Ju, S., Hu, S., Ohnishi, M., Nagai, N., Hirakawa, K., Shiomi, J.: Machine-learning-optimized aperiodic superlattice minimizes coherent phonon heat conduction. Phys. Rev. X **10**, 021050 (2020)
30. Narayana, S., Sato, Y.: DC magnetic cloak. Adv. Mater. **24**, 71 (2012)
31. Lan, C., Yang, Y., Geng, Z., Li, B., Zhou, J.: Electrostatic field invisibility cloak. Sci. Rep. **5**, 16416 (2015)
32. Huang, C.-W., Chen, Y.-C., Nishimura, Y.: Particle-in-cell simulation of plasma sheath dynamics with kinetic ions. IEEE T. Plasma Sci. **43**, 675 (2015)
33. Yu, Z.-Z., Xiong, G.-H., Zhang, L.-F.: A brief review of thermal transport in mesoscopic systems from nonequilibrium Green's function approach. Front. Phys. **16**, 43201 (2021)
34. Xing, G., Zhao, W., Hu, R., Luo, X.: Spatiotemporal modulation of thermal emission from thermal-hysteresis vanadium dioxide for multiplexing thermotronics functionalities. Chin. Phys. Lett. **38**, 124401 (2021)

Printed in the United States
by Baker & Taylor Publisher Services